THE PHYSICS OF OPTO-ELECTRONIC MATERIALS

THE PHYSICS OF OPTO-ELECTRONIC MATERIALS

Proceedings of the Symposium on the Physics of Opto-Electronic Materials held at the General Motors Research Laboratories in Warren, Michigan, on October 4-6, 1970

Edited by Walter A. Albers, Jr.

General Motors Research Laboratories
Warren, Michigan

SPRINGER SCIENCE+BUSINESS MEDIA, LLC 1971

Library of Congress Catalog Card Number 73-173832
ISBN 978-1-4684-1949-8 ISBN 978-1-4684-1947-4 (eBook)
DOI 10.1007/978-1-4684-1947-4

Originally published by Plenum Press, New York in 1971
Softcover reprint of the hardcover 1st edition 1971

PREFACE

The papers in this volume represent most of the contributions to the Symposium on the Physics of Opto-Electronic Materials held at the General Motors Research Laboratories in Warren, Michigan, on October 4, 5 and 6, 1970.

The purpose of this Symposium was to examine the current status of knowledge related to the controlled alteration of the optical properties of solids through externally-applied agencies, with the aim of assessing possible future directions of scientific effort to achieve efficient, practical control of light.

Since the advent of the laser, the scientific community has been motivated to explore, with a renewed vigor, methods of modulating light, and in the last decade several applications of the electrooptic effect in single crystal solids have been realized. During this same period of time the list of recognized optical modulation effects in solids (exclusive of the ordinary electrooptic effects) has grown rapidly, and recently dramatic demonstrations of light modulation by liquid crystal and ferroelectric ceramic materials have captured the attention of the scientific community. Unlike the single-crystal electrooptic effects which are quite suitable for modulation of coherent laser light, these latter materials promise relatively inexpensive approaches to the modulation of light from ordinary incoherent light sources. It was these new vistas of light modulation – and how they fit into our current understanding of the optical properties of solids – that the symposium addressed.

The papers appear in this volume in the same order that they were presented at the symposium. The first four were chosen to provide a fundamental theoretical and experimental framework of the optical properties of solids as they relate to light modulation, with an effort to include the most recent developments in our theoretical understanding. Subsequent papers are devoted to explicit light modulation approaches, each of which provides both a description of the underlying physics and experimental evaluations of the particular approach under consideration. I believe that this represents the first volume that provides the reader with a wide range of optical modulation physics under a single cover, and as such constitutes a potentially valuable reference source for workers in the field.

This symposium was the fourteenth in an annual series initiated in 1957 by the then vice president in charge of the General Motors Research Laboratories, Dr. L. R. Hafstad. The symposia have as their objective promotion of the interchange of knowledge between specialists from many allied disciplines in a rapidly developing or changing area of science or technology. Attendees characteristically represent the academic, government, and industrial institutions that are noted for their ongoing activities in the particular area of interest.

I am indebted to many people for their help in the planning, executing, and hosting of the symposium, as well as in the preparation of this volume. A special note of thanks goes to an advisory group consisting of Professors N. Bloembergen (Harvard), E. Burstein (Pennsylvania), M. L. Cohen (California), and P. Handler (Illinois), who were instrumental in organizing the program as well as actively participating either as speakers or as Session Chairmen; to all of the authors whose efforts are, after all, the sum and substance of the symposium; to Professor G. Pratt of MIT and Professor M. Balkanski of the University of Paris for their continuing valuable advise throughout the planning stages and their roles as Session Chairmen; and to the numerous other individuals with whom I consulted from time to time in evolving the symposium program. Of my many colleagues at the General Motors Research Laboratories who have been of assistance, I should like to especially thank R. L. Scott for providing an excellent environment for the symposium, and R. Thomson, J. Caplan, N. Muench, and F. Jamerson for their interest and encouragement. I extend my deepest gratitude to Mrs. E. Levinge and Miss J. Filinger for secretarial assistance and manuscript typing respectively. And finally, I want to extend my very special thanks to Dr. Robert Herman of the GM Research Laboratories for a most excellent talk, after the symposium banquet, which contributed substantially to the overall enjoyment of the symposium.

Walter A. Albers, Jr.

January 1971

ATTENDANCE LIST

(* Denotes Session Chairman)

D. Adler
Massachusetts Institute of Technology
Cambridge, Massachusetts

P. D. Agarwal
GM Research Laboratories
Warren, Michigan

W. G. Agnew
GM Research Laboratories
Warren, Michigan

W. A. Albers, Jr.*
GM Research Laboratories
Warren, Michigan

P. J. Alluzo
Chevrolet Motor Division
General Motors Corporation
Warren, Michigan

F. J. Arlinghaus
GM Research Laboratories
Warren, Michigan

D. E. Aspnes
Bell Telephone Laboratories
Murray Hill, New Jersey

M. Balkanski*
University of Paris
Paris, France

R. H. Bartel
Terex Division
General Motors Corporation
Hudson, Ohio

H. B. Bebb
Texas Instruments Corporation
Dallas, Texas

S. E. Beacom
GM Research Laboratories
Warren, Michigan

M. Berg
AC Spark Plug Division
General Motors Corporation
Flint, Michigan

C. E. Bleil
GM Research Laboratories
Warren, Michigan

N. Bloembergen*
Harvard University
Cambridge, Massachusetts

M. H. Brooks
GM Research Laboratories
Warren, Michigan

G. Burns
IBM Watson Research Center
Yorktown Heights, New York

E. Burstein
University of Pennsylvania
Philadelphia, Pennsylvania

A. V. Butterworth
GM Research Laboratories
Warren, Michigan

L. R. Buzan
GM Research Laboratories
Warren, Michigan

J. D. Caplan
GM Research Laboratories
Warren, Michigan

M. Cardona
Brown University
Providence, Rhode Island

J. A. Carol
AC Spark Plug Division
General Motors Corporation
Flint, Michigan

M. C. Chen
GM Research Laboratories
Warren, Michigan

M. L. Cohen*
University of California
Berkeley, California

J. M. Colucci
GM Research Laboratories
Warren, Michigan

W. D. Compton
Ford Scientific Laboratories
Dearborn, Michigan

W. Cornelius
GM Research Laboratories
Warren, Michigan

L. E. Cross
Pennsylvania State University
University Park, Pennsylvania

W. A. Daniel
GM Research Laboratories
Warren, Michigan

R. Davies
GM Research Laboratories
Warren, Michigan

P. J. Dean
Bell Telephone Laboratories
Murray Hill, New Jersey

M. Di Domenico, Jr.
Bell Telephone Laboratories
Murray Hill, New Jersey

J. O. Dimmock
Lincoln Laboratories
Massachusetts Institute of Technology
Lexington, Massachusetts

R. W. Dixon
Bell Telephone Laboratories
Murray Hill, New Jersey

T. Downs
Delco Moraine Division
General Motors Corporation
Dayton, Ohio

J. Ducuing
Institut d'Optique
Orsay, France

D. S. Eddy
GM Research Laboratories
Warren, Michigan

G. Fan
IBM Research Laboratory
San Jose, California

J. E. Fischer
Michelson Laboratory
China Lake, California

P. Franken
University of Michigan
Ann Arbor, Michigan

D. B. Fraser
Bell Telephone Laboratories
Murray Hill, New Jersey

A. Frova
University of Rome
Rome, Italy

D. L. Fry
GM Research Laboratories
Warren, Michigan

A. D. Gara
GM Research Laboratories
Warren, Michigan

Z. G. Gardlund
GM Research Laboratories
Warren, Michigan

J. G. Gay
GM Research Laboratories
Warren, Michigan

R. W. Gibson, Jr.
GM Research Laboratories
Warren, Michigan

J. A. Giordmaine
Bell Telephone Laboratories
Murray Hill, New Jersey

W. L. Grube
GM Research Laboratories
Warren, Michigan

P. Handler*
University of Illinois
Urbana, Illinois

J. L. Hartman
GM Research Laboratories
Warren, Michigan

D. J. Henry
GM Research Laboratories
Warren, Michigan

R. Herman
GM Research Laboratories
Warren, Michigan

R. Hickling
GM Research Laboratories
Warren, Michigan

J. W. Hile
GM Research Laboratories
Warren, Michigan

I. Hodes
Manufacturing Development
General Motors Corporation
Warren, Michigan

R. N. Hollyer
GM Research Laboratories
Warren, Michigan

J. C. Holzwarth
GM Research Laboratories
Warren, Michigan

J. J. Hopfield
Princeton University
Princeton, New Jersey

S. Iwasa
Energy Conversion Devices
Troy, Michigan

F. E. Jamerson
GM Research Laboratories
Warren, Michigan

W. J. Johnston
AC Spark Plug Division
General Motors Corporation
Flint, Michigan

T. Jones
Engineering Staff
General Motors Corporation
Warren, Michigan

I. P. Kaminov
Bell Telephone Laboratories
Holmdel, New Jersey

M. Kaplit
GM Research Laboratories
Warren, Michigan

J. C. Kent
GM Research Laboratories
Warren, Michigan

L. T. Klauder, Jr.
GM Research Laboratories
Warren, Michigan

D. A. Kleinman
Bell Telephone Laboratories
Murray Hill, New Jersey

S. K. Kurtz
Philips Laboratories
Briarcliff Manor, New York

W. E. Lamb
Yale University
New Haven, Connecticut

C. E. Land
Sandia Corporation
Albuquerque, New Mexico

D. Langer
Aerospace Research Laboratories
Dayton, Ohio

R. Landauer
IBM Watson Research Center
Yorktown Heights, New York

J. Lastovka
Bell Telephone Laboratories
Murray Hill, New Jersey

M. Lax
Bell Telephone Laboratories
Murray Hill, New Jersey

R. W. Lee
GM Research Laboratories
Murray Hill, New Jersey

L. L. Lewis
GM Research Laboratories
Warren, Michigan

T. N. Louckes
Oldsmobile Division
General Motors Corporation
Lansing, Michigan

G. Lucovsky
Xerox Corporation
Palo Alto, California

B. A. Maciver
GM Research Laboratories
Warren, Michigan

M. E. Mack
United Aircraft Research Laboratories
East Hartford, Connecticut

C. E. Manning
Packard Electric Division
General Motors Corporation
Warren, Ohio

R. J. McDonald
GM Research Laboratories
Warren, Michigan

J. C. McElhany
GM Research Laboratories
Warren, Michigan

M. E. Meyers, Jr.
GM Research Laboratories
Warren, Michigan

E. J. Miller
GM Research Laboratories
Warren, Michigan

K. T. Milne
Chevrolet Motor Division
General Motors Corporation
Warren, Michigan

J. H. Moran
Buick Motor Division
General Motors Corporation
Warren, Michigan

N. L. Muench
GM Research Laboratories
Warren, Michigan

W. Oakley
Delco Moraine Division
General Motors Corporation
Dayton, Ohio

C. K. N. Patel
Bell Telephone Laboratories
Holmdel, New Jersey

R. W. Perkins
Oldsmobile Division
General Motors Corporation
Lansing, Michigan

P. S. Pershan
Harvard University
Cambridge, Massachusetts

A. Pinczuk
University of Pennsylvania
Philadelphia, Pennsylvania

F. H. Pollak
Brown University
Providence, Rhode Island

G. W. Pratt, Jr.*
Massachusetts Institute of Technology
Cambridge, Massachusetts

C. F. Quate
Stanford University
Palo Alto, California

R. A. Reynolds
Texas Instruments Corporation
Dallas, Texas

J. F. Rhodes
GM Research Laboratories
Warren, Michigan

S. Robbins
Fisher Body Division
General Motors Corporation
Warren, Michigan

D. M. Roessler
GM Research Laboratories
Warren, Michigan

R. L. Saur
GM Research Laboratories
Warren, Michigan

M. Schaeffer
Fisher Body Division
General Motors Corporation
Warren, Michigan

T. P. Schreiber
GM Research Laboratories
Warren, Michigan

R. L. Scott
GM Research Laboratories
Warren, Michigan

B. Segall
Case Western Reserve University
Cleveland, Ohio

B. O. Seraphin
University of Arizona
Tucson, Arizona

D. Schenk
Delco Moraine Division
General Motors Corporation
Dayton, Ohio

Y. R. Shen
University of California
Berkeley, California

G. R. Smith
Engineering Staff
General Motors Corporation
Warren, Michigan

G. W. Smith
GM Research Laboratories
Warren, Michigan

W. V. Smith
IBM Corporation
Zurich, Switzerland

B. B. Snavely
Eastman Kodak Company
Rochester, New York

R. A. Soref
Sperry Rand Research Center
Sudbury, Massachusetts

Stadler, H. L.
Ford Scientific Laboratories
Dearborn, Michigan

T. Stapleton
Manufacturing Development
General Motors Corporation
Warren, Michigan

B. P. Stoicheff
University of Toronto
Toronto, Canada

J. Tauc
Brown University
Providence, Rhode Island

R. W. Terhune
Ford Scientific Laboratories
Dearborn, Michigan

P. Thacher
Sandia Corporation
Albuquerque, New Mexico

R. F. Thomson
GM Research Laboratories
Warren, Michigan

C. S. Tuesday
GM Research Laboratories
Warren, Michigan

J. A. Van Vechten
Naval Research Laboratories
Washington, D. C.

J. R. Wallace
Oldsmobile Division
General Motors Corporation
Lansing, Michigan

E. F. Weller, Jr.
GM Research Laboratories
Warren, Michigan

S. H. Wemple
Bell Telephone Laboratories
Murray Hill, New Jersey

R. J. Wilson
Detroit Diesel Allison Division
General Motors Corporation
Cleveland, Ohio

P. Wolff
Massachusetts Institute of Technology
Cambridge, Massachusetts

A. Yariv
California Institute of Technology
Pasadena, California

Y. Yoshida
Fisher Body Division
General Motors Corporation
Warren, Michigan

R. S. Zucker
Detroit Diesel Allison Division
General Motors Corporation
Cleveland, Ohio

CONTENTS

IV. LIGHT MODULATING MATERIALS – II

LIST OF SYMPOSIA

held at General Motors Research Laboratories

1957 *Friction and Wear* (edited by Robert Davies), published 1959.
1958 *Internal Stresses and Fatigue in Metals* (edited by Gerald M. Rassweiler and William L. Grube), published 1959.
1959 *Theory of Traffic Flow* (edited by Robert Herman), published 1961.
1960 *Rolling Contact Phenomena* (edited by Joseph D. Bidwell), published 1962.
1961 *Adhesion and Cohesion* (edited by Philip Weiss), published 1962.
1962 *Cavitation in Real Liquids* (edited by Robert Davies), published 1964.
1963 *Liquids: Structure, Properties, Solid Interactions* (edited by Thomas J. Hughel), published 1965.
1964 *Approximation of Functions* (edited by Henry L. Garabedian), published 1965.
1965 *Fluid Mechanics of Internal Flow* (edited by Gino Sovran), published 1967.
1966 *Ferroelectricity* (edited by Edward F. Weller), published 1967.
1967 *Interface Conversion for Polymer Coatings* (edited by Philip Weiss and G. Dale Cheever), published 1968.
1968 *Associative Information Techniques* (edited by Edwin L. Jacks), published 1971.
1969 *Chemical Reactions in Urban Atmospheres* (edited by Charles S. Tuesday), published 1971.
1970 *The Physics of Opto-electronic Materials* (edited by Walter A. Albers, Jr.), this volume.

THE CHEMISTRY OF THE OPTICAL CONSTANTS OF SOLIDS*

J. J. HOPFIELD

Department of Physics, Princeton University, Princeton, New Jersey

Most ideally, the theoretical solid state physicist would be able to calculate from first principles the linear and non-linear optical constants of practical use. In fact, there are very few calculations in all of solid state physics which can be done from first principles. Instead, there is normally an immense amount of feed-back between theory and experiment in which one evaluates from experiments the theoretical parameters whose meaning is simple but whose calculation is impossible. It is such quasi-first principles calculations which will be described. An alternative form of calculation is pure empiricism, in which one invents new constructs, and checks these constructs against experimental results to find out whether the constructs themselves are valid and useful or not. But the characteristic approach of physics begins at the beginning and goes as far as possible from first principles before using experimental numbers to conceal physical ignorance.

This paper starts by describing how optical constants are in principle to be obtained, and then describes how they are in fact calculated when they are calculated at all. It next investigates the relation of chemical thinking to linear optical constants and describes some of the work of Phillips[1], Van Vechten[2], Wemple and DiDomenico[3], Levine[4], and others[5]. This point of view leads to relations between chemistry and non-linear optical constants as well as the linear ones. In understanding the relation between chemistry and the optical constants there is also the attractive possibility of explaining the utility of *ad hoc* models by giving their parameters some first principles meaning.

For linear optical properties, one is interested in calculating the refractive index and the extinction coefficient n and k related to the complex dielectric constant ϵ by

References p. 16

$$(n + ik)^2 = \epsilon = \epsilon_1 + i\epsilon_2 . \tag{1}$$

To avoid the need for tensor indices we restrict ourselves to cubic crystals. If a uniform electric field $E_o \cos\omega t$ is applied, second order time dependent perturbation theory can be used to calculate the rate of making transitions and thus the rate of energy loss from the applied electric field. This rate is proportional to the square of a perturbation matrix element and to the density of states. The perturbation is

$$H_{pert} = [\Sigma_{electrons}\, eE \cdot p_i/m\omega] \sin\omega t. \tag{2}$$

For a single final state F the density of states can be written as a δ-function. The rate-of-transition calculation thus has the form

$$\text{Transition rate} = (\pi/2\hbar)\, |\langle 0|H_{pert}|F\rangle|^2\, \delta(E_o + \hbar\omega - E_F) . \tag{3}$$

State 0 is the exact ground state of the system and state F is an exact final state. The imaginary part of the dielectric function can be written in the form

$$\epsilon_2(\omega) = 2/E_o^{\,2}(4\pi\hbar/\Omega) \sum_F \text{(transition rate)} \tag{4}$$

where Ω is the volume of the crystal. All the non-linear (power law) optical coefficients can similarly be generated from perturbation theory.

The problem of the theory of optical constants is the actual evaluation of such sums. The difficulty is that the state 0 and the states F refer to the exact ground and exact excited states of the system. Conventional theory gives approximate descriptions of the ground state and the excited states. The question is then whether or not these approximate descriptions are good enough to obtain useful quantitative information about the optical constants of interest.

The simplest approach to the evaluation of optical constants is to substitute for the state 0 and the states F the Hartree-Fock approximation to these states. This involves calculating the one-electron bands in solids, calculating transition matrix elements between one-electron states, and adding up the results. In fact there are very few Hartree-Fock calculations for crystals. Instead, one-electron states are calculated for model potentials. These potentials contain parameters which cannot be calculated adequately from a theoretical point of view. When these parameters are evaluated empirically by fitting known band gaps and known masses in particular energy bands, fairly good agreement can be obtained between the calculations and the experimental optical properties of typical semiconductors. As an example, Fig. 1 shows a comparison of theoretical and experimental reflectivities for GaAs.

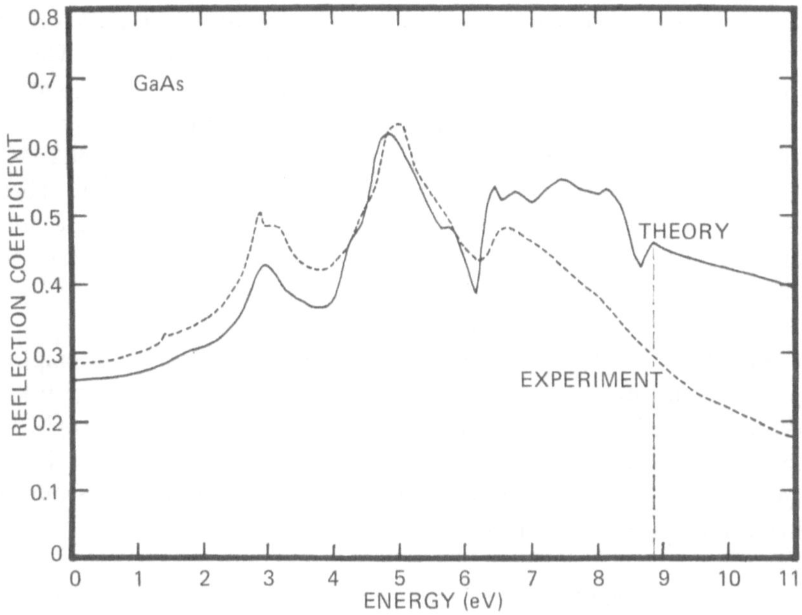

Fig. 1. A comparison of the experimental and calculated reflectivity of GaAs (after Walter and Cohen, ref. 5).

Coulomb interactions have been left out of such calculations in two important ways. First of all, such calculations contain no local field effects. If a solid were assembled of individual molecules (e.g., solid argon) one would write as a first approximation the polarizability of the crystal in terms of a Lorentz local-field expression for the polarizability of an assembly of atoms. In this way the polarization of each atom interacts with the next atom through the electric field due to the polarization on the atom in question. (It is possible to extend this molecular calculation to situations beyond that of simple argon. For example, one can calculate the polarizability of a Na^+ ion in the presence of the NaCl crystal environment as has recently been done by Graber[6].) Local field effects can be inserted in the ordinary Hartree-Fock calculation by complicating it. If a uniform electric field is applied, the electrons in the crystal polarize with a uniform and with a periodic part. The periodic part of the polarization will in turn itself produce an electric field. The problem of response to the total field can be solved self-consistently. In principle this involves calculating an ϵ which is a function of two reciprocal lattice vectors. This ϵ describes how the crystal polarization responds at one reciprocal lattice vector when an electric field of a different reciprocal lattice vector is applied. Although there has been much talk about making such calculations none has completely been carried out. While local field effects are not believed to be of major importance in semiconductors, they are expected to be important in insulators.

References p. 16

The second kind of Coulomb interaction neglected in the Hartree-Fock calculation are the interactions between the electrons and holes which bind them together in excitons. In the theoretical part of Fig. 1 there are no exciton states. There are two ways of introducing exciton states. One is to go back to the molecule point of view - molecules have discrete excitation levels - and the other is to look at the binding of a single electron-hole pair, as was originally done by Wannier and Mott. It is, however, difficult to fully introduce the effects into the calculations. These exciton effects create important details in the structure of $\epsilon_2(\omega)$. Similar Coulomb effects completely alter the nature of the thresholds which are present around band edges.

While the calculations just described contain a healthy admixture of empirical knowledge in the band structure, the theoretical constructs directly emerge from *ab initio* theory. One should note the distinction between this kind of empiricism and an empiricism which postulates, for example, bond polarizabilities. It should eventually be possible to evaluate the band structure empirical constants from theory because they can be precisely defined.

Properties of solids can be divided into two groups, accidental properties and chemical properties. The mobility of electrons in germanium is an example of an accidental property. If there were small changes in the band gap parameters, the k=0 conduction band valley in germanium would be the lower valley and the mobility of electrons would then be a factor of 10 to 100 times larger than it is in real germanium. At the same time the cohesive energy, elastic constants and similar *chemical* properties would not be changed appreciably. The dielectric constant would also not be changed. This insensitivity is expected of most low frequency linear and non-linear optical properties. On the other hand, the detailed behavior of the absorption spectrum of germanium is an accidental property, as are the optical constants very near the fundamental absorption edge.

The properties I have termed accidental are due to fine details of band structure. These can also be predicted only from a detailed calculation. On the other hand, the properties such as cohesive energy and the elastic constants have always been qualitatively understandable in terms of the binding of atoms together in a very chemical scheme. We will next look at the *chemical* behavior of the optical constants of solids by methods which are far simpler than detailed band structure calculations but which cannot predict "accidental" properties.

The relation between optical constants and chemical bonding has been recently emphasized by J. C. Phillips and J. Van Vechten. The bonding between two elements in the middle of the periodic table, for example the carbon-carbon bond or boron-boron bond, is termed covalent. The energy of binding of a boron-nitrogen bond is greater than that of either a boron-boron or a nitrogen-nitrogen bond.

Pauling attributed this increase in the binding of boron nitride over boron and nitrogen bonds to the ionic contribution in the bonding energy. It is due to a partial transfer of an electron from the nitrogen onto the boron atom, leaving each of the atoms partially charged and thus partially ionic. Pauling[7] made use of bond lengths and empirical rules for ionicity to understand the cohesive energy of a wide variety of partially covalent and partially ionic crystals, as well as of molecules. Fundamental to Pauling's description is the basic idea of bonding and anti-bonding orbitals. Given two atoms, each of which has, for example, a p-orbital about its center, the linear combination of these two orbitals with no node between the atoms has a lower energy than the linear combination with a node between the atoms. This linear combination with no node is called a bonding orbital and the linear combination with a node at the mid-point between atoms is called an antibonding orbital.

In crystals such as germanium, silicon and diamond, with four covalent bonds around a four-valent atom, the bonding states are filled and the anti-bonding states are empty. A pure ionic state on the other hand would occupy an orbital on one atom or the other, i.e., one of the original basis functions. In a partially ionic crystal the orbitals will be some compromise between single atom orbitals and the bonding orbitals. Thus in a partially ionic crystal the bonding and anti-bonding orbitals will be partly mixed. The energy difference between the bonding and anti-bonding orbitals, or more accurately, between the filled and the empty orbitals, can be directly measured in the optical properties of solids and in some sense is simply the typical excitation energy from a valence band to a conduction band[1].

If only two orbitals and two atoms are involved, the scheme can be simply described. Take the energy of a bonding orbital to be zero. In a purely covalent crystal the anti-bonding orbital would have some excitation energy E_c, and in terms of these two states the Hamiltonian for the electron is

$$\begin{pmatrix} 0 & 0 \\ 0 & E_c \end{pmatrix} \Leftrightarrow \begin{pmatrix} \text{bonding} \\ \text{antibonding} \end{pmatrix} \qquad (5)$$

The effect of partial ionicity is to provide a matrix element between the bonding and anti-bonding states, which can be written in terms of an off diagonal matrix element in each corner described by the parameter C/2. If E_c were zero, the states in the presence of C would turn out to be the single atom states. For a general crystal both parameters will be non-zero. The difference in energy between the occupied and the unoccupied state of this simple 2 x 2 Hamiltonian is given by

$$\Delta E = E_g = (E_o^2 + C^2)^{1/2} . \qquad (6)$$

References p. 16

Phillips has defined the ionicity as

$$f_i \equiv \text{ionicity} \equiv C^2/E_g^2 . \tag{7}$$

This is a *definition* and is useful only if this ionicity appears in expressions other than the defining one. The discussion given so far is for two orbitals. At many points of high symmetry in the zone, a similar kind of analysis can be made for the band gap separating some of the valence bands and some of the conduction bands in typical zinc-blende structure semiconductors. Thus the basic equations are defined to hold for semiconductors as well as for molecules.

To evaluate E_g for an experiment, Phillips has made use of a Kramers-Kronig relation. The real part of the dielectric constant at zero frequency can be written as

$$\epsilon_1(0) = 1 + (2/\pi)\int_0^\infty \epsilon_2(\omega)d\omega/\omega . \tag{8}$$

The f-sum rule is

$$\omega_p^2 = (2/\pi)\int_0^\infty \omega\epsilon_2(\omega)d\omega . \tag{9}$$

In the calculation of ω_p^2 it is presumed that an atom of valence four should be counted as having four electrons. Comparing these two integrals we see that

$$\epsilon_1(0) \equiv 1 + \omega_p^2/\omega_g^2 \tag{10}$$

where ω_g^2 is some average of the frequency squared. Phillips identified this mean squared frequency as $(E_g/\hbar)^2$. If $\epsilon_2(\omega)$ were all strongly concentrated at one single frequency, this identification would be unique. It represents only an approximate evaluation, however, given the actual form of the imaginary part of $\epsilon(\omega)$.

One has by definition obtained an experimental way of evaluating the parameter E_g^2 from the "static" dielectric constant (evaluated well above the ion motion frequencies). We would like to evaluate E_g and C separately. In order to do this, Phillips makes the additional supposition that E_c itself depends only on the lattice parameter. E_c exhibits a simple power-law dependence on lattice parameter for silicon, germanium and carbon, and it is presumed that this dependence will continue to hold in partially ionic crystals.

From the lattice constants and the measured E_g's (and thus E_c's) of the group four elements one can then determine the ionicities for all AB semiconductors, where the sum of the valences of A and B is 8. Probably the greatest triumph of this theory is its success in calculating whether a given material should have a covalent crystal structure (involving tetrahedral coordination) or should have an ionic crystal structure. It was found that all materials having ionicities less than 0.785 have tetrahedral coordination, while all having ionicities greater than this show typical ionic crystal structure.

Another interesting feature of this ionicity construct is the fact that some crystal properties do seem *empirically* to be related to this ionicity. For example, the cohesive energies of different AB semiconductors having tetrahedral coordination are a linear function of ionicity within a given row of the periodic table. There have been other attempts to correlate ionicity with physical properties, but most of these involve additional *ad hoc* suppositions. An example of these is the bond-charge model explanation of the values of the piezoelectric constants.

A strong pseudo-atom hypothesis underlies this discussion of the relationship between dielectric behavior and crystal binding. All details which take place in the core region of the atom have been ignored. For a physicist the basis for doing this might be taken to be the empirical observation that bromine and chlorine chemistry looks so similar. To the chemist, there is an interesting difference between bromine and chlorine, and that must be a core effect. The question is whether the core effect contains details of the core or whether it is simply a core size effect. One can model sodium as a one-electron atom whose electrostatic potential looks as shown in Fig. 2. If the cut-off distance is correctly chosen, one can fit the sodium valence electron s-states, p-states, and d-states amazingly well. If a different value for the parameter is chosen, the states of potassium will instead be generated. There are large differences in the wave functions of the sodium and potassium valence electrons in the core region. As far as the behavior of the elementary electronic excitations at low energies is concerned, however, the only property of the core which honestly enters is something having to do with the core size. This can be subsumed in a one-electron model of sodium or potassium. Similarly, in the case of the semiconductors under discussion, the use of a free electron mass and four electrons per silicon atom is tantamount to presuming that the core structure is ignorable and that silicon is really equivalent to a four-electron atom with a modified core potential.

I emphasize this hypothesis because it is very seldom stated in detail and because what will next be done is to use this hypothesis as strongly as possible. Granting this pseudo-atom hypothesis, there are still problems with the analysis of Phillips. There are too many suppositions in it which are both unproved and, because they lack precision, almost unprovable. On the other hand, there is much good physics present in the analysis. The next logical step is to reformulate the problem in such a way that Phillips' constructs come to the fore, but come from a

References p. 16

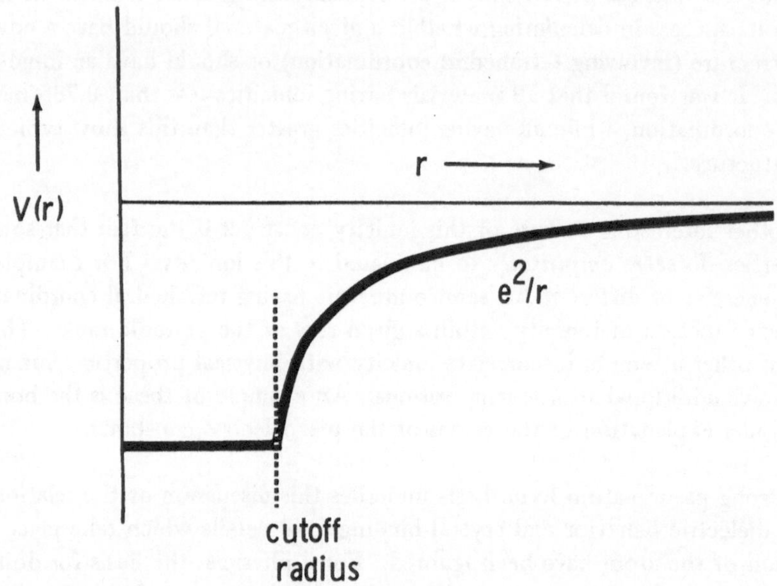

Fig. 2. A model local pseudopotential for alkali atoms. The longe-range part must be chosen to agree with actual potential.

well-defined mathematical procedure.

The ordinary f-sum rule can be viewed as arising from the fact that at high enough frequencies electrons do not notice the potential of the lattice. At high enough frequencies the charge distribution of the electron moves in an applied uniform electric field as a rigid unit, as though the electrons were free. In the next approximation there is a restoring force on this rigid motion of the charges due to the fact that the electron charge density peaks up around the ions. Displacement of the electron charge density peak from the ions produces a restoring force, described by a restoring force constant k. The high frequency limit of the dielectric constant is therefore[8]

$$\epsilon_1(\omega)_{\omega \to \infty} = 1 + \omega_p^2/(k/m - \omega^2) . \tag{11}$$

We can evaluate the high frequency behavior of $\epsilon_1(\omega)$ in two ways. One way is to use a Kramers-Kronig relation and write $\epsilon(\omega)$ in the limit $\omega \to \infty$ as an integral over $\epsilon_2(\omega)$. In the limit of large frequencies ω is beyond the range in which ϵ_2 exists. Expanding the denominator in powers of (ω'/ω), one obtains

$$\epsilon_1(\omega) \underset{\omega \to \infty}{\approx} 1 - (2/\pi)\omega^{-2} \int_0^\infty \omega'\epsilon_2(\omega')d\omega' - (2/\pi)\omega^{-4} \int_0^\infty \omega'^3\epsilon_2(\omega')d\omega' \ldots \tag{12}$$

By equating powers of $(1/\omega)$ of this equation with those of Eq. 11 two relations are obtained: The ordinary f-sum rule; and a relationship between the third moment of $\epsilon_2(\omega)$ and the constant k. The constant k is the restoring force due to the motion of a rigid charge distribution of electrons. This force can be obtained by integrating the electric field times the charge density over all the electrons. Since the electron wave function is itself rigid, the electron-electron interactions will not occur in the calculation, and only the electron-ion interactions are needed. As a result one can write k in terms of a sum over all the atoms in a cell of an integral of the deviation, $\delta\rho$, of the charge distribution from uniformity multiplied by the Laplacian of the potential due to the pseudo-atoms. At short ranges the pseudo-atom potentials may be slightly complicated. At long distances their potentials are Coulomb so the Laplacian of the potential will vanish at large distances. The final theorem is

$$(2/\pi) \int_0^\infty \omega^3\epsilon_2(\omega)d\omega = \omega_p^2 e/3mn \sum_{\substack{i \\ \text{atoms in} \\ \text{unit cell}}}^{n} \int_{\text{all space}} \delta\rho(r)\nabla^2 V_i(r)\ d^3r \tag{13}$$

The same theorem can be obtained by a rigorous calculation[8] without the argument about the rigid charge distribution.

This is a rigorous theorem for pseudo-atoms. Its connection with the point of view of Phillips can be described in the following way. According to Phillips there exists a squared frequency having a chemical interpretation. This frequency is obtainable from $\epsilon_2(\omega)$ (since the static dielectric constant can be found from $\epsilon_2(\omega)$). This squared frequency can be written as the sum of two terms[2], one relating to the symmetric part of the problem and the other to the anti-symmetric part (i.e., the covalent and ionic parts of the problem).

In the rigorous sum rule, there does exist a squared frequency which can be written

$$\omega_a^2 = (2/\pi\omega_p^2) \int_0^\infty \omega^3\epsilon_2(\omega)d\omega = e/3mn \sum_{\text{atoms}} \int \delta\rho(r)\nabla^2 V_i(r)\ d^3r \tag{14}$$

Given a two atom problem one can define the symmetric and the anti-symmetric parts of $\delta\rho$ and of the potential in the unit cell. Because the integral of a symmetric function times an anti-symmetric one vanishes, the right hand side becomes

References p. 16

$$\int \delta\rho_s \nabla^2 V_s d^3 r + \int \delta\rho_a \nabla^2 V_a d^3 r \, .$$

Thus ω_a^2 is rigourously the sum of two terms, one containing only symmetric quantities and the other containing only anti-symmetric ones.

This result is similar to that of Phillips and Van Vechten. However, one should note that ω_a^2 is not identically the same frequency as ω_g^2. Consider, for example, what would happen if silicon had a slightly overlapping band structure. The dielectric constant at zero frequency would be infinite. Phillips would have to make qualifying statements about how this situation should be interpreted in order to find the correct value of ω_g. The present calculation, on the other hand, would yield a finite value for this third moment and a well-behaved ω_a^2. Thus from a theoretical point of view the positive moments of $\epsilon_2(\omega)$, which do not give a heavy weighting to states near small band gaps, are a more satisfactory definition of an average appropriate to binding than is the static dielectric constant.

From a pragmatic point of view the static high-frequency dielectric constant and its frequency dependence are all that can easily be obtained from experiments. Thus these three moments

$$\epsilon_s - 1 = (2/\pi) \int_0^\infty \omega^{-1} \epsilon_2(\omega) d\omega \tag{15}$$

$$\omega_p^2 = (2/\pi) \int_0^\infty \omega \epsilon_2(\omega) d\omega \tag{16}$$

$$M_{-3} = (2/\pi) \int_0^\infty \omega^{-3} \epsilon_2(\omega) d\omega \tag{17}$$

are generally known, while the moment involved in calculating ω_a^2 is not. Worse, what one really needs is not actual experimental data from which to evaluate the integral for ω_a^2 but instead the value of $\epsilon_2(\omega)$ which would be present if experiments were done on pseudo-atoms instead of real atoms. There is unfortunately a difference at high frequencies due to the core wiggles in the valence band wave functions.

We must now look for a way of avoiding this evaluation difficulty by getting ω_a^2 from low frequency data. The way out of this dilemma is to recognize that these three moments can be thought of as the values of the function

$$\int_0^\infty \omega^N \epsilon_2(\omega) d\omega$$

for different values of the argument N. What we wish to do is to extrapolate this function from the values N = -3, N = -1, N = +1 to the value N = +3. If the function is well behaved such an extrapolation is possible, though the extrapolation will depend upon the nature of the pseudo-atom potential. For a purely Coulomb potential it is easily shown that this function of N will become logarithmically infinite for N = 3.5. This infinity is caused by the Coulomb potential singularity at the origin. Pseudopotentials which do not have a singularity as large as this will have smoother extrapolation properties. Given a potential, one can establish an extrapolation procedure for generating the third moment from the moments which are available experimentally.

A simple illustration of the use of the sum rule can be obtained using the dielectric model of Wemple and DiDomenico[3]. They proposed a model of $\epsilon_2(\omega)$ which represented an insulator with a band gap ω_t and upper cut-off frequency $b\omega_t$,

$$\epsilon_2(\omega) = (\pi/2)\,(\omega_p^2/\omega\omega_t)\,(b-1);\quad \omega_t<\omega<b\omega_t \qquad (18)$$

$$= 0 \qquad \text{otherwise}$$

This model obeys the f-sum rule and contains two adjustable parameters. We can use these two parameters to fit the measured dielectric constant and the measured moment M_{-3}. From this come two results of interest. One is that, within this model for typical semiconductors $\omega_g^2 = 1.4\,\omega_a^2$. The other is that a parameter which Wemple and DiDomenico found to be interesting from the point of view of empirical chemical classification turns out to be simply ω_p^4/ω_a^2. This is an elementary dielectric model, and the extrapolation procedure has not yet honestly been solved. But I view these results as encouraging.

If the antisymmetric potential is viewed as a small perturbation as one goes to the crystal GaAs from germanium, its effect can be evaluated in perturbation theory. The antisymmetric potential will cause an antisymmetric charge distribution which will be linear in the antisymmetric potential. Thus one will be able to write ω_g^2 for GaAs as that evaluated for germanium plus a term which is proportional to the square of the antisymmetric potential. The polarizability of the electron gas can be used as an approximate means of calculating $\delta\rho_G$ from V_G, and as a result one obtains a first-principles calculation of the dielectric constant of gallium arsenide (or f_i) knowing the dielectric constant of germanium. The inputs for this calculation are the pseudopotential coefficients V_G for gallium arsenide, which can be taken from a band structure calculation fit to experiment. This elementary evaluation generates the correct value of the difference in dielectric constant between Ge and GaAs with a precision of about 20%.

References p. 16

In summary, there is a rigorous sum rule about the high frequency dielectric behavior which can be used to derive theorems about a parameter which is closely related to, but is not exactly, the parameter E_c^2 used by Phillips. The advantage of this procedure is that all relations are well-defined, and the disadvantage of the procedure has been the strong use made of the pseudo-atom hypothesis. The extrapolation procedure of the dielectric constant function is important and can be best seen in Fig. 3. If $\epsilon_1(\omega)$ is some function such as that sketched and one knows the value of $\epsilon_1(\omega)$ and its derivative for small frequencies, the problem is to extrapolate up to high frequencies to obtain the high frequency behavior of $\epsilon_1(\omega)$. Knowing that there is a Kramers-Kronig relation, this procedure may be possible because the form of $\epsilon_2(\omega)$ can be fairly well described by its area (fixed by the f-sum rule) and a typical position determined from the low frequency $\epsilon_1(\omega)$.

Let us now look at the problem of non-linear optics. Generalizing the linear optics results, one can write theorems about non-linear optics at very high frequencies simply in terms of the charge distributions and the crystal potential. Such a theorem valid at high frequencies for the second harmonic generation coefficient ϵ_{xyz} is, for $\omega_1 + \omega_2 \longrightarrow \infty$

$$\epsilon_{xyz} = [e^2/(m^2\omega_1^2\omega_2^2)]\, i \sum_G G_x G_y G_z V_G \; \delta\rho_{-G}$$

or equivalently for the sum (19)

$$\sum_G G_x G_y G_z \; [V_G \delta\rho_G^*]_{\substack{\text{imag}\\\text{part}}}$$

This is a rigorous theorem.† The problem is the extrapolation of this theorem from the high frequency regime down to the low frequency regime where experiments are done. In this one expression a couple of interesting results immediately stand out. The lattice potential is written as the sum on G and the crystal charge distribution is similarly written. If the crystal has a center of inversion symmetry and that center is chosen as origin, V_G and $\delta\rho_G$ are real and ϵ_{xyz} thus vanishes (as it must for symmetry reasons). Suppose then that the crystal lacks inversion symmetry. In general V_G and $\delta\rho_G$ are then complex, but the second harmonic coefficient comes from the difference in the phase angles of the complex V and $\delta\rho$. In a simple metal where the crystal potential is small and the deviation of the charge density from uniform can be thought of as being linear in the crystal potential, V_G and $\delta\rho_G$ would have the same complex phase. Thus in a free-electron-like metal ϵ_{xyz} would vanish. It requires a non-linearity in the charge density at a particular crystal potential V_G to obtain a different complex phase for V and $\delta\rho$, and thus a non-vanishing second harmonic coefficient.

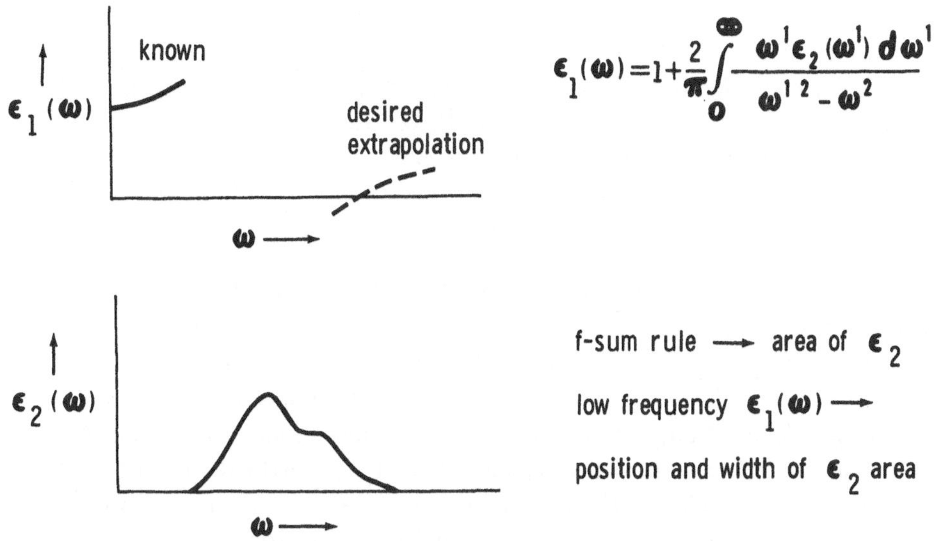

Fig. 3. A summary of the problem of the extrapolation of $\epsilon_1(\omega)$ to high frequencies.

The problem of the evaluation of the low-frequency electrooptic coefficient from the high frequency behavior is caused by the fact that ϵ_{xyz} does not have a well-determined sign. Consider for simplicity what would happen if one of the electric fields were static. There would then be a Kramers-Kronig relation relating the high-frequency behavior and the low-frequency behavior for the second field. Unfortunately the analog of $\epsilon_2(\omega)$ under those circumstances *would not necessarily be positive*. As a result it would be possible for the high-frequency behavior to have one sign and the low-frequency behavior to have the same sign. In contrast, the usual dielectric function (minus 1) changes sign between low frequencies and high frequencies. In a crystal with only one kind of bond - and crudely one kind of resonant oscillator - the sign ambiguity is unlikely to occur and an extrapolation from high frequency results to low frequency results should be possible. For crystals containing many different kinds of bonds and many different kinds of atoms, on the other hand, I do not see at this point any way of making the extrapolation.

If we assume one kind of bonding, then the elementary extrapolation for the low frequency behavior of ϵ_{xyz} will simply replace ω_1^2 and ω_2^2 by ω_g^2. The result is a chemical theory of Miller's Δ.[9] The only items necessary to evaluate this expression in a typical zinc-blende semiconductor are the actual bare pseudopotential coefficients and the charge densities in the crystal. The major contributor to this sum is the (1,1,1) direction. Reciprocal lattice vectors in the (1,0,0) and (1,1,0) directions do not contribute to the sum for symmetry reasons, while smallest (1,1,1) terms give a large and probably dominant contribution.

Levine[4] in a recent Physical Review Letter gave an expression for ϵ_{xyz} of tetrahedral semiconductors which was based on bond charges and a certain model of dielectric response. In present terms, his theory can be described as providing an *ansatz* in these materials for evaluating the sum. His *ansatz* is essentially equivalent to saying that, to a first approximation, the charge distribution does not change when the antisymmetric potential is turned on.

Recently Levine made his model more elaborate in such a fashion that, while it contained more hypotheses, it also contained the possibility of generating either sign of ϵ_{xyz}. From the general considerations above it is easy to understand why ϵ_{xyz} is likely to change in sign in going from covalently bonded materials toward more ionically bonded ones. In a crystal of zinc-blende structure the symmetric and the antisymmetric parts of the crystal potential for a particular reciprocal lattice vector are 90° out of phase. The real and imaginary parts of the potential or charge distribution are equivalent to the symmetric and antisymmetric parts of the coefficients. Fig. 4 is a representation of the phase angle of the potential and charge distribution coefficients. A purely covalent crystal has for symmetry reasons a zero

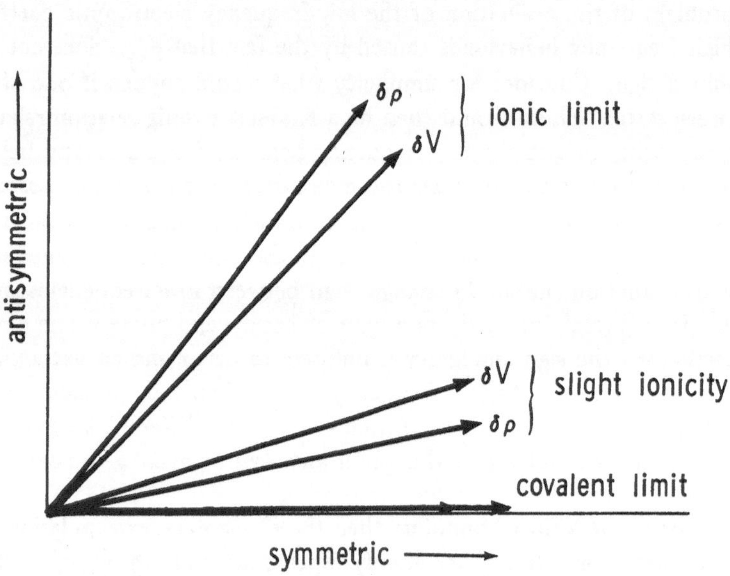

Fig. 4. The complex values of V and $\delta\rho$ for the (111) direction in typical zinc-blende crystals ranging from purely covalent to ionic. Note the expected change in sign of the difference in phase angle between $\delta\rho$ and V.

phase angle of both the charge distribution and the electrostatic potential. When a small antisymmetric potential is turned on, the crystal remains in the Pauling limit, with very little redistribution of charge. This produces a second harmonic coefficient of well-defined sign, and in fact of the observed sign. The sign will be reversed if the phase angle between the potential and the charge distribution is reversed. In the ionic crystal limit, the charge distribution is essentially all on one site and is an equal mixture of symmetric and antisymmetric parts. However, because the ion which has no electron on it still has a potential, the potential has a larger symmetric part than antisymmetric part thus making a phase angle which is less than 45° in Fig. 4. The reversal in phase angle difference should produce a change in sign of ϵ_{xyz} in going from chiefly covalent to chiefly ionic crystals.

I am not sure of the extent to which this theorem about non-linear optical coefficients is going to be useful. The extrapolation problem seems enormous. On the other hand, problems of model building without exact theorems to help also seem enormous, so this theorem may be useful in two ways. First of all, one may be able to interpret the models which correctly describe the non-linear parameters in terms of what they are saying about $\delta\rho$ and δV. Second, in *simple* systems like the zinc-blende semiconductors, there is real hope of evaluating V and $\delta\rho$ from information about electron wave functions at a few points of the Brillouin zone. As a result it should be possible to directly check this calculation to see whether or not the extrapolation works in zinc-blende systems. It must be emphasized that in complicated systems I do not see that this theorem can possibly be useful. What it can do is aid in the understanding of simple systems which in turn may aid in understanding more complicated ones. (Of course it may also be true that the best non-linear optical materials turn out to be the best for accidental reasons, not for chemical ones).

To recapitulate, since solids are held together by electrical forces it is not surprising that the electromagnetic response of a solid yields a wealth of information about the bonding and about the wave functions of the solid. If we know the wave functions we can obviously calculate the non-linear optical response of the system. The question which I am trying to ask is: Can one get from understanding of the bonding of the solid directly to information about the "chemical" optical properties without the necessity of calculating an infinite number of wave functions in between, making the Hartree-Fock approximation and filling this connecting link with myriads of details of theory? The theoretical phenomenology of Phillips and co-workers in this field suggests that you can. What I am suggesting is that there is a rigorous means of making such a connection, at least in some cases, through the use of sum rules and the use of dispersion relations to go from the frequency region where the sum rules are correct to the frequency region of experimental interest.

References p. 16

REFERENCES

* *Research supported in part by the U.S. Air Force OFSR under contract AF 49 (638) 1545.*

1. *J. C. Phillips, Phys. Rev. Lett. 20, 550(1969); 22, 645(1969); Covalent Bonding in Crystals, Molecules, and Polymers, (The University of Chicago Press, Chicago, 1969); Physics Today, Feb. 1970, p. 23.*
2. *J. A. Van Vechten, Phys. Rev. 182, 891(1969); 187, 1007(1969).*
3. *S. H. Wemple and M. DiDomenico, Phys. Rev. Lett. 23, 1156(1969).*
4. *B. F. Levine, Phys. Rev. Lett. 22, 787(1969); Bull. Amer. Phys. Soc., Series II, 15, 335 (1970).*
5. *J. P. Walter and M. L. Cohen, Phys. Rev. 183, 763(1969).*
6. *M. A. Graber, Optical Absorption of the Alkali Halides, PhD Thesis, Princeton University, 1970 (to be published).*
7. *See for example L. Pauling, The Nature of the Chemical Bond (Cornell University Press, Ithaca, 1960), pp. 65 - 105.*
8. *J. J. Hopfield, Phys. Rev. B2, 973(1970).*
9. *R. C. Miller, Appl. Phys. Lett. 5, 17(1964).*

† *Note added in proof: the proof of this conjecture was worked out by T. C. McGill (private communication).*

NONLINEAR OPTICAL EFFECTS

Y. R. SHEN

Department of Physics, University of California and Inorganic Materials Research Division, Lawrence Radiation Laboratory, Berkeley, California

ABSTRACT

Third-harmonic generation in cholesteric liquid crystals, far-infrared generation by ultrashort pulses, and self-focusing of light in nonlinear media are taken as examples for illustration of nonlinear optical effects. Depending on the problems, different approximations are used in solving the nonlinear wave equations. Connection between nonlinear optical effects and properties of the media is briefly discussed.

INTRODUCTION

By definition, nonlinear optics deals with problems involving nonlinear interaction of light with matter.[1] It arises because the response of a medium to the fields is generally nonlinear, as indicated by the nonlinear constituitive equation $\mathbf{D} = \epsilon(\mathbf{E}) \cdot \mathbf{E}$. Only when the field dependence in the dielectric tensor $\epsilon(\mathbf{E})$ is negligible, can the medium be approximated as linear. If, however, the fields are sufficiently strong, then the field dependence in ϵ can no longer be neglected. The nonlinear response of the medium changes drastically the characteristics of wave propagation in the medium, and gives rise to many interesting nonlinear optical phenomena.

As an example, suppose we can expand $\epsilon(\mathbf{E})$ into a power series of E.

$$\epsilon(\mathbf{E}) = \epsilon_o + \epsilon^{(2)} :\mathbf{E} + - - - . \quad (1)$$

The second term gives rise to a nonlinear polarization $\mathbf{P}^{NL} = \epsilon^{(2)} :\mathbf{EE}/4\pi$. If E is oscillating at a frequency ω, then P^{NL} is oscillating at a frequency 2ω. Since oscillating dipoles radiate, the nonlinear polarization P^{NL} now acts as a source for the generation of the second-harmonic field at 2ω. Usually in the visible, ϵ_o is around 3 esu, and $\epsilon^{(2)}$ ranges from 10^{-4} - 10^{-7} esu. For an input beam of 1 Mwatts/cm^2 corresponding to E ≈ 20 esu, the second-harmonic generated from a 1-cm $LiNbO_3$ crystal ($\epsilon^{(2)}_{zxx} \approx 9 \times 10^{-6}$ esu) can have a maximum intensity around 100 Kwatts/cm^2 which is certainly non-negligible.

Just as in linear optics, there are two types of problems in nonlinear optics. One is to relate the nonlinear optical constants to the microscopic properties of a given medium. Because of the higher-rank tensors, a nonlinear dielectric tensor usually contains more independent elements than the linear one. We can, at least in principle, learn more about the properties of the medium by studying the nonlinear case. For this type of problem, what we often do is to measure the nonlinear optical constants on the one hand and, on the other hand, to calculate them from some microscopic theory. By comparing the theoretical and experimental results, we hope to learn something about the properties of the medium. Recently, there has been considerable interest in developing simple microscopic theories for evaluating lower-order nonlinear optical constants. In this respect, the chemical bond theory developed by Phillips, Van Vechten, Levine, etc.[2] has turned out to be most successful. Here, since we already have several papers on the subject in the conference proceedings, we shall not dwell on it further.

The other type of problem in nonlinear optics is to study how the light waves propagate in a nonlinear medium with given nonlinear optical constants. This type of problem is basically more important. For example, in order to deduce the nonlinear optical constants of a medium from measurements, one must first understand nonlinear wave propagation in such a medium. In principle, these problems are fairly simple. All we have to do is to solve the nonlinear wave equations. Unfortunately, nonlinear wave equations are generally difficult to solve. Specific approximations must be used for specific problems. Physics actually comes in when we try to use various approximations to find a solution. In this paper, we shall give three examples to illustrate the point. These are the problems which we are presently working on in our laboratory. Therefore, they also serve as examples for a section of the current nonlinear optical research.

PHASE-MATCHED THIRD-HARMONIC GENERATION IN LIQUID CRYSTALS

Harmonic generation in a nonlinear medium is one of the oldest problems in nonlinear optics. Phase-matched third-harmonic generation has been achieved in anisotropic crystals[3] and in dye solutions.[4] Here, we would like to show that it can also be achieved in cholesteric liquid crystals which have some unusual characteristics.[5]

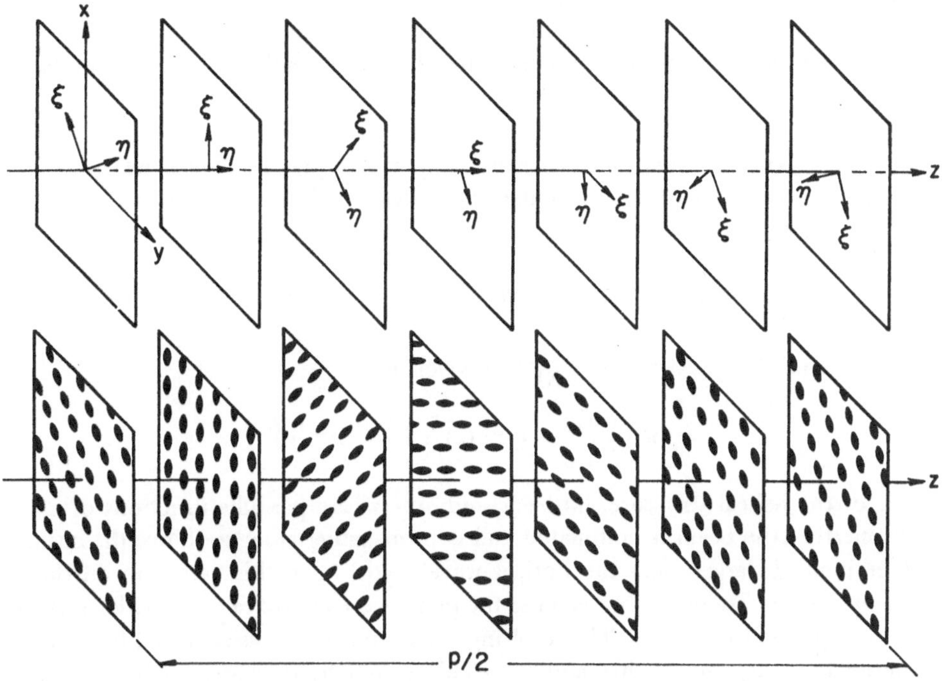

Fig. 1. Arrangement of moldcules in a cholesteric liquid crystal. (Taken from I. G. Chistyakov, Usp. Fiz. Nauk 89, 563(1966) [Translation: Uspekhi 9, 551(1967)]).

For convenience in later discussion, let us first review briefly the general characteristics of a cholesteric liquid crystal.[6] A cholesteric liquid crystal is usually composed of rod-like molecules. Fig. 1 shows the arrangement of molecules in such a medium. In each plane perpendicular to the z-axis, the molecules are all aligned with the principal molecular axes along $\hat{\xi}$ and $\hat{\eta}$. As the plane advances along the z-axis, the direction of molecular alignment gradually rotates, so that the medium acquires an overall helical structure. Referred to the coordinates $\hat{x}$ and $\hat{y}$ in the lab system, the molecular axes $\hat{\xi}$ and $\hat{\eta}$ can be written as

$$\begin{aligned} \hat{\xi} &= \hat{x}\cos(2\pi z/p) + \hat{y}\sin(2\pi z/p) \\ \hat{\eta} &= -\hat{x}\sin(2\pi z/p) + \hat{y}\cos(2\pi z/p) \end{aligned} \tag{2}$$

where p is the helical pitch, which can be varied by varying temperature, composition, external fields, etc. Because of its helical structure, a cholesteric liquid crystal has some interesting optical properties. It behaves like a grating in diffracting light and possesses strong optical activity. Since the pitch is comparable with optical wavelengths, the macroscopic optical dielectric constants now vary periodically along z. In each plane of molecular alignment, the dielectric constants ϵ_ξ and ϵ_η along $\hat{\xi}$ and $\hat{\eta}$ respectively are clearly different. As a result of these characteristic properties, we have found some unusual phase-matching conditions for third-harmonic generation in a cholesteric liquid crystal as we shall see.

From the Maxwell equations, we can easily derive the nonlinear wave equation for generation of the third-harmonic field $\mathbf{E}^{(3\omega)}\exp(-i3\omega t)$ in a cholesteric liquid crystal.

$$[\nabla \times (\nabla \times) - (3\omega/c)^2 \epsilon(z)]\ \mathbf{E}^{(3\omega)} = (3\omega/c)^2\ 4\pi\ \mathbf{P}_{NL}^{(3\omega)} \tag{3}$$

where the nonlinear polarization $\mathbf{P}_{NL}^{(3\omega)}$ is given by

$$4\pi\ \mathbf{P}_{NL}^{(3\omega)} = \epsilon^{(3)}(z) : \mathbf{E}^{(\omega)}\mathbf{E}^{(\omega)}\mathbf{E}^{(\omega)}. \tag{4}$$

Because of the helical structure, both $\epsilon(z)$ and $\epsilon^{(3)}(z)$ are periodic functions of z. We assume that the beam is collimated and is propagating along $\hat{z}$. Since the beam diameter is much larger than the optical wavelength, λ, in this case, the diffraction effect is negligible. Therefore, we can use the plane-wave approximation with $\nabla \times (\nabla \times)$ in Eq. (3) replaced by $-\partial^2/\partial z^2$. The resulting equation is most easily solved by first eliminating the z-dependence in ϵ and $\epsilon^{(3)}$. This is done by transforming to a coordinate system rotating around $\hat{z}$ with a rotational transformation $\mathbf{R}(\theta = 2\pi z/p)$.[5,7] Physically, the helical structure can be considered as a twisted birefringent medium, and the rotational transformation is designed to untwist the helical structure.

In this rotating coordinate system, Eq. (3) now becomes

$$[\partial^2/\partial z^2 - (2\pi/p)^2 + (4\pi/p)\sigma\partial/\partial z + (3\omega/c)^2 \epsilon_T]\mathbf{E}_T^{(3\omega)}$$

$$= -(3\omega/c)^2 \epsilon_T^{(3)} :\ \mathbf{E}_T^{(\omega)}\mathbf{E}_T^{(\omega)}\mathbf{E}_T^{(\omega)} \tag{5}$$

where $\mathbf{E}_T = \mathbf{R}\cdot\mathbf{E}$, $\sigma = \mathbf{R}(3\pi/2)$, ϵ_T and $\epsilon_T^{(3)}$ are now independent of z, and ϵ_T is diagonalized with the principal values ϵ_ξ and ϵ_η along $\hat{\xi}$ and $\hat{\eta}$ respectively. Consider first the linear wave propagation in the medium. The solution can be obtained by neglecting the nonlinear term in the wave equation, and is in the form[7]

$$\mathbf{E}_T^{(\omega)} = \sum_{j=\pm} \mathcal{E}_{Tj}^{(\omega)} \exp[i\kappa_j z - i\omega t] \tag{6}$$

with $\mathcal{E}_{Tj}^{(\omega)} = (\hat{\xi}\,\mathcal{E}_\xi + \hat{\eta}\,\mathcal{E}_\eta)$

$$|\kappa_\pm^{(\omega)}| = (\omega/c)\left\{(\epsilon + \lambda'^2)^2 \pm [(\epsilon+\lambda'^2)^2 - (\epsilon-\lambda'^2)^2 + \alpha^2]^{1/2}\right\}^{1/2}$$

$$\left(\mathcal{E}_\eta^{(\omega)}/\mathcal{E}_\xi^{(\omega)}\right) = if_j^{(\omega)} = i[\kappa_j^2 + (\omega/c)^2(\lambda'^2 + \alpha - \epsilon)]/2\kappa_j(\omega/c)\lambda'$$

where $\epsilon \equiv (\epsilon_\xi+\epsilon_\eta)/2$, $\alpha \equiv (\epsilon_\xi-\epsilon_\eta)/2$, $\lambda' \equiv \lambda/p$, κ_- is negative for $\lambda'^2 > \epsilon_\xi$ and $\mathcal{E}^{(\omega)}_{Tj}$ is independent of z.

If the nonlinear term becomes non-negligible, there is then a continuous transfer of energy from the fundamental to the third-harmonics through nonlinear coupling. As a result, the field amplitudes $\mathcal{E}^{(\omega)}$ and $\mathcal{E}^{(3\omega)}$ should now depend on z. However, physically, since the nonlinear coupling is weak, we expect that the rate of energy transfer is small, and hence

$$|\partial \mathcal{E}^{(3\omega)}_T/\partial z| << |\kappa^{(3\omega)}\mathcal{E}^{(3\omega)}| \text{ and } |\partial^2 \mathcal{E}^{(3\omega)}_T/\partial z^2| << |\kappa^{(3\omega)}\partial \mathcal{E}^{(3\omega)}_T/\partial z|.$$

Equation (5) can therefore be approximated by

$$\begin{aligned}[2i\kappa_j + (4\pi/p)\sigma]\ (\partial/\partial z)\mathcal{E}^{(3\omega)}_{Tj} &= -(3\omega/c)^2 \sum_{k,\ell,m} \epsilon_T^{(3)}: \\ &: \mathcal{E}^{(\omega)}_{Tk}\mathcal{E}^{(\omega)}_{T}\mathcal{E}^{(\omega)}_{Tm}\exp(i\Delta\kappa_{jk\ell m}z)\end{aligned} \tag{7}$$

where $\Delta\kappa_{jk\ell m} = \kappa^{(\omega)}_k + \kappa^{(\omega)}_\ell + \kappa^{(\omega)}_m - \kappa^{(3\omega)}_j$.

Usually, the depletion of the fundamental pump energy by third-harmonic generation is also small. As a good approximation, the fundamental field amplitue $\mathcal{E}^{(\omega)}_T$ can be taken as a constant. It is then easy to solve Eq. (7). We find,[5] assuming $E^{(3\omega)}_T = 0$ at $z = 0$,

$$\begin{aligned}E^{(3\omega)} &= \sum_{j,k,\ell,m} A_j(\epsilon^{(3)}_T)_{jk\ell m}\mathcal{E}^{(\omega)}_{Tk}\mathcal{E}^{(\omega)}_{Tk}\mathcal{E}^{(\omega)}_{Tm}[\sin(\Delta\kappa_{jk\ell m}z/2)]/[\Delta\kappa_{jk\ell m}/2]\ \cdot \\ &\cdot \exp[i(\kappa^{(3\omega)}_j + i\Delta\kappa_{jk\ell m}/2)z]\end{aligned} \tag{8}$$

where $(\epsilon^{(3)}_T)_{jk\ell m} = \mathcal{E}^{(3\omega)}_{Tj} \cdot \epsilon^{(3)}_T : \mathcal{E}^{(\omega)}_{Tk}\hat{\epsilon}^{(\omega)}_{T\ell}\hat{\epsilon}^{(\omega)}_{Tm}$

$$A_j = i(3\omega/c)^2/2[\kappa^{(3\omega)}_j - 4\pi f_j^{(3\omega)}/p(1+f_j^2)].$$

It is clear from the above equation that the third-harmonic intensity will be a maximum if the momentum or phase matching condition $\Delta\kappa_{jk\ell m} = 0$ is satisfied. This is possible in a cholesteric liquid crystal for several combinations of j, k, ℓ, and m, since $\kappa_\pm$ in Eq. (6) can be varied by varying the helical pitch p. Near phase-matching, $\Delta\kappa_{jk\ell m} \approx 0$, the generated third-harmonic power is given by $(c/2\pi n) \int |E^{(3\omega)}|^2\, dx\, dy$, where

$$\begin{aligned}|E^{(3\omega)}|^2 &= |E_T^{(3\omega)}|^2 \\ &\cong |A_j(\epsilon^{(3)}_T)_{jk\ell m}E^{(\omega)}_{Tk}E^{(\omega)}_{T\ell}E^{(\omega)}_{Tm}|^2 [\sin^2(\Delta\kappa_{jk\ell m}z/2)]/[\Delta\kappa_{jk\ell m}/2]^2.\end{aligned} \tag{9}$$

If $\epsilon_T^{(3)}$ for the medium is known, we can predict the third-harmonic power generated at phase-matching by a given input beam. Conversely, measurements of the third-harmonic output near phase-matching should, in principle, enable us to determine the nonlinear dielectric constant $(\epsilon_T^{(3)})_{jk\ell m}$ from Eq. (9).

The experimental setup for harmonic generation experiments is fairly standard.[8] In our case, a mode-locked Nd-glass laser beam was normally incident on a liquid crystal sample whose helical axis was made parallel to the direction of beam propagation. The third-harmonic output was then detected by a photomultiplier with proper filtering. In order to achieve phase-matching, the helical pitch of the sample was tuned by adjusting the sample temperature. We have observed in our experiments most of the predicted phase-matching conditions.[9] Here, we shall mention only two of them.

In the first case, a cholesteric mixture composed of 1.75 parts of cholesteryl chloride and 1 part of cholesteryl myristate was used.[5] Knowing ϵ and α in Eq. (6) from measurements, we calculated $\kappa_\pm$ as functions of the pitch p. We then predicted that the phase-matching condition $\Delta\kappa_{++++} = 0$ can be satisfied with $p = \pm 17.3\ \mu m$ where + and - refer to right and left helixities. The sample temperatures corresponding

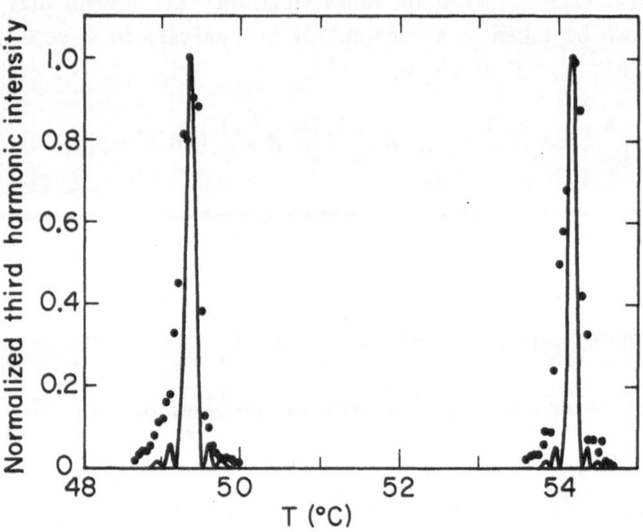

Fig. 2. Normalized third-harmonic intensity vs temperature near the phase matching temperatures for the mixture of 1.75 cholesteryl chloride and 1 cholesteryl myristate, in a cell 130 μm thick. The peak at the lower temperature (corresponding to left helica structure) is generated by left-circularly polarized fundamental waves and the one at the higher temperature by right-circularly polarized fundamental waves. The solid line is the theoretical phase-matching curve and the dots are experimental data points. The uncertainty in the experimental third-harmonic intensity is about 20%.

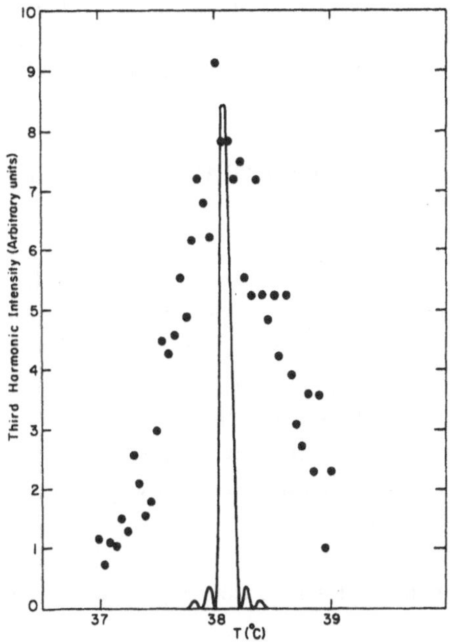

Fig. 3. Backward-propagating third-harmonic intensity vs temperature near the phase-matching temperature for the mixture of 30% cholesteryl chloride, 35% cholesteryl nonanoate, and 35% cholesteryl oleyl carbonate, in a cell of 130 μm thick. The solid curve is the theoretical phase-matching curve assuming a monochromatic incoming field.

to these pitch values are 49.4 and 54.2°C respectively. The third-harmonic waves in this case propagate in the same direction as the fundamental waves. In Fig. 2, the observed third-harmonic intensity is plotted as a function of temperature. It is seen that the phase-matching peaks indeed appear at the predicted temperatures.

In the second case, a mixture of 30% cholesteryl chloride, 35% cholesteryl nonanoate, and 35% cholesteryl oleyl carbonate was used.[9] The phase-matching condition $\Delta\kappa_{----} = 0$ can now be satisfied with $p = 0.47$ μm at a sample temperature of 38.2°C. However, in this case, $\kappa_-^{(\omega)}$ and $\kappa_-^{(3\omega)}$ correspond to fundamental and third-harmonic waves propagating in opposite directions in the laboratory frame. The observed third-harmonic phase-matching peak again appeared at the predicted temperature, as shown in Fig. 3. Here, the observed peak is much broader than the theoretical one calculated under the assumption that the fundamental laser beam is monochromatic. In the experiments, the mode-locked laser pulses had roughly a 10 cm^{-1} band of frequencies. Since different frequency components lead to slightly different phase-matching temperatures here, the overall phase-matching curve was now appreciably broadened. One may have noticed that in this case, with the fundamental and the third-harmonic waves propagating in opposite directions, the linear momentum does not seem to be conserved in the laboratory frame. However,

References p. 31

the momentum difference is actually compensated by the lattice momentum.[9,10] This is equivalent to an umklapp process in a crystal. The unit lattice momentum here is given by $2\pi/p$, as a result of the helical structure of the liquid crystals.

FAR-INFRARED GENERATION BY ULTRASHORT PULSES IN NONLINEAR CRYSTALS

Far-infrared radiation can be generated by the difference-frequency generation process in a nonlinear crystal.[11] Two laser beams with two different frequencies can be used as the primary pumping fields.[12] An ultrashort pulse, however, contains a broad band of frequencies. For example, a 1-psec. pulse has a bandwidth of about 15 cm^{-1}. In a nonlinear crystal, the different frequency components in the band can beat with one another to generate difference-frequency fields in the far-infrared. From the practical point of view, this nonlinear optical process has the potential of providing a powerful, far-infrared source.

The wave equation for the far-infrared field **E** generated by beating of the laser field $\mathbf{E}_\ell$ has the form

$$\nabla \times (\nabla \times \mathbf{E}) + (\partial^2/c^2\partial t^2)\mathbf{D} = -4\pi \mathbf{P}_{NL}^{(2)} \tag{10}$$

where $\mathbf{D} = \epsilon \cdot \mathbf{E}$ and $4\pi \mathbf{P}_{NL}^{(2)} = \epsilon^{(2)} \cdot \mathbf{E}_\ell \mathbf{E}_\ell$ if we neglect dispersion at the far infrared. One would expect that the solution of the above equation is quite similar to the case of harmonic generation. However, since the far-infrared frequencies are now comparable with the inverse of the ultrashort pulse width, we can no longer neglect the time-derivative of the field amplitude in the wave equation as we did in the case of harmonic generation. The approximation of replacing $\nabla \times (\nabla \times \mathbf{E})$ by $-(\partial^2/\partial z^2)\mathbf{E}$ also breaks down because the far-infrared wave-lengths are comparable with the transverse dimension of the beam. Diffraction of the far-infrared now becomes important, especially in the low-frequency limit. As a result, the wave vectors of the far-infrared radiation generated in the nonlinear crystal span over a large cone. Because the refraction index at the far-infrared is usually large (6.6 for ordinary ray in $LiNbO_3$) reflection and refraction at the boundaries are also important. For a crystal slab with the beam propagating perpendicular to the plane surfaces, the far infrared generated with a wave vector at a large angle is totally reflected at the boundary surfaces and can never get out of the crystal. The far infrared generated at small angles sees a strong Fabry-Perot interference. It also suffers strong refraction at the boundaries, so that the far-infrared radiation is spread into a broad cone outside the crystal.

We then realize that, in the present case, we must solve the complete 4-dimensional equation in (10) with proper boundary conditions. The calculation is clearly much more difficult than that for harmonic generation. However, with appropriate approximations, we can greatly simplify the solution.[13] Because of the limited space here, we shall not go into the discussion of how we actually solve the equation. We have calculated far-infrared spectra generated in $LiNbO_3$ by Nd mode-locked pulses under different conditions. Figure 4 gives one example.[13]

In Fig. 4, the far-infrared is generated in a 1 mm slab of $LiNbO_3$ through the dielectric constant $\epsilon^{(2)}_{zzz}$ by pulses with a 2-psec. pulse width. This spectrum is quite easy to understand physically. The dashed curve is calculated without boundary conditions. The three peaks at 5, 8.4, and 11.8 cm^{-1} come in as the secondary peaks of the phase-matching curve with phase matching occurring at zero frequency. They correspond to $\Delta k = 3\pi/2$, $5\pi/2$, and $7\pi/2$ respectively, where Δk is the momentum mismatch in the difference-frequency generation process. There would be the major phase-matching peak at zero frequency, if it were not for the fact that low-frequency fields do not radiate efficiently. The cutoff of the phase-matching curve at the low-frequency end due to radiation and diffraction effects gives rise to the first peak at 2 cm^{-1} in Fig. 4. With the boundary conditions included, the curve is now modified by the Fabry-Perot interference pattern, as shown by the solid curve with spikes. However, in actual experiments, the spectrometer has limited resolution. If the resolution is larger than the width of the spikes, then after convoluting the solid curve in Fig. 4 with the slit function, the spectrum would again have the form of the dashed

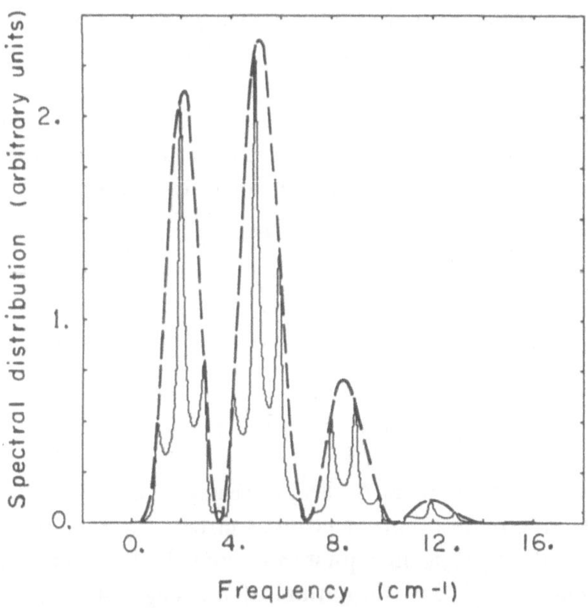

Fig. 4. Calculated spectrum of the far-infrared generated by a 2-psec. short Nd laser pulse in a 1-mm slab of $LiNbO_3$, oriented with the z-axis parallel both to the laser polarization and to the surfaces of the slab. The dashed and the solid curves were calculated with and without the boundary conditions respectively.

References p. 31

curve. This has been shown to agree well with the experimental results.[14] A Nd mode-locked laser generating pulses with peak power of a few Gwatts was used in the experiments. The spectrum of the far-infrared output was analyzed by a Michelson interferometer. Normalization against fluctuations was achieved by the usual method of splitting the far-infrared output into two beams. Details of the experiments will be published elsewhere.[14]

Instead of $\epsilon^{(2)}_{zzz}$, we can also use the dielectric constant $\epsilon^{(2)}_{yyz}$ of $LiNbO_3$ by properly orienting the crystal with respect to the laser polarization. Phase matching of the collinear difference-frequency generation now occurs at different frequencies depending on the orientation of the crystal. Our calculations show that, in this case, the far-infrared spectra consist of sharp phase-matching curves. The positions of the phase-matching peaks can be tuned by rotating the crystal. Experimental results are again in agreement with theoretical prediction.

SELF-FOCUSING OF LIGHT IN LIQUIDS

As a last example, let us discuss how the propagation of a laser beam in a medium can be vastly different from the linear wave propagation because the refractive index of the medium depends on the laser intensity. This is one of the most fascinating, but also the most difficult problems in nonlinear optics. Only after many years of struggle, we now begin to gain insight in the solution of the problem.

Let us first have a qualitative discussion on the self-focusing phenomenon. Consider an isotropic medium, such as a liquid, as an example. Since the medium has an inversion symmetry, its refractive index (or dielectric constant) can be written as

$$n(E) = n_o + \Delta n \tag{11}$$

where we have, in the steady-state,

$$\Delta n = \Delta n_o = n_2|E|^2 + n_4|E|^4 + \cdots .$$

Physically, the induced refractive index Δn arises because of molecular reorientation and redistribution, electrostriction, and electronics effects. In experiments with Q-switched laser pulses propagating in liquids, the contribution from molecular reorientation often dominates.[15] The magnitude of n_2 ranges from 10^{-11} to 10^{-13} esu. In most media, the induced Δn is a positive quantity.

We now propagate a laser beam into such a medium. Assume that the collimated laser beam has initially a plane wavefront, but has a finite cross-section with a

Gaussian intensity profile. Then, as the beam propagates into the medium, the central part of the beam sees a larger refractive index than the edge, and therefore travels at a slower speed than the edge. As a result, the wavefront of the beam gets distorted more and more as the beam propagates (see Fig. 5). Since propagation of energy must be normal to the wavefront, the beam now appears focused by itself. The beam distortion, being a nonlinear, cumulative effect, makes the beam focus rather suddenly to a small spot of a few microns in radius.

From the above description, it is seen that the finite beam cross-section is important for self-focusing, and in the focusing region, both the intensity and the phase of the optical field change rapidly, so that they cannot be considered as subject to small perturbation. The wave equation for this problem can be written as

$$[\nabla \times (\nabla \times) + (\partial^2/c^2\partial t^2)(n_0+\Delta n)^2]\mathbf{E} = 0 \qquad (12)$$

where Δn, if it is mainly due to molecular orientation, can be assumed to obey a simple relaxation equation

$$(\partial/\partial t + 1/\tau)\Delta n(t) = \Delta n_0/\tau. \qquad (13)$$

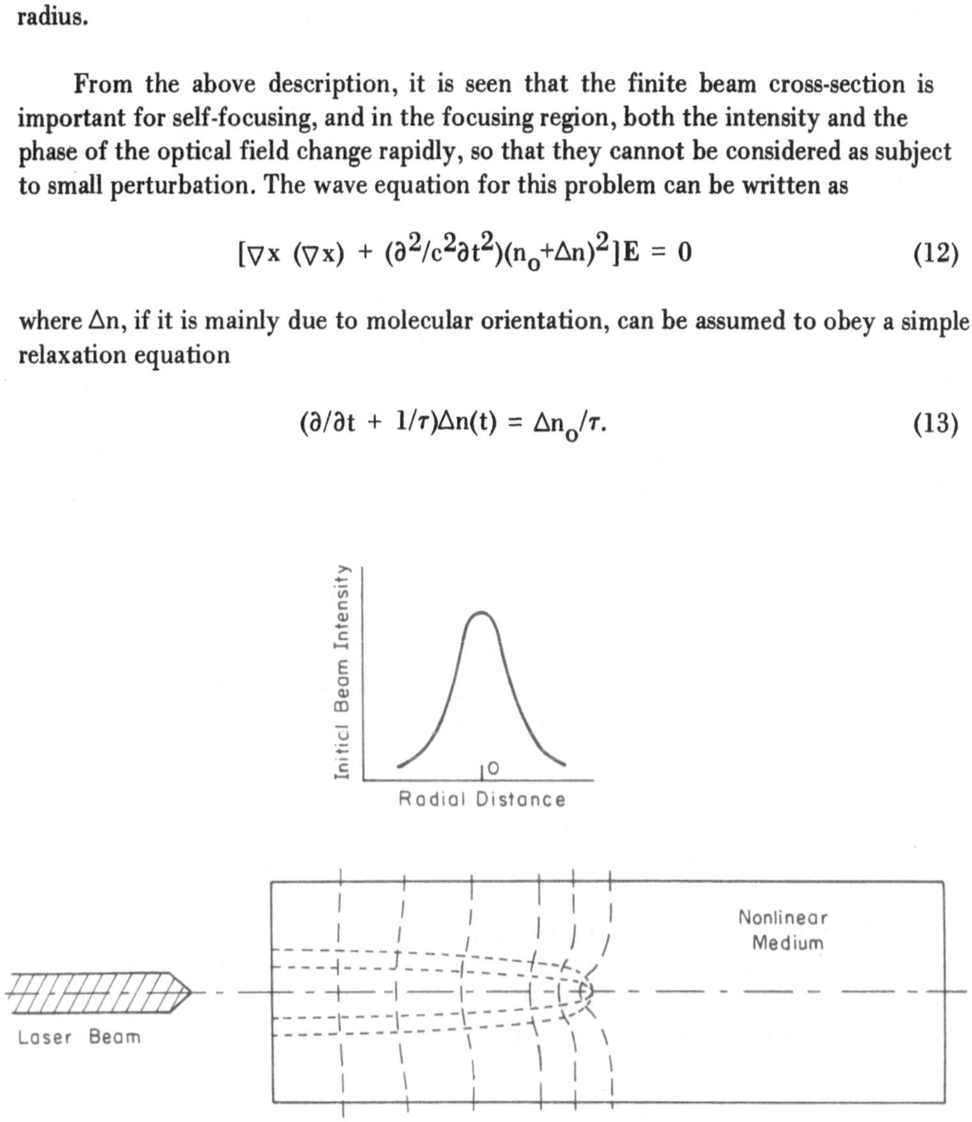

Fig. 5. Description of self-focusing of a laser beam in a nonlinear medium.

References p. 31

For ordinary, non-viscous liquids, the orientational relaxation time τ is of the order of a few picoseconds. The solution of the above equations, being extremely nonlinear, is clearly very difficult. In order to facilitate the solution by approximations, we must rely very much on our physical judgment.

When the laser pulse is much longer than the relaxation time, the time derivative in Eq. (13) can be neglected. If the beam intensity is not exceptionally high, then we can write

$$\Delta n(t) \cong n_2 |E(t)|^2.$$

Assuming a steady-state case with the above nonlinearity, Kelley[16] has found that a beam with power P would self-focus to a point at

$$z_f \cong K/(P^{1/2} - P_{cr}^{1/2}) \qquad (14)$$

$$K = (n/4)(a^2/f)(c/n_2)^{1/2}$$

where P_{cr} is the critical power for self-trapping,[17] a is the beam radius, and f is a parameter of the order of 1.[18] As seen from Eq. (14), there is a critical power P_{cr} below which the beam cannot self-focus ($z_f < 0$). This happens because diffraction overcomes the self-focusing action. Eq. (14) has actually been verified by expieriments.[18]

However, for a pulsed laser beam, the input power P is a function of time. Consequently, from Eq. (14), z_f should also be a function of time, and hence, the focal spot should appear moving along the axis.[19] If we take a time-integrated photograph of the self-focused beam from the side, the continuous series of focal spots would then appear as a bright filament. This is the so-called small-scale filament many research groups have observed.[20]

Let us see more quantitatively how the focal spot moves along the axis. Suppose we have a 1 -nsec. input pulse with a given pulse shape as shown in Fig. 6. At t_A, the light beam with a power $P(t_A)$ enters the medium. It then propagates in the medium with the light velocity c/n_o (along the dashed line in Fig. 6), and finally self-focuses at z_A. At an earlier time, the light beam entering the medium has a smaller power, and it therefore self-focuses at a distance farther away. With the help of Eq. (14), we can now plot out the whole curve describing the position of the focal spot as a function of time.[21]

If ℓ is the length of the medium, the focal spot should first appear inside the medium at z_D. It then immediately splits into two. One first moves backward towards the laser to z_B and then turns around and moves forward. The other simply moves all

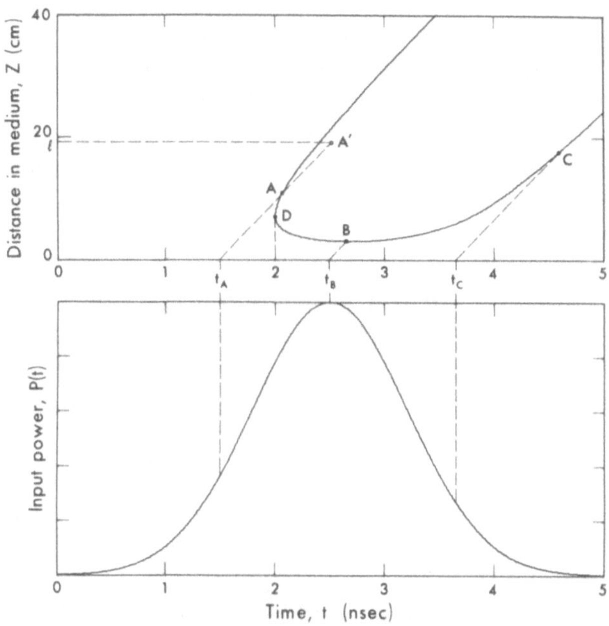

Fig. 6. The lower trace describes the input power P(t) as a function of time t. The peak power is 42.5 Kwatts and the half width at the e^{-1} point is 1 nsec. The upper trace, calculated from Eq. (1), describes the position of the focal spot as a function of time. The values of P_{cr} and K used are 8 Kwatts and 11.6 cm/(KW)$^{1/2}$ respectively, which correspond roughly to an input beam of 400 μ in diameter propagating in CS_2. The dotted lines, with the slope equal to the light velocity in CS_2, indicate how the light propagates along the z-axis at various times.

the way to the end of the medium. Here, let us consider only the second branch since it is more interesting, although experiments have proved the existence of both branches.[21-23] As shown in Fig. 5, the forward-moving branch has a slope larger than c/n_0, and therefore the focal spot must have a velocity faster than the velocity of light in the medium. This can be demonstrated quite easily by experiments. We can use liquid as the nonlinear medium, and immerse two beam splitters in it. Each beam splitter couples out a short light pulse when the focal spot strikes the beam splitter. Knowing the distance between the two beam splitters, we can find, by measuring the time separation between the two pulses, the average velocity for the focal spot to travel from one beam splitter to the other. This was actually done.[23] In one case, for example, the two beam splitters immersed in toluene were separated by 15 cm. It would take 0.75 nsec. for light to travel such a distance, but we measured a time of 0.25 nsec. between the two pulses. That the focal spot can move faster than light does not violate the law of relativity, since the continuous series of focal spots is actually formed by continuous focusing of different parts of the input pulse. A direct consequence of the fact that the focal spot moves faster than light is the possible formation of trapped filaments of light.[23] Because of its high intensity, the moving focal spot leaves a track of induced Δn, which lasts over a period of the order of the relaxation time τ. This channel of Δn forms an optical waveguide traveling along with the focal spot. The self-focused light trailing after the forward-moving focal spot can then be partially trapped in the waveguide for a certain distance. The trapping length depends on the relative velocity of the focal spot with respect to light. If the focal spot has a velocity close to the light velocity, so that it keeps in step with the light for a long distance, then a long trapped filament can result. Since the trapping length varies with time as a result of the focal-spot movement, light emitted from the filament at the end of the medium is strongly phase-modulated, and hence shows a remarkable spectral broadening with semi-periodical structure (a broadening of a few hundred to a few thousand cm^{-1} against a linewidth of less than 1 cm^{-1} for the input laser beam).[24] The model is also successful in explaining many anomalous observations on small-scale filaments.[23] However, we shall not go into detailed discussion here.

CONCLUSION

The above three examples illustrate some general aspects of nonlinear optical research. There are, of course, many other intersting nonlinear optical problems, such as stimulated light scattering, nonlinear propagation of ultrashort pulses, self-induced transparency effect, etc. The field of nonlinear optics is still growing. So far, research on nonlinear optics has been mainly on investigation of basic phenomena. However, quite a few important applications of nonlinear optics have already been found. The coherent, tunable light source is just one example. It is hopeful that someday we may find nonlinear optics at least as useful as linear optics both in practical applications and in studying properties of matter.

ACKNOWLEDGMENTS

This work was performed under the auspices of the U.S. Atomic Energy Commission.

REFERENCES

1. *For detailed discussion see, for example. N. Bloembergen, Nonlinear Optics (W. A. Benjamin, Inc., New York 1965).*
2. *J. C. Phillips, Phys. Rev. Letters 20, 550(1968).*
   J. C. Phillips and J. A. Van Vechten, Phys. Rev. 183, 709(1969).
   J. A. Van Vechten, Phys. Rev. 182, 891; 187, 1007(1969).
   B. F. Levine, Phys. Rev. Letters 22, 787(1969); 25, 440(1970).
3. *P. D. Maker and R. W. Terhune, Phys. Rev. 137, A801(1965).*
4. *P. P. Bey, J. F. Giuliani, and H. Rabin, Phys. Rev. Letters 19, 819(1967).*
5. *J. W. Shelton and Y. R. Shen, Phys. Rev. Letters 25, 23(1970).*
6. *See, for example, G. W. Gray, Molecular Structure and the Properties of Liquid Crystals (Academic Press, New York, 1962) p. 39.*
7. *H. N. de Vries, Acta Cryst. 4, 219(1951).*
8. *See, for example, R. K. Chang and L. K. Galbraith, Phys. Rev. 171, 993(1968).*
9. *J. W. Shelton and Y. R. Shen, presented at the Sixth International Quantum Electronics Conference, Kyoto, Japan, 1970) paper 11-3; Phys. Rev. Letters 26, 538 (1971)*
10. *N. Bloembergen (private communication).*
11. *F. Zernike and P. R. Berman, Phys. Rev. Letters 15, 999(1965).*
12, *D. W. Faries, K. A. Gehring, P. L. Richards, and Y. R. Shen, Phys. Rev. 180, 363(1969).*
   N. Van Tran and C. K. N. Patel, Phys. Rev. Letters 22, 463(1969).
   T. Yajima and K. Inoue, IEEE J. Quantum Electron. QE-5, 140(1969).
13. *J. R. Morris and Y. R. Shen, Optics Comm. (to be published).*
14. *K. H. Yang, Y. R. Shen, J. W. Shelton, and P. L. Richards (to be published). A preliminary report was presented at the Sixth International Quantum Electronic Conference at Kyoto, Japan, 1970; paper 8-6.*
15. *Y. R. Shen, Phys. Letters 20, 378(1966).*
16. *P. L. Kelley, Phys. Rev. Letters 15, 1005(1965).*
17. *R. Y. Chiao, E. Garmire, and C. H. Townes, Phys. Rev. Letters 13, 479(1964).*
18. *C. C. Wang, Phys. Rev. Letters 16, 344(1966).*
19. *A. L. Dyshko, V. N. Lugovoi, and A. M. Prokhorov, Zh. Eksperim. i Theo. Fiz. Pis'ma Redakt. 6, 655(1967), [Translation: JETP Letters 6, 146(1967)]; V. N. Lugovoi and A. M. Prokhorov, ibid. 7, 153(1968) [Translation: JETP Letters 7, 344(1966)].*
20. *R. Y. Chiao, E. Garmire, and C. H. Townes, Phys. Rev. Letters 13, 479(1964); R. Y. Chiao, M. A. Johnson, S. Krinsky, H. A. Smith, C. H. Townes, and E. Garmire IEEE J. Quantum Electron. QE-2, 467(1966); R. G. Brewer, J. R. Lifshitz, E. Garmire, R. Y. Chiao, and C. H. Townes, Phys. Rev. 166, 326(1968).*
21. *M. M. T. Loy and Y. R. Shen, Phys. Rev. Letters 22, 994(1969).*
22. *V. V. Korobkin, A. M. Prokhorov, R. V. Serov, and M. Ya. Shchelev, ZHETF. Pis. Red. 11, 153(1970); [Translation: JETP Letters 11, 94(1970)].*
23. *M. M. T. Loy and Y. R. Shen; presented at the Sixth International Quantum Electronic Conference, Kyoto, Japan, 1970, paper 9-7; Phys. Rev. Letters 25, 1333 (1970).*
24. *F. Shimizu, Phys. Rev. Letters 19, 1097(1967);*
   R. G. Brewer, Phys. Rev. Letters 19, 8(1967);
   A. C. Cheung, D. M. Rank, R. Y. Chiao, and C. H. Townes, Phys. Rev. Letters 20, 786(1968).

REFERENCES

1. [illegible] Pseudopotentials [illegible] (W. A. Benjamin, New York, 1966).
2. J. C. Slater, Phys. [illegible] 36, 57 (1930).
3. J. C. Phillips and L. Kleinman, Phys. Rev. 116, 287 (1959).
4. [illegible] Phys. Rev. 122, [illegible] (1961).
5. [illegible] Phys. Rev. Letters [illegible] (19[illegible]).
6. [illegible] Solid State Physics 24, 249 (1970).
7. [illegible] Phys. Rev. Letters [illegible] (19[illegible]).
8. [illegible] Phys. Rev. Letters [illegible] (19[illegible]).
9. [illegible]
10. [illegible] Phil. Mag. [illegible] (19[illegible]).
11. [illegible] Phys. Rev. [illegible] (19[illegible]).
12. [illegible] Phys. Rev. [illegible] (19[illegible]).
13. [illegible] Phys. Rev. Letters [illegible] (19[illegible]).
14. [illegible] (private communication).
15. [illegible] Phys. Rev. [illegible] (19[illegible]).
16. [illegible] Phys. Rev. [illegible] (19[illegible]).
17. [illegible] Phys. Rev. Letters [illegible] (19[illegible]).
18. [illegible] (19[illegible]).
19. [illegible] (to be published).
20. [illegible] report was presented at [illegible] (19[illegible]).
21. [illegible] Phys. Letters [illegible] (19[illegible]).
22. [illegible] (19[illegible]).
23. [illegible] Phys. Rev. Letters [illegible] (19[illegible]).
24. [illegible] Phys. Rev. Letters [illegible] (19[illegible]).

LIGHT SCATTERING BY COLLECTIVE EXCITATIONS IN DIELECTRICS AND SEMICONDUCTORS

E. BURSTEIN and A. PINCZUK

Physics Department and Laboratory for Research on the Structure of Matter,

University of Pennsylvania, Philadelphia, Pennsylvania 19104*

INTRODUCTION

The early experimental and theoretical investigations of spontaneous Raman scattering in solids were concerned with first and second order Raman scattering by lattice vibrations which provided information about the frequencies and symmetries of the long wavelength ($q \approx 0$) modes as well as information about modes at two phonon density of states "critical points" in the Brillouin zone. Largely, as a result of the availability of cw lasers, covering a range of frequencies extending from the infrared to the ultraviolet, and improved photo-detectors and associated electronics, these investigations have now been extended to essentially all of the other "low energy" excitations in solids, namely to plasmons and their linearly coupled modes with LO phonons, polaritons (coupled TO phonon-photon modes), excitons, magnons and their linearly coupled modes with acoustic phonons, single particle electron-hole pair excitations, and vibrational and electronic excitation of impurities.[1] Furthermore, they have been extended to opaque solids, i.e., metals and narrow gap semiconductors.

The scattering (diffraction) of light by the temporal and spatial fluctuations of the electric susceptibility which are induced by elementary excitations[2] provides information about the symmetry and wave vector dependent frequency of the excitation as well as about their non-linear interaction with EM waves. We present here a review of the macroscopic (phenomenological) and microscopic theories of spontaneous Raman scattering of light by optical phonons and their linearly coupled modes with plasmons and photons and a discussion of recent experimental and theoretical developments dealing with the dependence of the scattering intensity on scattering wave vector (*spatial dispersion*), on frequency of the incident EM radiation

References pp. 77-79

(*resonance enhancement*) and on externally applied forces (*morphic effects*). In doing so, we will be concerned more with the nature of the non-linear interactions of the collective excitation with EM modes, than with the information which the Raman scattering spectra provide about the excitations themselves.

PHENOMENOLOGICAL THEORY OF LIGHT SCATTERING

For our purposes we shall consider the incident EM radiation, the scattered EM radiation, and the elementary excitations to be plane waves. The incident radiation is characterized by frequency ω_i, wave vector $\mathbf{k}_i$, polarization vector $\hat{e}_i$ and thermal occupation number $\bar{n}(\omega_i) = [\exp(h\omega_i/kT) - 1)^{-1}$; the scattered radiation is characterized by ω_s, $\mathbf{k}_s$, $\hat{e}_s$ and $\bar{n}_s(\omega_s)$; and the elementary excitation is characterized by ω_j, $\mathbf{k}_j$, $\hat{e}_j$, and $\bar{n}_j(\omega_j)$ [Fig. (1)]. The elementary scattering process involves the annihilation of an incident photon, the emission or annihilation of one quantum of elementary excitation and the emission of a scattered photon. It is characterized by a *scattering frequency* $\omega_i - \omega_s = \Omega$ and a *scattering wave vector* $\mathbf{q} = \mathbf{k}_i - \mathbf{k}_s$. The requirement that energy (frequency) and momentum (wave vector) be conserved in the scattering process is expressed by $\Omega = \pm\omega_j$ and by the Bragg relation $\mathbf{q} = \pm\mathbf{k}_j$, where the (-) sign corresponds to a Stokes scattering process in

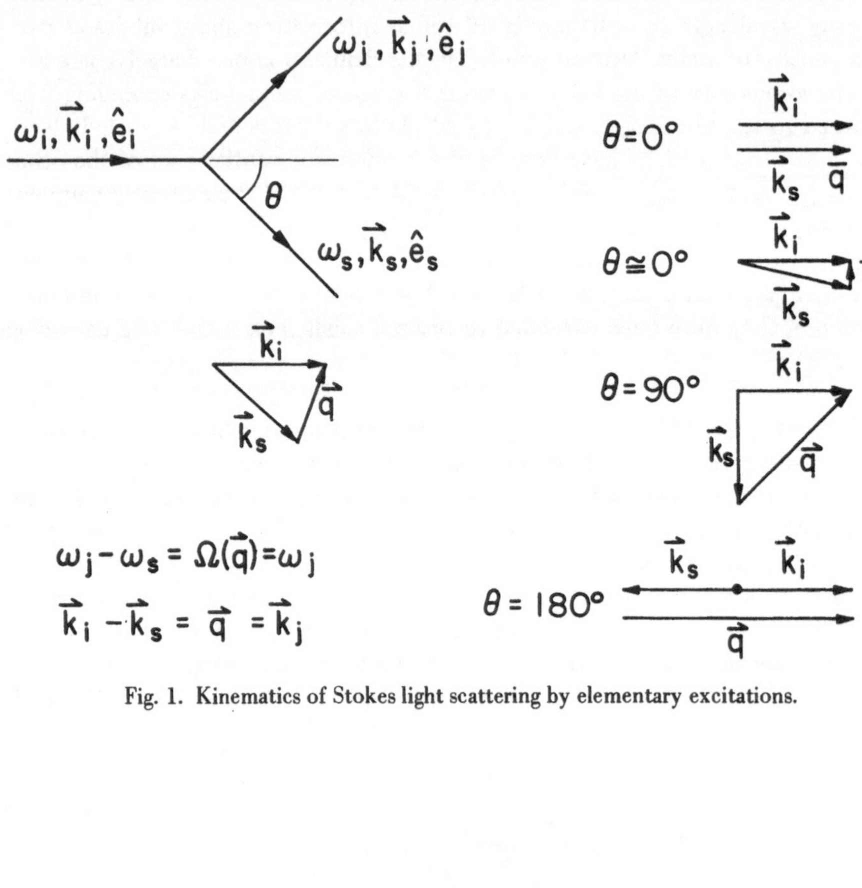

Fig. 1. Kinematics of Stokes light scattering by elementary excitations.

which a quantum of excitation is emitted and the (+) sign corresponds to an anti-Stokes scattering process in which a quantum of excitation is annihilated. By varying the scattering angle θ between 0° (forward scattering) and 180° (backward scattering), it is possible to vary the magnitude of $\mathbf{q}$ from 0 in forward scattering to 10^6 cm^{-1} in backward scattering. It is therefore possible under conditions of wave vector conservation ($\mathbf{k}_j = \mathbf{q}$) to observe scattering by excitations with wave vectors in the range $0 \ll k_j \ll 10^6$ cm^{-1}. Although this range of wave vector is small compared to a reciprocal lattice vector ($K \approx 10^8$ cm^{-1}), it represents an important range for a number of elementary excitations which is not otherwise readily accessible.

It should be noted that there are spontaneous scattering phenomena, such as impurity induced Raman scattering by optical phonons,[3] in which momentum is not conserved and $\mathbf{k}_j \neq \mathbf{q}$. In such cases the excitations that participate in the scattering can have wave vectors ranging from $\mathbf{k}_j = 0$, to $\mathbf{k}_j = \mathbf{K}$.

Higher order scattering processes are also observed in which two or more quanta of elementary excitation participate in the elementary scattering process.[4] In the case of second order (two quanta) scattering, the conservation of frequency and wave vector is given by $\Omega = \pm (\omega_j \pm \omega_j')$ and $\mathbf{q} = (\mathbf{k}_j \pm \mathbf{k}_j')$. Accordingly, the magnitude of $\mathbf{k}_j$ (and $\mathbf{k}_j'$) of the two elementary excitation participating in the scattering process can range from $\mathbf{k}_j \approx 0$ to $\mathbf{k}_j = \mathbf{K}$.

We consider next the factors that determine the scattering efficiency of a type j elementary excitation (ω_j, $\mathbf{k}_j$). The "first-order" spontaneous scattering efficiency (defined as the ratio of the number of EM modes scattered into unit solid angle per unit time per unit frequency interval, to the number of incident EM modes crossing unit area of the scattering volume in unit time) can be expressed phenomenologically in the form[5]

$$S_j^{(1)}(\mathbf{q}) = (\omega_i/\omega_s A)\, d^2\sigma_j^{(1)}(\mathbf{q})/d\omega_s d\Theta = (\omega_s/c)^4 VL(\hat{e}_i \cdot \underset{\sim}{\chi}_j^{(1)}(\mathbf{q}) \cdot \hat{e}_s)^2 \quad (2.1)$$

where $d^2\sigma_j^{(1)}(\mathbf{q})/d\omega_s d\Theta$ is the differential scattering cross-section; $V = AL$ is the scattering volume; L is the scattering length;[6] and $\underset{\sim}{\chi}_j^{(1)}(\mathbf{q})$ is the first-order *transition electric susceptibility* tensor for the scattering process. In the case of collective excitations, the "transition" electric susceptibility $\underset{\sim}{\chi}(\omega_i,\omega_s,\mathbf{q})$ can be expressed as a Taylor expansion in powers of the normal coordinates $\underset{\sim}{Q}_j(\mathbf{k}_j)$, of the collective excitations as follows:

$$\underset{\sim}{\chi}(\omega_i,\omega_s,\mathbf{q}) = \underset{\sim}{\chi}(\omega_i{=}\omega_s,\ \mathbf{k}_i{=}\mathbf{q}) + \sum_j \left[d\chi_{is}/dQ_j(\mathbf{k}_j)\right]_{\mathbf{k}_j=\mathbf{q}}\ \underset{\sim}{Q}_j(\mathbf{k}_j)$$

$$+ \sum_{jj'} \left[d^2\chi_{is}/dQ_j(\mathbf{k}_j)dQ_{j'}(\mathbf{k}_{j'})\right]_{\mathbf{k}_j \pm \mathbf{k}_j' = \mathbf{q}}\ \underset{\sim}{Q}_j(\mathbf{k}_j)\underset{\sim}{Q}_{j'}(\mathbf{k}_{j'}) + \ldots$$

$$= \underset{\sim}{\chi}(\omega_i{=}\omega_s,\ \mathbf{k}_i{=}\mathbf{q}) + \underset{\sim}{\chi}_j^{(1)}(\mathbf{q}) + \underset{\sim}{\chi}_{jj'}^{(2)}(\mathbf{q}) + \ldots \qquad (2.2)$$

$\chi_j^{(1)}(\mathbf{q})$ represents the first-order change in the electric susceptibility induced by type j collective excitations, and $\chi_{jj'}^{(2)}(\mathbf{q})$ represents the second order change induced by type j and j′ collective excitations. The first and second order *transition electric susceptibility* tensors can also be expressed, in terms of the Hamiltonian of the medium $\mathcal{H}$, as

$$\chi_j^{(1)}(\mathbf{q}) = \left[\partial^3\mathcal{H}/\partial E_i(\omega_i)\partial E_s(\omega_s)\partial Q_j(\mathbf{k}_j)\right]_{\mathbf{k}_j=\mathbf{q}}\ Q_j(\mathbf{k}_j)$$

$$\chi_{jj'}^{(2)}(\mathbf{q}) = \left[\partial^4\mathcal{H}/\partial E_i(\omega_i)\partial E_s(\omega_s)\partial Q_j(\mathbf{k}_j)\partial Q_{j'})\right]_{\mathbf{k}_j \pm \mathbf{k}_j' = \mathbf{q}}\ Q_j(\mathbf{k}_j)Q_{j'}(\mathbf{k}_{j'}) \qquad (2.3)$$

Since the frequencies of the two electric fields involved, ω_i and ω_s, are different $\underset{\sim}{\chi}_j^{(1)}(\mathbf{q})$ and $\underset{\sim}{\chi}_{jj'}^{(2)}(\mathbf{q})$ are not symmetrical. However, as noted by Kleinman[7] in connection with the non-linear susceptibility of a medium, the transition electric susceptibility tensors are to a good approximation symmetrical in frequency regions where the medium is transparent.

The polarization selection rules for first-order light scattering are determined by

$$\hat{e}_i \cdot \underset{\sim}{\chi}_j^{(1)}(\mathbf{q}) \cdot \hat{e}_s = \sum_{\mu\lambda\nu\sigma} \hat{e}_{i\mu}\ [\chi_j^{(1)}(\mathbf{q})]_{\lambda\nu}\ \hat{e}_{s\sigma}\ \delta_{\mu\lambda}\ \delta_{\nu\sigma} \neq 0 \qquad (2.4)$$

where the subscripts $\mu\lambda\nu\sigma$ designate x, y, z. The symmetry of the collective excitation determines the non-zero components of $\underset{\sim}{\chi}_j^{(1)}(\mathbf{q})$ and these together with the polarization selection rules determine the components of $\hat{e}_i$ and $\hat{e}_s$ for which the scattering efficiency is non-zero.

It should be noted that the scattering of light by a collective excitation mode, the diffraction of light by a collective excitation mode, and the parametric mixing of two EM modes with a collective excitation mode involve essentially the same transition electric susceptibility and associated polarization selection rules. They differ only in the thermal occupation factors and in the fact that it is possible to observe diffraction and parametric mixing processes under conditions in which momentum is not conserved.

In the case of optical phonons, the normal coordinate $\tilde{Q}_j(\mathbf{q})$ corresponds to the relative atomic displacement coordinate $\mathbf{u}_{jo}(\mathbf{q})$; in the case of acoustic phonons it corresponds to the elastic displacement coördinate $\mathbf{w}_{ja}(\mathbf{q})$; and in the case of plasmons it corresponds to the free carrier displacement coordinate $\mathbf{x}_p(\mathbf{q})$. We note however that in the limit $q \to 0$, $\mathbf{w}_{ja}(\mathbf{q})$ represents a uniform translation of the medium which does not affect the electric susceptibility. At finite $\mathbf{q}$ the change in the electric susceptibility, which is induced by acoustic phonons, comes from the displacement gradient $\tilde{\Sigma}_{ja}(\mathbf{q}) = \nabla\, \mathbf{w}_{ja}(\mathbf{q}) = i\mathbf{q}\, \mathbf{w}_{ja}(\mathbf{q})$. A similar situation exists in the case of plasmons since, in the limit $q \to 0$, $\mathbf{x}_P(\mathbf{q})$ represents a uniform translation of the free carriers relative to the fixed charges, which also has no effect on the electric susceptibility. At finite $\mathbf{q}$, the change in the electric susceptibility, which is induced by plasmons, comes from the electron displacement gradient $\tilde{\gamma}_P(\mathbf{q}) = \nabla\, \mathbf{x}_P(\mathbf{q}) = i\mathbf{q}\cdot\mathbf{x}_P(\mathbf{q})$.[8] In cubic crystals $\tilde{\gamma}_P(\mathbf{q}) = \rho_P(\mathbf{q})/n$ where $\rho_P(\mathbf{q})$ is the charge density fluctuation, and n is the free carrier density.

In the case of optical phonons the first-order change in the electric susceptibility $\tilde{\chi}^{(1)}_{jo}(\mathbf{q})$ involves contributions from the macroscopic field and the elastic displacement gradients of the mode as well as from the relative atomic displacements.[5] Thus $\tilde{\chi}^{(1)}_{jo}(\mathbf{q})$ can be expressed in the form

$$\begin{aligned}\tilde{\chi}^{(1)}_{jo}(\mathbf{q}) &= [\partial\chi/\partial u_j(\mathbf{q})]_{E,\Sigma}\, u_{jo}(\mathbf{q}) + [\partial\chi/\partial E(\mathbf{q})]_{u_{jo},\Sigma}\, \tilde{E}_{jo}(\mathbf{q}) + [\partial\chi/\partial\Sigma(\mathbf{q})]_{u_{jo},E}\, \tilde{\Sigma}_{jo}(\mathbf{q}) \\ &= \tilde{a}_j(\mathbf{q})\, \mathbf{u}_{jo}(\mathbf{q}) + \tilde{b}(\mathbf{q})\, \mathbf{E}_{jo}(\mathbf{q}) + \tilde{f}(\mathbf{q})\, \tilde{\Sigma}_{jo}(\mathbf{q}) \qquad (2.5)\end{aligned}$$

where $\mathbf{E}_{jo}(\mathbf{q}) = (dE/d\, u_j)\, \mathbf{u}_{jo}(\mathbf{q})$ and $\tilde{\Sigma}_{jo}(\mathbf{q}) = (d\Sigma/d\, u_j)\, \mathbf{u}_{jo}(\mathbf{q})$ are the electric field and the elastic displacement gradients of the type j optical phonons; $\tilde{a}_j(\mathbf{q})$ is the atomic displacement Raman tensor of the optical phonon at constant $\tilde{E}$ and $\tilde{\Sigma}$, $\tilde{b}(\mathbf{q})$ is the electro-optic tensor at constant $\tilde{u}_{jo}$ and $\tilde{\Sigma}$; and $\tilde{f}(\mathbf{q})$ is the elasto-optic tensor at constant $\tilde{u}_{jo}$ and $\tilde{E}$. We note that $\tilde{E}_{jo}(\mathbf{q})$ is zero for $q \approx 0$ TO phonons. It is also zero for $q \approx 0$ LO phonons which have a zero effective charge. In centrosymmetric crystals $\underset{\sim}{\Sigma}_{jo}(q)$ is zero for odd parity optical phonons. However, even in case of even parity phonons in centrosymmetric crystals and mixed parity optical phonon in crystals lacking a center of symmetry, $\underset{\sim}{\Sigma}_{jo}(q)$ is generally quite small, particularly when the optical phonon frequency is considerably greater than the frequencies of the acoustical phonons having the same wave vector, and can generally be neglected.

On substituting $\underset{\sim}{\chi}^{(1)}_{jo}(q)$ into Eq. (2.1) one obtains the following expression for the Stokes scattering efficiency of type j optical phonons:

$$S^{(1)}_{jo}(q) = (\omega_s^4 VL/c^4)\,|\hat{e}_i \cdot \{\underset{\sim}{a}_j(q) + \underset{\sim}{b}(q)\,(dE/du_{jo}(q))\} \cdot \hat{e}_s|^2\,|\underset{\sim}{u}_{jo}(q)|^2 \qquad (2.6)$$

In the absence of optical phonon damping $|\underset{\sim}{u}_{jo}(q)|^2$ is given by[9]

$$|\underset{\sim}{u}_{jo}(q)|^2 = (\hbar/2NV\overline{m}_{jo}\,\omega_{jo})\,[\overline{n}_{jo}(\omega_{jo}) + 1] \qquad (2.7)$$

where N is the number of unit cells per unit volume; $\overline{m}_{jo}$ is the reduced mass and $\overline{n}_{jo}(\omega_{jo})$ is the thermal occupation number of the optical phonon.

It should be pointed out that $\underset{\sim}{\chi}^{(1)}_{jo}(q)$ can also be expressed in terms of the local electric field, $E^{loc}_{jo}(q)$ of the optical phonon rather than in terms of the macroscopic electric field.[5] Thus, $\chi^{(1)}_{jo}(q)$ can be written in the form

$$\chi^{(1)}_{jo}(q) = [\partial\chi/\partial u_j(q)]_{E_j^{loc}}\, \underset{\sim}{u}_{jo}(q) + [\partial\chi/\partial E_j(q)]_{u_j}\, E_j^{loc}(q) \qquad (2.8)$$

$$= \underset{\sim}{a}_j^{loc}(q)\, u_j(q) + \underset{\sim}{b}_j^{loc}(q) E^{loc}_{jo}(q)$$

where $\underset{\sim}{a}_j^{loc}(q)$ is the atomic displacement Raman tensor at constant local field, and $\underset{\sim}{b}_j^{loc}(q)$ is an electro-optic type Raman tensor characteristic of the type j optical phonons. (We have omitted the contribution to $\underset{\sim}{\chi}^{(1)}_{jo}(q)$ from the elastic strain of phonons.) This expression for $\underset{\sim}{\chi}^{(1)}_{jo}(q)$ is equivalent to the one in which $\underset{\sim}{\chi}^{(1)}_{jo}(q)$ is expressed in terms of the macroscopic field. However, it is not a particularly useful one, since apart from the problem of defining $\underset{\sim}{E}_j^{loc}(q)$, a detailed knowledge of the microscopic properties of the medium is needed to establish its magnitude.

In the case of acoustic phonon $\underset{\sim}{\chi}_{ja}^{(1)}(q)$ involves contributions from the macroscopic electric field and from the relative atomic displacements, as well as from the elastic displacement gradients of the acoustic phonons, and is given by[5]

$$\underset{\sim}{\chi}_{ja}^{(1)}(q) = [\partial\chi/\partial\Sigma(q)]_{E,u_j}\, \underset{\sim}{\Sigma}_{ja}(q) + [\partial\chi/\partial E(q)]_{\Sigma,u_j}\, E_{ja}(q) + \sum_k [\partial\chi/\partial u_k(q)]_{E,\Sigma}\, u_{ka}(q)$$

$$= \underset{\sim}{f}(q)\, \underset{\sim}{\Sigma}_{ja}(q) + \underset{\sim}{b}(q) E_{ja}(q) + \sum_k \underset{\sim}{a}_k(q) u_{ka}(q) \qquad (2.9)$$

where $u_{ka}(q) = [du_k/d\Sigma(q)]\,\underset{\sim}{\Sigma}_{ja}(q)$; $\underset{\sim}{u}_k(q)$ is the atomic displacement coordinate of type k optical phonons; and $E_{ja}(q) = [dE/d\Sigma(q)]\,\underset{\sim}{\Sigma}_{ja}(q)$. The inner displacement coordinate $u_{ka}(q)$ is zero when the corresponding optical phonons have odd parity. The macroscopic electric field $E_{ja}(q)$ is zero in non-piezoelectric crystals. In piezoelectric crystals, $E_{ja}(q)$ is predominantly longitudinal.

The scattering efficiency of type j acoustic phonons is given by

$$S_{ja}^{(1)}(q) = (\omega_s^4 VL/c^4)[\hat{e}_i \cdot \{ \underset{\sim}{f}(q) + \underset{\sim}{b}(q)\, [dE/d\Sigma(q)] + \sum_k \underset{\sim}{a}_k(q)\, [du_k/d\Sigma(q)] \} \cdot \hat{e}_s]^2 \, |\underset{\sim}{\Sigma}_{ja}(q)|^2,$$

$$|\Sigma_{ja}(q)|^2 = q^2\, (\hbar/2\,\rho\, V\, \omega_{ja})\, (\bar{n}_{ja}(\omega_{ja}) + 1) \qquad (2.10)$$

where ρ is the density of the medium.

In the case of polariton (coupled TO phonon-photon) modes (Ω_π, q) in crystals with a single infrared active optical phonon, $\underset{\sim}{\chi}_j^{(1)}(q)$ involves contributions from the macroscopic transverse electric field $E_\pi(q)$ and the macroscopic magnetic field, $H_\pi(q)$ associated with the photon content of the coupled modes and from the relative atomic displacements $u_{j\pi}(q)$ associated with the type j TO phonon content of the coupled modes[10,11]

$$\underset{\sim}{\chi}_{\pi}^{(1)}(q) = [\partial\chi/\partial u_j(q)]_{E,H}\, u_{j\pi}(q) + [\partial\chi/\partial E_T(q)]_{u_j H}\, E_{\pi}(q) + [\partial\chi/\partial H(q)]_{u_j E}\, H_{\pi}(q)$$

$$= \underset{\sim}{a}_j(q) u_{j\pi}(q) + \underset{\sim}{b}_T(q)\, E_{\pi}(q) + \underset{\sim}{h}(q) H_{\pi}(q) \tag{2.11}$$

where $\underset{\sim}{b}_T(q)$ is the transverse electro-optic Raman tensor, and $\underset{\sim}{h}(q)$ is the magneto-optical Raman tensor at constant u_j and E. It should be noted that the transverse electro-optic tensor $\underset{\sim}{b}_T(q) = [\partial\chi/\partial E_T(q)]$ and the longitudinal electro-optic tensor $\underset{\sim}{b}_L(q) = [\partial\chi/\partial E_L(q)]$ are equivalent only in the limit $q \to 0$.

The scattering efficiency of the polaritons is given by

$$S_{\pi}^{(1)}(q) = (\omega_s^4 VL/c^4)\, |\hat{e}_i \cdot \{ \underset{\sim}{a}_j(q)[(e_j^*/\overline{m}_j)/(\omega_j^2(q) - \omega^2)] + \underset{\sim}{b}_T(q) +$$

$$\underset{\sim}{h}(q)\, \epsilon(\omega_{\pi}, q)^{1/2} \} \cdot \hat{e}_s|^2 \;\; |\underset{\sim}{E}_{\pi}(q)|^2 \tag{2.12}$$

where e_j^* and $\overline{m}_j$ are the transverse effective charge and the reduced mass of the type j optical phonon, and $\epsilon(\omega_{\pi}, q)$ is the dielectric constant of the medium at $\omega = \omega_{\pi}$. In the absence of damping, $|\underset{\sim}{E}_{\pi}(q)|^2$ is given by[12,13]

$$|\underset{\sim}{E}_{\pi}(q)|^2 = [4\pi\hbar\omega_{\pi}(q)\, v_g(\omega_{\pi}, q)/c\epsilon^{1/2}(\omega_{\pi}, q)]\, (\overline{n}_{\pi}(\omega_{\pi}) + 1) \tag{2.13}$$

where $v_g(\omega_{\pi}, q)$ is the group velocity of the polariton.

Finally, we note that in the case of coupled LO phonon-plasmon modes (Ω_{λ}, q) in crystals with a single infrared active optical phonon mode, $\underset{\sim}{\chi}_j^{(1)}(q)$ involves contributions from the free carrier displacement gradient, the macroscopic longitudinal electric field content of the mode, and the relative atomic displacements of the coupled modes,[5,14,15] and is given by

$$\underset{\sim}{\chi}_{\lambda}^{(1)}(q) = [\partial\chi/\partial\gamma(q)]_{E,u_j}\, \underset{\sim}{\gamma}_{\lambda}(q) + [\partial\chi/\partial E_L(q)]_{\gamma,u_j}\, E_{\lambda}(q) + [\partial\chi/\partial u_j(q)]_{\gamma,E}\, u_{j\lambda}(q)$$

$$= \underset{\sim}{c}(q)\gamma_{\lambda}(q) + \underset{\sim}{b}_L(q) E_{\lambda}(q) + \underset{\sim}{a}_j(q) u_{j\lambda}(q) \tag{2.14}$$

The tensor coefficients $\underset{\sim}{a}_j(q)$, $\underset{\sim}{b}(q)$, $\underset{\sim}{c}(q)$, $\underset{\sim}{f}(q)$ and $\underset{\sim}{h}(q)$ are dependent on the frequencies of the incident and scattered radiation and on the scattering wave vector. In the long wavelength $(q = 0)$ limit, the electro-optic tensor $\underset{\sim}{b}(q)$ is a polar third rank tensor which has non-zero components only in piezoelectric crystals; the magneto-optic tensor $\underset{\sim}{h}(q)$ is a pseudo third rank tensor which has non-zero components in all crystal classes; the free carrier displacement gradient tensor $\underset{\sim}{c}(q)$ and the strain-optic tensor $\underset{\sim}{f}(q)$ are fourth rank tensors whose non-zero components are determined by the point group symmetry of the medium. The form of the atomic displacement Raman tensor $\underset{\sim}{a}_j(q)$ of an optical phonon is determined by the symmetry of the optical phonon.

The atomic displacement Raman tensor $\underset{\sim}{a}_j(q)$ is, in the limit $q \rightarrow 0$, non-zero only for even parity optical phonons (i.e., only even parity $q \approx 0$ optical phonons can be Raman active). This symmetry selection rule for Raman scattering by optical phonons in centrosymmetric crystals is complementary to the symmetry selection rule for infrared absorption by optical phonons, that the effective charge tensor of optical phonons, $\underset{\sim}{e}_j^*(q)$, is in the limit $q \rightarrow 0$, non-zero only for odd parity optical phonons (i.e., only odd parity optical phonons can be infrared active). In centrosymmetric crystals having all atomic sites at centers of symmetry, such as NaCl and cubic $BaTiO_3$ type crystals, the optical phonons have only odd parity and are Raman inactive. In diamond type crystals, the optical phonons have even parity and are infrared inactive but Raman active. On the other hand, even parity optical phonons are not necessarily Raman active, and odd parity optical phonons are not necessarily infrared active. In crystals lacking a center of inversion, the $q \approx 0$ optical phonons do not have well defined parity and can therefore be both Raman and infrared active.

The q-dependence of the Raman scattering efficiency of a given collective excitation manifests itself most strikingly in situations where one or more components of $\underset{\sim}{\chi}_j^{(1)}(\mathbf{q})$ is zero in the limit $q \rightarrow 0$ but is non-zero at finite $\mathbf{q}$, as in the case of F_{1u} symmetry optical phonons in the alkali halides and in the cubic perovskites, which are Raman inactive in the limit $q \rightarrow 0$ but are Raman active at finite q.[16] The dependence of $\underset{\sim}{\chi}_j^{(1)}(\mathbf{q})$ on the scattering wave vector can be expressed in the form of a Taylor expansion in powers of $\mathbf{q}$ as follows:

$$\underset{\sim}{\chi}_j^{(1)}(\mathbf{q}) = \underset{\sim}{\chi}_j^{(1)}(q{=}0) + (d\chi_j^{(1)}/dq)\mathbf{q} + (d^2\chi_j^{(1)}/dq^2)\mathbf{q}^2 + \ldots \tag{2.15}$$

$$= \underset{\sim}{\chi}_j^{(1)} + \underset{\sim}{\chi}_{jq}^{(1)} + \underset{\sim}{\chi}_{jq^2}^{(1)} + \ldots \quad .$$

Thus in the case of optical phonons $\chi^{(1)}_{\sim jq}$ is given by

$$\chi^{(1)}_{\sim joq} = \chi^{(1)}_{\sim jo} = [\partial^2\chi/\partial u_{jo}\partial q]_E \, \underset{\sim}{q}\underset{\sim}{u}_{jo}(q) + [\partial^2\chi/\partial E\partial q]_{u_j} \underset{\sim}{q}\underset{\sim}{E}_{jo}(q) \qquad (2.16)$$

We note that second order derivative terms $(\partial^2\chi/\partial u_{jo}\partial q)\underset{\sim}{q}\underset{\sim}{u}_{jo}$ and $(\partial^2\chi/\partial E\partial q)\,\underset{\sim}{q}\,\underset{\sim}{E}_{jo}$ are equivalent to $[\partial_\chi/\partial \nabla u_{jo}(q)] \nabla \underset{\sim}{u}_{jo}(q)$ and $[\partial_\chi/\partial \nabla E_{jo}(q)] \nabla \underset{\sim}{E}_{jo}(q)$ where $\nabla \underset{\sim}{u}_{jo}(q) = i\underset{\sim}{q}\underset{\sim}{u}_{jo}(q)$ and $\nabla \underset{\sim}{E}_{jo}(q) = i\underset{\sim}{q}\underset{\sim}{E}_{jo}(q)$. Since **q** is a polar vector the coefficient $[\partial^2\chi/\partial E\partial q]_{u_j} = \underset{\sim}{b}_q$ is a polar fourth rank tensor having non-zero components for all crystals classes. Thus, although $\underset{\sim}{b}(q)$ is zero in centrosymmetric crystals, in the limit $q \to 0$, it has non-zero components when **q** is finite. The coefficient $[\partial^2\chi/\partial u_j\partial q] = \underset{\sim}{a}_{jq}$ is a tensor whose functional form is determined by the symmetry character of $\underset{\sim}{u}_j(q\approx 0)$. In the case of the odd parity F_{1u} symmetry optical phonons in NaCl and cubic $BaTiO_3$ type crystals $\underset{\sim}{u}_j(q\approx 0)$ corresponds to a polar vector and therefore $\underset{\sim}{a}_{jq}$ is a non-zero fourth rank tensor.

The form of the atomic displacement Raman tensors, $\underset{\sim}{a}_j$, and $\underset{\sim}{a}_{jq}$ for optical phonons in crystals having NaCl, zincblende (ZnS) and wurtzite (ZnO) structures are given in Tables I and II. We note in the case of the F_2 symmetry LO phonons in zincblende type crystals and the E_1 symmetry LO phonons in wurtzite type crystals which are Raman active in the limit $q \to 0$, that the q-dependent contributions to $\chi^{(1)}_j(q)$ have non-zero diagonal components, whereas the q-independent contributions only have off-diagonal components (with respect to cubic reference axes). Thus the q-dependent and q-independent scattering by the LO phonons involve different polarization selection rules and can be readily distinguished from one another.

MICROSCOPIC THEORY OF LIGHT SCATTERING

The elementary scattering processes in dielectrics and semiconductors, which play a role in the first-order scattering of EM radiation by phonons, and their coupled modes with plasmons and photons, involve interband and intraband electronic transitions, in which an "incident" photon $(\omega_i,\underset{\sim}{k}_i)$ is annihilated, a quantum of the

TABLE I

Atomic Displacement Raman Tensor, $a_j^{\dagger(4)}$

Cubic				
		$\begin{bmatrix} - & - & - \\ - & - & - \\ - & - & - \end{bmatrix}$	$\begin{bmatrix} - & - & - \\ - & - & - \\ - & - & - \end{bmatrix}$	$\begin{bmatrix} - & - & - \\ - & - & - \\ - & - & - \end{bmatrix}$
NaCl	O_h	$F_{1u}(x)$	$F_{1u}(y)$	$F_{1u}(z)$
		$\begin{bmatrix} - & - & - \\ - & - & a \\ - & a & - \end{bmatrix}$	$\begin{bmatrix} - & - & a \\ - & - & - \\ a & - & - \end{bmatrix}$	$\begin{bmatrix} - & a & - \\ a & - & - \\ - & - & - \end{bmatrix}$
ZnS	T_d	$F_2(x)$	$F_2(y)$	$F_2(z)$
Hexagonal				
		$\begin{bmatrix} - & - & a_E \\ - & - & - \\ a_E & - & - \end{bmatrix}$	$\begin{bmatrix} - & - & - \\ - & - & a_E \\ - & a_E & - \end{bmatrix}$	$\begin{bmatrix} b & - & - \\ - & b & - \\ - & - & c \end{bmatrix}$
CdS	C_{6v}	$E_1(x)$	$E_1(y)$	$A_1(z)$

† *Referred to cubic axes.*

TABLE II

Linear q-dependent Atomic Displacement Raman Tensor, a_{jq} *† (16)

Cubic		$\begin{bmatrix} \cdot & \cdot & a_q \\ \cdot & \cdot & \cdot \\ a_q & \cdot & \cdot \end{bmatrix}$	$\begin{bmatrix} \cdot & \cdot & \cdot \\ \cdot & \cdot & a_q \\ \cdot & a_q & \cdot \end{bmatrix}$	$\begin{bmatrix} a_q' & \cdot & \cdot \\ \cdot & a_q' & \cdot \\ \cdot & \cdot & a_q'' \end{bmatrix}$
NaCl	$O_h(q_z)$	$F_{1u}(x)$	$F_{1u}(y)$	$F_{1u}(z)$
ZnS	$T_d(q_z)$	$F_{2u}(x)$	$F_2(y)$	$F_2(z)$
Hexagonal		$\begin{bmatrix} \cdot & \cdot & A_q \\ \cdot & \cdot & \cdot \\ A_q & \cdot & \cdot \end{bmatrix}$	$\begin{bmatrix} \cdot & \cdot & \cdot \\ \cdot & \cdot & A_q \\ \cdot & A_q & \cdot \end{bmatrix}$	$\begin{bmatrix} B_q & \cdot & \cdot \\ \cdot & B_q & \cdot \\ \cdot & \cdot & C_q \end{bmatrix}$
CdS	$C_{6v}(q_x)$	$E_1(x)$	$E_1(y)$	$A_1(z)$
		$\begin{bmatrix} D_q & \cdot & \cdot \\ \cdot & E_q & \cdot \\ \cdot & \cdot & F_q \end{bmatrix}$	$\begin{bmatrix} \cdot & G_q & \cdot \\ G_q & \cdot & \cdot \\ \cdot & \cdot & \cdot \end{bmatrix}$	$\begin{bmatrix} \cdot & \cdot & H_q \\ \cdot & \cdot & \cdot \\ H_q & \cdot & \cdot \end{bmatrix}$
CdS	$C_{6v}\, q_x$	$E_1(x)$	$E_1(y)$	$A_1(z)$

* *The linear electric field dependent atomic displacement Raman tensor a_{jE} has the same form as a_{jq}.*

† *Referred to cubic axes.*

collective excitation ($\omega_j,\mathbf{k}_j$) is annihilated or emitted, and a "scattered" photon ($\omega_s,\mathbf{k}_s$) is emitted.[4,9] Scattering processes in which the EM radiation interacts directly with the collective excitation also occur, but are generally less important than those in which the EM radiation interacts through the intermediary of the electronic excitation, particularly when $\omega_j \ll \omega_i$.

We consider first the type of electronic transitions which take place in the absence of free carriers. These are illustrated in Figs. (2), (3) and (4). The numbers ①,② and ③ designate the order in which the electronic transitions take place. The scattering medium is initially in its electronic ground state $|0\rangle$ with all valence band states filled and all conduction band states empty. Transition ① involves a direct interband excitation of an electron from a valence band state to a conduction band state to form a continuum electron-hole pair intermediate state $|\alpha\rangle$. Transition ② involves either an intraband excitation of the electron (or hole) as shown in Fig. (2) or a direct interband transition as shown in Fig. (3), to form a second continuum electron-hole pair intermediate state $|\beta\rangle$. Transition ③ involves the recombination of the electron-hole pair which returns the medium to its electric ground state. Momentum is conserved in each of the three transitions, whereas energy is conserved only for the overall scattering process. Scattering processes in which electronic transition ② involves an intraband excitation are designated as *two band* scattering processes, and those in which electronic transition ② involves an interband excitation are designated as *three band* scattering processes.

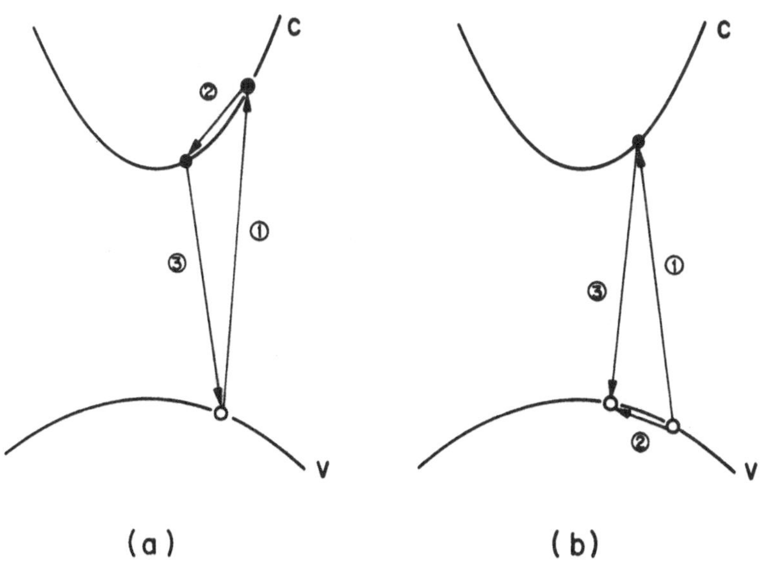

Fig. 2. Schematic diagram of the interband and intraband transitions which play a role in *two band* scattering processes. The numbers 1 , 2 , and 3 indicate the time order of the electronic transitions.

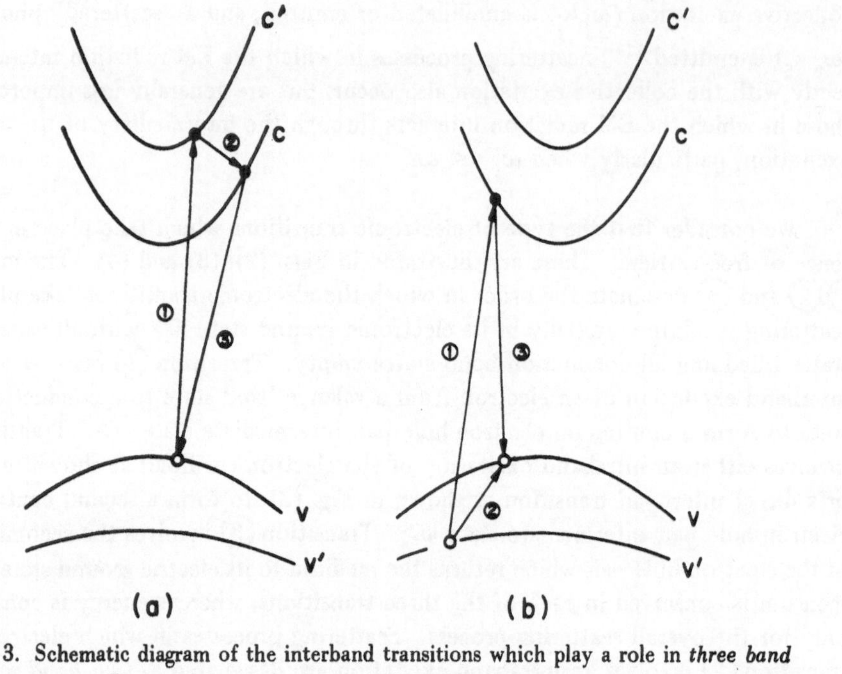

Fig. 3. Schematic diagram of the interband transitions which play a role in *three band* scattering processes. The numbers 1 , 2 , and 3 indicate the time order of the electronic transitions.

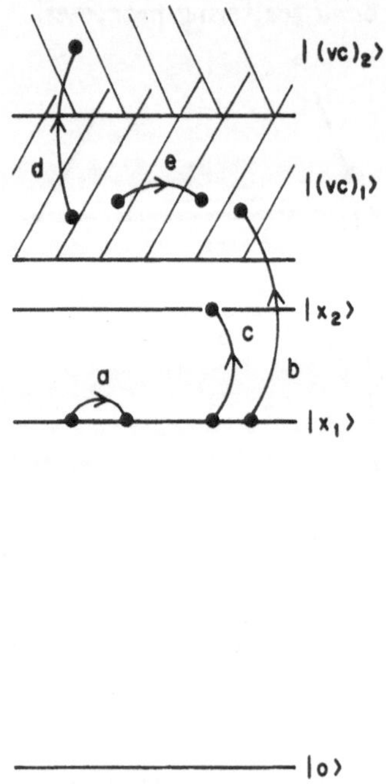

Fig. 4. Schematic diagram of the transitions between discrete and continuum exciton intermediate states.

The interband excitation in transition ① may also form a discrete (or a continuum) exciton as intermediate state [Fig. (4)]. Transition ② will then involve either an "intraband" or "interband" excitation of the exciton, and transition ③ will involve the recombination of the exciton electron-hole pair.

The absorption and emission of the photons and the collective excitation quantum can take place in any time order. There are accordingly six Stokes scattering processes and six anti-Stokes scattering processes. The diagramatic representation of the Stokes scattering process in which transition ① is accompanied by the annihilation of an incident photon; transition ② is accompanied by the emission of a collective excitation quantum; and transition ③ is accompanied by the emission of a scattered photon is shown in Fig. (5).

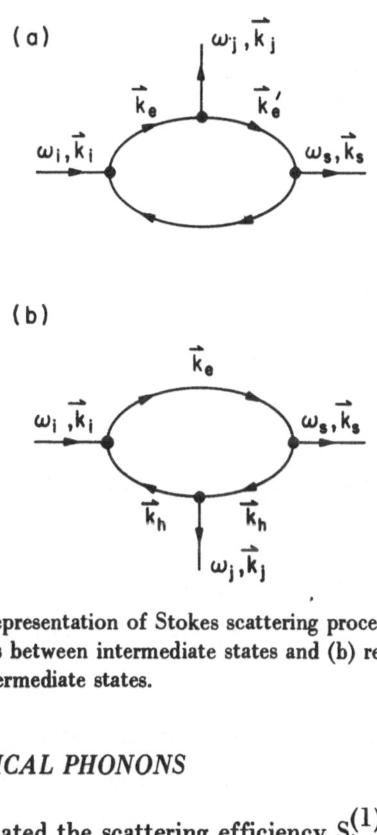

Fig. 5. Diagramatic representation of Stokes scattering processes: (a) represents processes involving electron-transitions between intermediate states and (b) represents processes involving hole-transitions between intermediate states.

SCATTERING BY OPTICAL PHONONS

Loudon[9] has calculated the scattering efficiency $S_j^{(1)}(q)$ of optical phonons in diamond and zincblende type crystals on the basis of the three step scattering processes, using third order perturbation theory. His result for $S_j^{(1)}(q)$ (which is also applicable to scattering by polaritons and by coupled LO phonon-plasmon modes) can be expressed in the form

$$S_j^{(1)}(q) = (e/\hbar mc)^4\, LV^2\, |R_j^{(1)}(q)|^2$$

$$R_{jo}^{(1)}(q) = V^{-1} \sum_{\alpha\beta} (p^s_{o\beta})(H^j_{\beta\alpha})(p^i_{\alpha o})/[(\omega_{o\beta}(k) - \omega_s)(\omega_{\alpha o}(k) - \omega_i)] \tag{3.2}$$

$$+ (p^i_{o\beta})(H^j_{\beta\alpha})(p^s_{\alpha o})/[(\omega_{o\beta}(k) + \omega_i)(\omega_{\alpha o}(k) + \omega_s)]$$

$$+ 4 \text{ other terms}$$

where $(p^s_{o\beta}) = \langle 0|\mathbf{p}\cdot\hat{e}_s|\beta\rangle$, $(p^i_{\alpha o}) = \langle\alpha|\mathbf{p}\cdot\hat{e}_i|0\rangle$, etc. are the momentum matrix elements which result from the electron-radiation interaction, H_{eR}; $(H^j_{\beta\alpha}) = \langle\beta|H_{eQj}|\alpha\rangle$ is the matrix elements of the electron-collective excitation interaction, H_{eQj}; $\hbar\omega_{\alpha o}(k) = \mathcal{E}_\alpha(k) - \mathcal{E}_o$ and $\hbar\omega_{o\beta}(k') = \mathcal{E}_o - \mathcal{E}_\beta(k')$; $\mathcal{E}_\alpha(k)$ and $\mathcal{E}_\beta(k')$ are the energies of the electron-hole (e-h) pair states, $|\alpha\rangle$ and $|\beta\rangle$ to which the virtual intermediate states correspond; and the summation is taken over all e-h pair states $|\alpha\rangle$ and $|\beta\rangle$. In the case of *two band* Stokes scattering processes involving intraband electron and interband hole transitions $\omega_{o\beta}(k')$ and $\omega_{o\alpha}(k)$ are related as follows:

$$\omega_{o\beta}(k') \approx \omega_{o\alpha}(k) - \mathbf{q}\cdot\mathbf{v}_e \qquad \text{electron scattering}$$

$$\omega_{o\beta}(k') \approx \omega_{o\alpha}(k) - \mathbf{q}\cdot\mathbf{v}_h \qquad \text{hole scattering}$$

where $v_e(k) = \hbar k/m_e$ and $v_h(k) = \hbar k/m_h$ are the velocities of the electron and hole respectively in state $|\alpha\rangle$, and m_e and m_h are the effective masses of the electron and hole.

As shown by Loudon, $R_j^{(1)}(q)$, the "Raman scattering tensor" of the optical phonon, is related to $\chi_j^{(1)}(q)$ the first-order *transition electric susceptibility* as follows:

$$R_j^{(1)}(q) = (m^2\hbar^2\omega_s^2/c^2)\,\chi_j^{(1)}(q) \tag{3.4}$$

In semiconductors, $\omega_{\alpha o}(\mathbf{k})$ and $\omega_{\beta o}(\mathbf{k})$ are generally much larger than ω_j and, as ω_j is generally small compared to ω_i, the major contribution to $R_j^{(1)}(\mathbf{q})$ will generally come from the terms involving $(H^j_{\beta\alpha})$, displayed in Eq. (3.2), corresponding to *two band* and *three band* scattering processes in which the electronic transitions between the intermediate states $|\alpha\rangle$ and $|\beta\rangle$ take place with the emission of a collective excitation quantum.

The matrix elements of H_{eQ} involve contributions from the elastic displacement gradient $\underset{\sim}{\Sigma}_j(\mathbf{q})$, the relative atomic displacements $\mathbf{u}_j(\mathbf{q})$ and the macroscopic (longitudinal or transverse) electric field $\mathbf{E}_j(\mathbf{q})$ of the collective excitation. Thus, $(H^j_{\beta\alpha})$, the matrix element of H_{eQj} for electronic transitions between intermediate states $|\alpha\rangle$ and $|\beta\rangle$ is given by

$$(H^j_{\beta\alpha}) = \Xi^{\Sigma}_{\beta\alpha}(\mathbf{q})\,\Sigma_j(\mathbf{q}) + \Xi^{u_j}_{\beta\alpha}(\mathbf{q})\,u_j(\mathbf{q}) + (F^j_{\beta\alpha}) + (A^j_{\beta\alpha}) \tag{3.5}$$

where $\Xi^{\Sigma}_{\beta\alpha}(\mathbf{q})$ is the elastic deformation potential; $\Xi^{u_j}_{\beta\alpha}(\mathbf{q})$ is the atomic displacement deformation potential; $(F^j_{\beta\alpha})$ is the matrix element of the Frohlich (electron-macroscopic longitudinal electric field) interaction and $(A^j_{\beta\alpha})$ is the momentum matrix element. $(F^j_{\beta\alpha})$ and $(A^j_{\beta\alpha})$ are given by

$$\begin{aligned}(F^j_{\beta\alpha}) &= eV_j(\mathbf{q})\,\langle\beta|e^{i\mathbf{q}\cdot\mathbf{r}}|\alpha\rangle \\ (A^j_{\beta\alpha}) &= [e\,A_j(\mathbf{q})/mc]\,\langle\beta|\mathbf{p}\cdot\mathbf{e}_j\,e^{i\mathbf{q}\cdot\mathbf{r}}|\alpha\rangle\end{aligned} \tag{3.6}$$

where $V_j(\mathbf{q}) = iE_{jL}(\mathbf{q})/q$ and $\underset{\sim}{A}_j(\mathbf{q}) = (ic/\omega)\,\underset{\sim}{E}_{jT}(\mathbf{q})$. When $|\alpha\rangle$ and $|\beta\rangle$ involve continuum e-h pair states in the same pair of bands, $(F^j_{\beta\alpha})$ and $(A^j_{\beta\alpha})$ correspond to intraband matrix elements and, in the case of simple energy bands, are given by

$$\begin{aligned}(F^j_{\beta\alpha})_{e,h} &= \pm\, i\, e\,[E_{jL}(\mathbf{q})/q]\;\delta\,(\mathbf{k}' - \mathbf{k} - \mathbf{q}) \\ (A^j_{\beta\alpha})_{e,h} &= \pm\, i\, e\,[E_{jT}(\mathbf{q})/q]\,(\hbar\mathbf{k}/mc)\;\delta\,(\mathbf{k}' - \mathbf{k} - \mathbf{q})\end{aligned} \tag{3.7}$$

where the (+) sign corresponds to intraband hole transitions and the (-) sign corresponds to intraband electron transitions, i.e., the electron and hole intraband matrix elements have the same magnitude but are opposite in sign. We note that for $E_{jL}(\underset{\sim}{q}) = E_{jT}(\underset{\sim}{q})$, $(A^j_{\beta a}) = \hbar k/mc\ (F^j_{\beta a})$, i.e., $(A^j_{\beta a})$ is much smaller than $(F^j_{\beta a})$.

When $|a\rangle$ and $|\beta\rangle$ represent continuum e-h pair states in different pairs of valence and conduction bands, $(F^j_{\beta a})$ and $(A^j_{\beta a})$ correspond to electron (or hole) interband matrix elements and are given by[9]

$$(F^j_{\beta a})_{vv',cc'} = i\,e\,[E_j(q)(p^j_{\beta a})/\omega_{\beta a}]\ \delta\,(k' - k - q) = (A^j_{\beta a})_{vv',cc'} \tag{3.8}$$

where $\hbar\omega_{\beta a} = \mathcal{E}_\beta(k') - \mathcal{E}_a(k)$. The electron and hole interband matrix elements are different in magnitude since $(p^j_{\beta a})_{cc'} \neq (p^j_{\beta a})_{vv'}$ and $(\omega_{\beta a})_{cc'} \neq (\omega_{\beta a})_{vv'}$(where cc′ designates electron interband transitions and vv′ designate hole interband transitions).

When the intermediate e-h pair states $|a\rangle$ and $|\beta\rangle$ correspond to discrete or continuum exciton states, $(F^j_{\beta a})$ and $(A^j_{\beta a})$ are given by[17]

$$\begin{aligned}(F^j_{\beta a}) &= e\,V_j(q)\ \langle x_{n'}|\exp[i(m_h/M)q\cdot r] - \exp[-i(m_e/M)q\cdot r]\,|x_n\rangle \\ (A^j_{\beta a}) &= (e\,A_j(q)/mc)\ \langle x_{n'}\,|\ \exp[i(m_h/M)q\cdot r] - \exp[-i(m_e/M)q\cdot r]\ \ p\cdot\hat{e}_j\,|x_n\rangle\end{aligned} \tag{3.9}$$

where $M = m_e + m_h$; $r = r_e - r_h$ is the relative coordinate of the electron-hole pair; and $|x_{n'}\rangle$ and $|x_n\rangle$ are s-type exciton states when the interband transitions $|0\rangle$ to $|a\rangle$ and $|\beta\rangle$ to $|0\rangle$ are optically allowed.

The matrix elements of the Frohlich interaction for transitions between s-type exciton states $(F^j_{xs'xs})$ are equal to zero in the limit $q \rightarrow 0$, but are non-zero and diagonal at finite **q** (providing $m_e \neq m_h$).[18,19] This applies to transitions involving continuum exciton states as well as to transitions involving discrete exciton states. On the other hand $(A^j_{xs'xs})$, the momentum matrix element for transitions between s states are zero independent of q.

When $|x_{n'}\rangle$ and $|x_n\rangle$ correspond to hydrogenic ls exciton states, $(F^j_{ls\,ls})$ is given by

$$(F^j_{ls\,ls}) = i\,e\,[E_j(\underset{\sim}{q})/q]\ \left[1 - \left\{1 + \tfrac{1}{4}q^2a_x^2\right\}^{-2}\right] \tag{3.10}$$

where a_x is the radius of the exciton in the ls state. The magnitude of $(F^j_{ls\,ls})$ is a maximum at $qa_x = 2$ and, for $qa_x << 1$, it is equal to $eE_j qa_x^2/2$.

An interesting situation arises when the intermediate states correspond to impurity-bound exciton states, since the impurity states destroy the translational symmetry of the crystal and thereby the requirement that wave vector be conserved, i.e., the requirement that $\mathbf{k}_j = \mathbf{k}_i - \mathbf{k}_s$. In this situation, LO phonons with wave vector $\mathbf{k}_j$ different from the scattering wave vector $\mathbf{q} = \mathbf{k}_i - \mathbf{k}_s$ and, in particular, phonons with $k_j \approx 2/a_x$, can participate in the scattering process and the corresponding Frohlich interaction matrix element can be quite large.[20]

The elastic strain deformation potentials $\Xi^{\Sigma}_{\beta a}(\mathbf{q})$ are non-zero in all crystal classes. The atomic displacement deformation potentials $\Xi^{u_j}_{\beta a}(\mathbf{q})$ are, in the limit $q \to 0$, non-zero only for even parity or mixed parity optical phonons. Both types of deformation potentials have magnitudes which are characteristic of the energy band involved and therefore have different magnitudes for electron and hole transitions, i.e., $[\Xi^{\Sigma}_{\beta a}(\mathbf{q})]_e \neq [\Xi^{\Sigma}_{\beta a}(\mathbf{q})]_h$ and $[\Xi^{u_j}_{\beta a}(\mathbf{q})]_e \neq [\Xi^{u_j}_{\beta a}(\mathbf{q})]_h$. Furthermore, since interband deformation potentials are generally much smaller in magnitude than the intraband deformation potentials, the major contributions to the elasto-optic and atomic displacement Raman tensors generally comes from *two band* scattering processes.

In high symmetry crystals, such as those with zincblende and NaCl type structures, the Frohlich interaction matrix element for intraband electron and intraband hole transitions are equal in magnitude but opposite in sign. Accordingly, since the scattering processes involving intraband electron transitions and the scattering corresponding to intraband hole transitions from the same intermediate state $|a\rangle$ have in the limit $q \to 0$ the same energy denominators, their contributions cancel.[9] The same situation holds for the terms involving the momentum matrix element of the electron-radiation interaction for intraband electron and intraband hole transitions which are also equal in magnitude but opposite in sign. Thus *two band* scattering processes involving continuum e-h pairs as intermediate states do not contribute to the electro-optic Raman tensor in zincblende and NaCl type crystals, in the limit $q \to 0$. However, as first pointed out by Hamilton[21] the energy denominators are not the same when q is non-zero, and the electron and hole terms do not cancel. The resulting q-dependent contributions to the Raman scattering tensor become appreciable when ω_i is close to $\omega_a(\mathbf{k})$. Since the intraband matrix elements $(F^j_{\beta a})$ and $(A^j_{\beta a})$ are diagonal in the electron states, the q-dependent contribution to the electro-optic Raman tensor from *two band* scattering processes will have diagonal components.

The interband matrix elements of the Frohlich interaction (and the interband momentum matrix elements of the electron-radiation interaction which are equal to one another in the electric dipole approximation) have different magnitudes for electron and hole transitions and the corresponding energy denominators are also different. Consequently, the electron and hole contributions do not cancel. However, because of parity selection rules, *three band* scattering processes involving these matrix elements do not contribute to the electro-optic Raman tensor in centrosymmetric crystals in the limit $q \rightarrow 0$, but do so in crystals lacking a center of inversion in which parity selection rules are not applicable.

Three band processes can contribute to the electro-optic Raman tensor in centrosymmetric crystals when $q \neq 0$. The q-dependent contributions from *three band* processes, which of course also occur in crystals lacking a center of inversion, can have either diagonal or non-diagonal components dependent on the point group symmetry of the crystal and on the directions of **q**.

SCATTERING BY PLASMONS

We next consider the "three step" elementary scattering processes which play a role in the scattering of EM radiation by plasmons and by coupled LO phonon-plasmon modes in semiconductors. For illustrative purposes we will consider scattering by plasmons (ω_p,k_p) in "degenerate" n-type semiconductors. Apart from occupation factors arising from the presence of electrons in the conduction band, which modify the contributions from the various elementary scattering processes discussed above, the presence of the electron plasma allows two additional types of processes to participate. In one type shown in Fig. (6a), an intraband [or interband (not illustrated)] transition ① of a *free* electron constitutes the first step. It is followed by an interband transition ② of an electron from the valence band to the state left empty by the intraband transition. In transition ③, the excited electron in the conduction band recombines with the hole in the valence band, returning the crystal to the electronic "ground" state. In the second type of scattering process shown in Fig. (6b) the first step is an interband transition ① of an electron from the valence band to an empty state in one of the conduction bands. This is followed by an interband transition ② in which another electron in the conduction band combines with the hole in the valence band. Transition ③ involves the intraband or interband (not shown) transition of the excited electron into the state in the conduction band left empty by the second transition, thereby returning the crystal to its electronic ground state. The annihilation and emission of photons and collective excitation quanta can take place in any time order. However, for $\omega_i \approx \omega_s >> \omega_j$, the important contributions to the Raman scattering tensor of the collective excitations (i.e., plasmons or coupled LO phonon-plasmons) arise from *two band* scattering processes in which the intraband transitions are accompanied by the annihilation or emission of

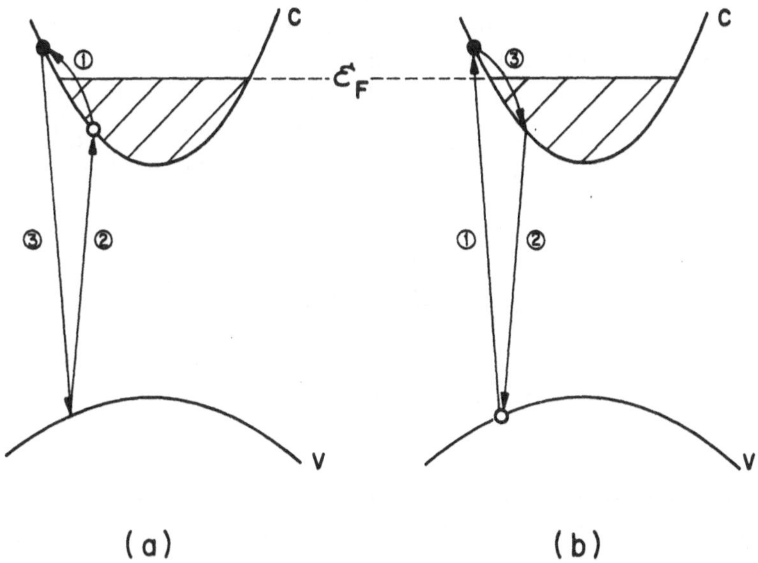

Fig. 6. Schematic diagram of interband and intraband transitions which play a role in two band scattering by plasmons in n-type semiconductors. The numbers 1 , 2 , and 3 indicate the time order of the electronic transitions.

collective excitation quanta. This is so because the energy denominators for these scattering processes are much smaller than the energy denominator for processes in which the intraband transitions are accompanied by the annihilation or emission of photons. (*Three band* scattering processes may be important when the separation between the conduction bands is relatively small.) Since the annihilation or emission of the photons can take place in any order, there are two Stokes scattering processes of the first type and two Stokes scattering processes of the second type, making a total of four Stokes scattering processes. Using third order perturbation theory, one obtains the following expression for $R_P^{(1)}(\mathbf{q})$, the *two band* electro-optic contribution to the Raman scattering tensor from the four Stokes scattering processes

$$R_j^{(1)}(\mathbf{q}) = V^{-1} \sum_{\mathbf{k}} \left\{ (p^i_{o\alpha})(p^s_{\beta o})(F^j_{\beta\alpha}) / [\omega_P + \mathcal{E}_c(\mathbf{k}+\mathbf{q}) - \mathcal{E}_c(\mathbf{k})] \right\} \mathrm{x}$$

$$\left\{ 2\omega_\alpha(\mathbf{k}) / [\omega_\alpha(\mathbf{k}) - \omega_1^2] \right\} [f(\mathbf{k}) - f(\mathbf{k}+\mathbf{q})], \qquad (3.11)$$

$$(F^j_{\beta\alpha}) = e\, V(\mathbf{q})\, \delta(\mathbf{k}' - \mathbf{k} + \mathbf{q}) = i\, e\, [E_P(\mathbf{q})/q]\, \delta(\mathbf{k}' - \mathbf{k} + \mathbf{q})$$

where $\mathcal{E}_c(\mathbf{k})$ and $\mathcal{E}_c(\mathbf{k}+\mathbf{q})$, f(k) and f(k + q) are the energies and the Fermi occupation factors of the conduction band states corresponding to $|\alpha\rangle$ and $|\beta\rangle$. (In the case of scattering by coupled LO phonon-plasmon modes one would also have an atomic displacement contribution involving the intraband deformation potential interaction.) When ω_i is not very close to resonance, we can replace $\omega_\alpha(\mathbf{k})$ by ω_G. If we then assume that $(p^i_{o\alpha})$ and $(p^s_{\beta o})$ are independent of k, $R^{(1)}_P(\mathbf{q})$ is, to a good approximation, given by

$$R^{(1)}_P(q) = [i\, e\, |p_{o\alpha}|^2\, E_P(q)/q]\, [2\omega_G/(\omega_G^2 - \omega_i^2)]\, x$$

$$\sum_{\mathbf{k}} [f(\mathbf{k}) - f(\mathbf{k+q})]/[\hbar\omega_p + \mathcal{E}(\mathbf{k+q}) - \mathcal{E}(\mathbf{k})] \tag{3.12}$$

We then make use of the fact that the frequency and wave vector dependent longitudinal dielectric constant of the electron plasma is given by

$$\epsilon_L(\omega,\mathbf{q}) = \epsilon_{Lo}(\omega,\mathbf{q}) - (4\pi e^2/q^2) \sum_{\mathbf{k}} [f(\mathbf{k}) - f(\mathbf{k+q})]/[\hbar\omega + \mathcal{E}(\mathbf{k+q}) - \mathcal{E}(\mathbf{k})] \tag{3.13}$$

where $\epsilon_{Lo}(\omega,\mathbf{q})$ is the contribution to the dielectric constant of the medium from other electric dipole excitations. $\epsilon_L(\omega,\mathbf{q})$ is zero at $\omega = \omega_P$ and, therefore,

$$\sum_{\mathbf{k}} [f(\mathbf{k}) - f(\mathbf{k+q})]/[\hbar\omega_P + \mathcal{E}(\mathbf{k+q}) - \mathcal{E}(\mathbf{k})] = +\, q^2\, \epsilon_{Lo}(\omega,\mathbf{q})/4\pi e^2 \tag{3.14}$$

On introducing this expression into Eq. (3.12) one obtains

$$R^{(1)}_P(\mathbf{q}) = i|(p_{o\alpha})|^2\; [2\omega_G/(\omega_G^2 - \omega_i^2)]\; q\;\; E_P(\mathbf{q})\epsilon_{Lo}(\omega,\mathbf{q})/4\pi e\;. \tag{3.15}$$

We note that the *two band* electro-optic contribution to $R^{(1)}_j(\mathbf{q})$ from the two additional types of scattering processes is proportional to $\nabla\, \underset{\sim}{E}_P(\mathbf{q}) = i\mathbf{q}\underset{\sim}{E}_P(\mathbf{q})$. One must also add to this contribution the q-independent and q-dependent electro-optic contributions to $R^{(1)}_P(\mathbf{q})$ coming from *two band* and *three band* elementary scattering processes of the type discussed in the section on *Scattering by Optical Phonons*

FREQUENCY DEPENDENCE OF THE RAMAN SCATTERING EFFICIENCY

The magnitude of the first-order *transition electric susceptibility* of optical phonons $\underset{\sim}{\chi}_j^{(1)}(\mathbf{q})$, which is determined by the superposition of the frequency dependent atomic displacement and electro-optic contributions from *two bands* and *three bands* scattering processes throughout the Brillouin zone, may be expected to exhibit strong resonance enhancements when ω_i approaches resonances in the interband electronic transitions. In semiconductors such resonance enhancements occur at the fundamental ("direct") absorption edges and at frequencies corresponding to critical points in the combined density of states for interband transitions. Thus, in the case of zincblends type semiconductors, resonance enhancements may be expected to occur at the E_o (fundamental) energy gaps, and at the E_1 and the E_2 energy gaps (Fig. 7). The E_2 gap makes the dominant contribution to the low frequency electric susceptibility, and we may also expect the scattering processes at the E_2 gap to make the dominant contribution to $\underset{\sim}{\chi}_j^{(1)}(\mathbf{q})$ when ω_i is far below ω_G, the frequency of the fundamental absorption edge. Although the E_o gap makes the smallest contribution to the low frequency susceptibility, and very likely makes the smallest contributions to $\underset{\sim}{\chi}_j^{(1)}(\mathbf{q})$ when $\omega_i << \omega_G$, we may in general expect the scattering processes at the E_o gap to make the dominant contribution to $\underset{\sim}{\chi}_j^{(1)}(\mathbf{q})$

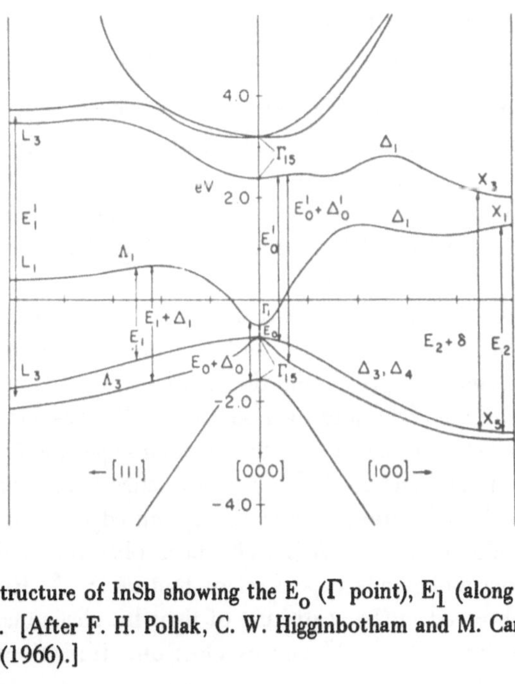

Fig. 7. Band structure of InSb showing the E_o (Γ point), E_1 (along <111>) and E_2 (X point) critical points. [After F. H. Pollak, C. W. Higginbotham and M. Cardona, J. Phys. Soc. Japan, Suppl. 21, 20(1966).]

when ω_i is close to ω_G. We may similarly expect the contribution from scattering processes at the E_1 gap to be the dominant contribution when ω_i is close to $E_1/\hbar$.

$\underset{\sim}{\chi}_j^{(1)}(q)$ can, to a good approximation, be expressed as a superposition of the contribution from the E_o, E_1 and E_2 gaps as follows:

$$\underset{\sim}{\chi}_j^{(1)}(q) = \underset{\sim}{\chi}_j^{(1)E_o}(q) + \underset{\sim}{\chi}_j^{(1)E_1}(q) + \underset{\sim}{\chi}_j^{(1)E_2}(q) \tag{4.1}$$

We may also write each term in this expression as the sum of the contributions from scattering processes involving discrete exciton and continuum e-h pair intermediate states.[22] Thus, in the case of the contribution from the E_o gap we may write

$$\underset{\sim}{\chi}_j^{(1)E_o}(q) = \underset{\sim}{\chi}_{j\,ex}^{(1)E_o}(q) + \underset{\sim}{\chi}_{j\,cont}^{(1)E_o}(q) \tag{4.2}$$

As the oscillator strengths of exciton bands are generally very small, $\underset{\sim}{\chi}_{j\,ex}^{(1)E_o}(q)$ may be expected to be important only when ω_i is in the vicinity of the exciton resonances, i.e., when $\omega_i \approx \omega_{ex}$. On the other hand, $\underset{\sim}{\chi}_{j\,cont}^{(1)E_o}(q)$ may be expected to make an appreciable contribution even when $\omega_i << \omega_g$.

The contributions to $\underset{\sim}{\chi}_j^{(1)}(q)$ from the various gaps may have different signs, and furthermore the atomic displacement and the electro-optic contributions of scattering processes at a given gap may have opposite signs. Consequently, $\underset{\sim}{\chi}_j^{(1)}(q)$, and therefore the scattering efficiency of optical phonons and their coupled modes with photons and plasmons, may exhibit minima and maxima as the frequency of the incident radiation approaches the E_o and E_1 gaps and the magnitudes of the contributions from E_o and E_1, relative to that from E_2, changes.

The resonance enhancement of light scattering by optical vibration modes was first observed experimentally in liquids,[23] and a theoretical treatment of resonance enhancement of scattering by optical vibration modes in liquids and solids was given by Ovander and co-workers.[24,25] and by Loudon.[26] The resonance enhanced light scattering by LO and TO phonons at the fundamental edge was first observed by Leite and co-workers in CdS and ZnSe.[27,28] Resonance enhanced light scattering has since been observed at the fundamental absorption edge of other zincblende and wurtzite type semiconductors.[29-31] It has also been observed at the E_1 energy gap in InSb, InAs and other zincblende type semiconductors.[32-35] Resonance enhancement has also been observed in the scattering of light by plasmons and single particle excitations in semiconductors[14,36,37] and by electronic transitions in paramagnetic crystals.[38] A very strong resonance enhancement of spin flip Raman scattering by impurities has also been observed in CdS.[39]

RESONANT RAMAN SCATTERING BY TO PHONONS IN OPAQUE SEMICONDUCTORS[41]

Under resonance conditions, the Raman scattering tensor of TO phonons can be approximated by the first term in Eq. (3.2), which exhibits the strongest resonance enhancement.

$$R_j^{(1)}(q) \approx V^{-1} \sum_{\alpha\beta} [(p^s_{o\beta})(H^j_{\beta\alpha})(p^i_{\alpha o})] / \left\{ [\omega_\beta(k') - \omega_s][\omega_\alpha(k) - \omega_i] \right\} \tag{4.3}$$

Furthermore, the largest contribution to $R_j^{(1)}(q)$ from this term will come from *two band* scattering processes in which $\omega_\alpha(k) \approx \omega_\beta(k') \approx \omega_i \approx \omega_s$. If one assumes that the interband momentum matrix elements and the intraband matrix elements of the electron-phonon interaction are constant for the two bands under consideration, the resonant Raman scattering tensor for TO phonons can, in the limit $q \to 0$, be written in the form

$$R_j^{(1)} = V^{-1} \left\{ [(\Xi^{u_j}_{\beta\alpha})_e + (\Xi^{u_j}_{\beta\alpha})_h]/\omega_j \right\} (p^s_{o\beta})(p^i_{\alpha o}) \times$$

$$\sum_{\mathbf{k}} [(\omega_\alpha(k) - \omega_i)^{-1} - (\omega_\beta(k') - \omega_s)^{-1}] \tag{4.4}$$

where $\mathbf{k}$ is the wave vector of the electron (and hole) in the intermediate state, and we have omitted the designation of $\mathbf{q}$ dependence for convenience.

In the case of *two band* scattering processes,

$$(\Xi^{u_j}_{\beta\alpha})_e + (\Xi^{u_j}_{\beta\alpha})_h = (d\mathcal{E}_G/du_j)\, d\underset{\sim}{u}_j \tag{4.5}$$

where $\mathcal{E}_G$ is the energy gap between the two bands involved. We note that the contribution to the electric susceptibility of the crystal for the interband transitions between the two bands is given by[9]

$$\chi(\omega_i) = (e^2/\hbar V m\omega^2)(p^s_{o\beta})(p^i_{\alpha o}) \sum_{\mathbf{k}} (\omega_k - \omega_i)^{-1} \tag{4.6}$$

where the summation over **k** is the same as in Eq. (4.4). In the case of cubic crystals, equations (4.4) and (4.6) can be combined to obtain the following expression for the Raman scattering tensor of TO phonons and for the deformation potential contribution to the Raman scattering tensor of LO phonons:

$$R_j^{(1)} \approx (m\omega_i^2/e^2\hbar)\left\{[\chi(\omega_i) - \chi(\omega_s)]/(\omega_i - \omega_s)\right\}(d\mathcal{E}_G/du_j)\underset{\sim}{u}_j \tag{4.7}$$

where $\chi(\omega_i)$ and $\chi(\omega_s)$ are the complex electric susceptibilities at ω_i and ω_s. When $\chi(\omega)$ is changing monotonically, Eq. (4.7) can be written in the form

$$R_j^{(1)} \approx (m\omega_i^2/e^2\hbar)(d\chi/d\omega)(d\mathcal{E}_G/du_j)\,\underset{\sim}{u}_j\,. \tag{4.8}$$

A similar expression involving the *intraband* matrix elements of the Frohlich interaction can be derived for the q-dependent resonance-enhanced scattering tensor by LO phonons. Such an expression would be applicable in situations where the Raman scattering by LO phonons has a contribution from the two-band scattering process, arising either from q-dependent terms in the energy denominator, or from q-dependent exciton-LO phonon matrix elements.

The expression for $R_j^{(1)}$ given in Eq. (4.8) can be used to estimate the relative strengths of the resonance enhanced Raman scattering tensor of TO phonons at different energy gaps. The magnitudes of $(d\chi/d\omega)$ are much larger in the vicinity of the E_1 gaps than in the vicinity of the E_o gaps. This reflects the fact that the interband density of states at the E_1 critical point is much larger than at the E_o critical point. Accordingly, if the atomic displacement deformation potentials at the E_1 and E_o gaps are not very different in magnitude, the resonance enhanced Raman scattering tensor for TO phonons will be larger in the vicinity of the E_1 gap than in the vicinity of the E_o gap. This accounts for the weak TO phonon scattering intensities when ω_i is in the vicinity of an $E_o + \Delta_o$ gap[37] and the large scattering intensities when ω_i is in the vicinity of an E_1 gap.[31]

RESONANCE ENHANCED LIGHT SCATTERING AT THE FUNDAMENTAL EDGE

In their initial investigations, Leite and Porto[27] observed a marked enhancement of the Raman scattering by LO phonons in CdS at low temperatures at frequencies near the absorption edge and interpreted their observations in terms of Loudon's theory of resonant enhanced scattering via continuum e-h pair intermediate states. In addition, they observed 2 - LO phonon peaks at twice the frequency of the single LO phonon peaks, which also exhibited a resonance enhancement. Birman and Ganguly[41] suggested that the dominant mechanism for the resonance scattering by optical phonons in CdS involved excitons as intermediate states and formulated a theory of resonance enhanced light scattering by TO and LO phonons on this basis. Further data obtained by Leite et al.[28] in CdS and ZnSe indicated that TO phonons also exhibited a marked enhancement in scattering efficiency in the resonance region. Their data indicated that the intensity of the TO phonon peak initially exhibits a greater enhancement than the LO phonon peak, but then saturates as resonance is approached, whereas that of the LO phonon peak continues to increase. The large increase of the ratio of the LO phonon to the TO phonon scattering intensities, I_{LO}/I_{TO}, in the resonance region was attributed by Burstein and co-workers[18,42] to the fact that the exciton-optical phonon coupling via the Frohlich interaction is much stronger than the corresponding coupling via the deformation potential. They also concluded that the contributions to the Raman scattering tensor of the LO phonons in the resonance region is dominated by scattering processes involving excitons as intermediate states, whereas the contributions to the Raman scattering tensor of the TO phonons is dominated by scattering processes involving continuum e-h pairs as intermediate states.

More recent studies of the scattering intensities of TO phonons in CdS over a somewhat wider range of frequencies by Ralston *et al*[43] showed that, as ω_i approached ω_G, the scattering efficiencies of the A_1 and E_1 symmetry TO phonons in CdS exhibit a pronounced dip prior to the onset of the resonance enhancement (Fig. 8). They attributed the dip to the destructive interference between resonant and non-resonant contributions to the Raman scattering tensor.

In subsequent studies, carried out at $\omega_i > \omega_G$, Leite *et al.*[44] and Klein and Porto[45] independently observed multiple-LO phonon peaks in CdS, at integral multiples of the frequency of the single LO phonon peaks, superimposed on a broad luminescence band. Multiple-LO phonon peaks have since been observed in other zincblende and wurtzite type crystals and Scott *et al.*[30] have noted that the number of multiple-LO phonon peaks observed in different semiconductors varies monotonically with the polaron coupling coefficient, a, ranging from n = 2 in InSb (a = 0.02) to n = 0 in CdS (a = 0.71). It should be noted that multi-TO phonon peaks have not been observed, thus far, which is consistent with the fact that the TO phonon-exciton coupling is small.

In their investigation of the multi-LO phonon spectrum, Klein and Porto noted that both the fluorescence and Raman scattering intensities increased as ω_i approached ω_{ex}. Furthermore they found that the polarization selection rules for the first-order (one phonon) resonant LO phonon scattering were different from the polarization selection rules for the non-resonant LO phonon scattering. Thus, at $\omega_i \approx \omega_{ex}$ they observed a strong normally forbidden first-order scattering by E_1 symmetry

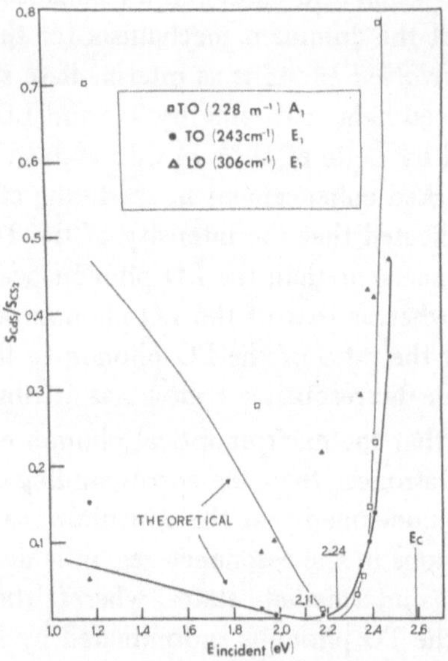

Fig. 8. The Raman scattering efficiencies of CdS (80°K) as a function of incident photon energy. $\mathcal{E}_G$ is the electron-energy gap of CdS. The solid lines are theoretical curves. [After J. M. Ralston, R. L. Wadsack and R. K. Chang, Phys. Rev. Letters 25, 814(1970).]

LO phonons for $\hat{e}_i \| \hat{e}_s$, whereas at ω_i far below ω_{ex} they observed the usual allowed scattering by the LO phonons for $\hat{e}_i \perp \hat{e}_s$.

Damen *et al.*[46] have carried out an experimental study of the half width and shapes of the multi-LO phonon peaks in CdS, ZnO and ZnS. They find a striking dependence of the 2-LO phonon line shape and width on the frequency of the incident radiation; namely, as ω_i increases toward ω_{ex}, the peaks narrow and shift toward somewhat high frequencies. They attribute this behavior to the fact that far away from ω_{ex}, LO phonons of all wave vectors participate in 2-LO phonon scattering processes and consequently the shape of the 2-LO phonon peak is in part determined by the dispersion (ω_j vs. k_j) curve of the LO phonons. As ω_i approaches ω_{ex}, the exciton intermediate states begin playing a more dominant role and there is an increased contribution from LO phonons with (small) wave vectors corresponding to $k_j = k_{j'} \approx 2/a_x$. The resulting peaks are narrower and the position of the peaks correspond to twice the frequency (ω_L) of the $q \approx 0$ LO phonons. They find that the intensity of the 2-LO phonon peak remains strong and also remains narrow as ω_i or ω_s becomes greater than ω_G. They also find that widths of the higher multi-phonon-peaks are comparable to those of the 2-LO phonon peak and suggest that the scattering process involves LO phonons of different symmetry, i.e., combinations of A and E symmetry phonons, rather than the combinations of phonons having the same symmetry.

Martin and Damen[19] have more recently investigated the strong exciton-enhanced forbidden ($\hat{e}_i \parallel \hat{e}_s$) 1-LO phonon scattering at incident radiation frequencies $\omega_i < \omega_{ex}$ where optical absorption is not very strong and scattering is a bulk property. Their data on the scattering by E_1 symmetry LO phonons in high purity CdS crystals indicate that the forbidden 1-LO phonon scattering is weak away from resonance but increases very rapidly near resonance, becoming much stronger than any allowed one phonon peak, whereas the allowed ($\hat{e}_i \perp \hat{e}_s$) 1-LO phonon scattering is strong away from resonance, but is quite weak near resonance (Fig. 9). Their results show that the intensity of the forbidden 1-LO phonon peak, relative to that of the TO phonon peak, increases as ω_i approaches ω_{ex}, while that of the allowed 1-LO phonon peak actually decreases. (There is actually no real contradiction between these results and those reported by Leite *et al*[28] which were based on unpolarized scattering measurements.) They also report that the 2-LO phonon peak observed for $\hat{e}_i \parallel \hat{e}_s$ is actually stronger than any of the one-phonon peaks. They attribute the large intensity of the 2-LO phonon peak to scattering by LO phonons with wave vectors $\underset{\sim}{k}_j \approx \underset{\sim}{k}_{j'} \approx 2/a_x$.

Martin and Damen attribute the resonance-enhanced forbidden ($\hat{e}_i \parallel \hat{e}_s$) 1-LO phonon peak to the q-dependent diagonal contribution to the electro-optic Raman tensor from scattering processes involving discrete and continuum exciton intermediate states. They discarded the possibility that the intermediate states involved

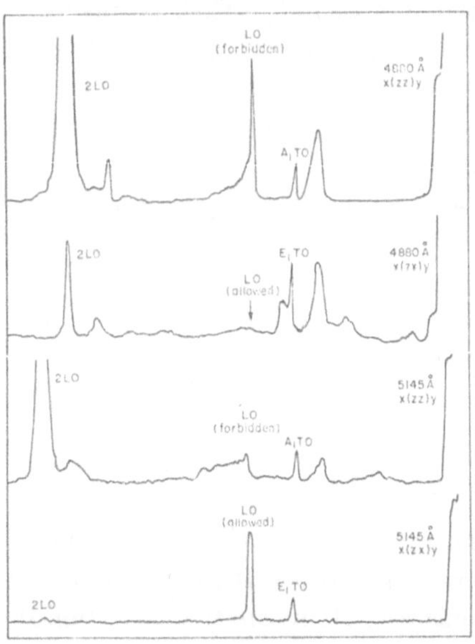

Fig. 9. Uncorrected experimental trace of the Raman scattered light near resonance in CdS for two scattering configurations: x(zz)y, corresponding to $\underset{\sim}{k}_i \parallel X$, $\underset{\sim}{k}_s \parallel X$ and $\hat{e}_i \parallel \hat{e}_s \parallel Z$, and x(zx)y, corresponding to $\underset{\sim}{k}_i \parallel X$, $\underset{\sim}{k}_s \parallel Y$ and $\hat{e}_i \parallel Z$ and $\hat{e}_i \parallel X$. [From R. M. Martin and T. C. Damen, Phys. Rev. Lett. (in press).]

impurity-bound excitons since the observed width of the forbidden 1-LO phonon peak is independent of ω_i and is the same as that for the allowed 1-LO phonon peak. Martin[47] has carried out a theoretical calculation of the q-dependent resonance enhanced scattering by E_1 symmetry LO phonons for $(\hat{e}_i \parallel \hat{e}_s)$ assuming a hydrogenic exciton model with $m_e/m_h = 0.2$, appropriate to the B series exciton in CdS, and summing the matrix elements over all discrete and continuum exciton intermediate states. Martin finds that the contribution from the continuum exciton states is about one half that from the discrete exciton states, and moreover has the opposite sign. His calculated values of the scattering efficiency are in rather good agreement with the experimental data. Thus, his calculated value for the scattering efficiency at $(\omega_x^{(1)} - \omega_i)/\omega_j = 1.8$ is 1.3×10^{-7} cm/str and the observed value is 6.6×10^{-7} cm/str.

Fontana and Mulazzi[48] have recently reported the observation of an exciton enhanced 2-LO phonon peak in TlBr. They did not observe any q-dependent first-order scattering by the odd parity LO phonons, in part because ω_i was not sufficiently close to ω_{ex}, and in part because of the small magnitudes of the scattering wave vector.

A detailed investigation of the scattering wave vector dependence of the forbidden scattering intensity of the E_1symmetry LO phonons in CdS has recently also been carried out by Colwell and Klein.[20] Their results based on a quantitative comparison of the forward and back scattering intensities for a CdS crystal which had been sulfur treated at 800°K to quench the luminescence indicated no dependence of the scattering intensity on the scattering wave vector **q**. They concluded from these results that momentum was not conserved in the scattering process and that the intermediate states therefore involved impurity bound excitons. Colwell and Klein also obtained a theoretical estimate of the q-independent forbidden scattering efficiency based on an impurity bound exciton model, which when combined with the data on the observed efficiency yielded a reasonable value for the concentration of impurity bound exciton sites.

POLARITON THEORY OF LIGHT SCATTERING BY OPTICAL PHONONS

In the Loudon[26] and the Birman and Ganguly[41] theories of light scattering by optical phonons, the interaction between the incident and scattered EM radiation and the interband electronic excitations are treated within the framework of perturbation theory. However, when the frequency of the EM radiation is close to resonance in the electronic transitions, as is the case when the frequency is close to an exciton absorption line, a large fraction of the energy of the EM modes is contained in the electronic excitation. Under these circumstances, it is advantageous to treat the incident and scattered radiation modes as polaritons, i.e. as coupled photon-electronic excitation modes.[49]

The polariton modes of a dielectric medium are characterized by the dispersion relation $k^2c^2 = \Omega^2\,\epsilon(\Omega,k)$ where $\epsilon(\Omega,k)$ is the frequency and wave vector dependent dielectric constant of the medium. For simplicity we will ignore spatial dispersion effects. The dispersion relation for the polariton modes $\Omega(k)$ in a crystal having one or more exciton levels which couple with EM radiation and one or more infrared active optical phonons has the form[18,49]

$$k^2c^2 = \Omega^2\,\epsilon(\omega) = \Omega^2[1 + \epsilon_{cont}(\Omega) + \epsilon_{ex}(\Omega) + \epsilon_{lat}(\Omega)] \qquad (4.9)$$

$$\epsilon_{cont}(\Omega) = \sum_{nn'}\sum_{m}(4\pi\beta_m\omega_m^2)/(\omega_m^2 - \Omega^2 - i\Omega\gamma_m)$$

$$\epsilon_{ex}(\Omega) = \sum_{j}(4\pi\beta_{xj}\omega_{xj}^2)/(\omega_{xj}^2 - \Omega^2 - i\Omega\gamma_{xj})$$

$$\epsilon_{lat}(\Omega) = \sum_{j}(4\pi\beta_{jo}\omega_{jo})/(\omega_{jo}^2 - \Omega^2 - i\Omega\gamma_{jo})$$

where $\epsilon_{cont}(\Omega)$, $\epsilon_{ex}(\Omega)$ and $\epsilon_{lat}(\Omega)$ are the contributions to $\epsilon(\Omega)$ from continuum e-h pair excitations, exciton excitations and TO phonons respectively; ω_m and γ_m are frequency and damping constants of the mth continuum e-h pair excitation of a given pair of valence and conduction bands nn′; and $4\pi\beta_m$ is the contribution of that excitation to the low frequency ($\Omega = 0$) dielectric constant ϵ_s; ω_{xj} and γ_{xj} are the frequency and damping constant of the jth exciton excitation and $4\pi\beta_{xj}$ is the contribution of that excitation to ϵ_s; ω_{jo} and γ_{jo} are the frequency and damping constant of the type j infrared active TO phonon and $4\pi\beta_{jo}$ is the jth TO phonon contribution to ϵ_s. One notes that $\epsilon_{ex}(\Omega)$ and $\epsilon_{lat}(\Omega)$ exhibit resonance peaks at ω_{xj} and ω_{jo} respectively. In real situations the sharpness of the resonance is limited by the damping constants of the excitation and, in the case of excitons, by spatial dispersion effects. The frequency dependence of $\epsilon_{cont}(\Omega)$ depends on the character of the energy bands and the nature of the critical points in the combined density of states of the pair of energy bands. In the case of parabolic energy bands at a "direct gap", which have an M_o type critical point, the contribution to $\epsilon_{cont}(\Omega)$ in the vicinity of the absorption edge has the approximate form,[50]

$$\epsilon^{Mo}_{cont}(\Omega) = \begin{cases} \epsilon^{Mo}_{s\,cont}\Omega^{-2}\,[2\omega_g^{1/2} - (\omega_g + \Omega)^{1/2}]; \; \Omega > \omega_g \\ \epsilon^{Mo}_{s\,cont}\Omega^{-2}\,[2\omega_g^{1/2} - (\omega_g + \Omega)^{1/2} - (\omega_g - \Omega)^{1/2}]; \; \Omega < \omega_g \end{cases} \quad (4.10)$$

where $\epsilon^{Mo}_{s\,cont} = \epsilon^{Mo}_{cont}(\Omega = 0)$. Thus, $\epsilon^{Mo}_{cont}(\Omega)$ exhibits a small peak at ω_g (with an infinite slope on the low frequency side and a finite slope on the high frequency side), but is otherwise a relatively slowly varying function at $\omega < \omega_g$, even in the absence of damping.

The polariton dispersion curves in the vicinity of the absorption edge in ZnSe are sketched in Fig. 10. The polariton modes $\Omega(\mathbf{k})$ are an admixture of photon, exciton, continuum e-h pairs and TO phonons whose strengths are determined by the oscillator strengths of the excitations, and by the factors $[\omega_m^2 - \Omega^2(\mathbf{k})]$, $[\omega_{xj}^2 - \Omega^2(\mathbf{k})]$ and $[\omega_{jo}^2 - \Omega^2(\mathbf{k})]$ which appear in the expression for $\epsilon(\Omega)$. One notes that even though the TO phonon-EM radiation coupling constant may be large,

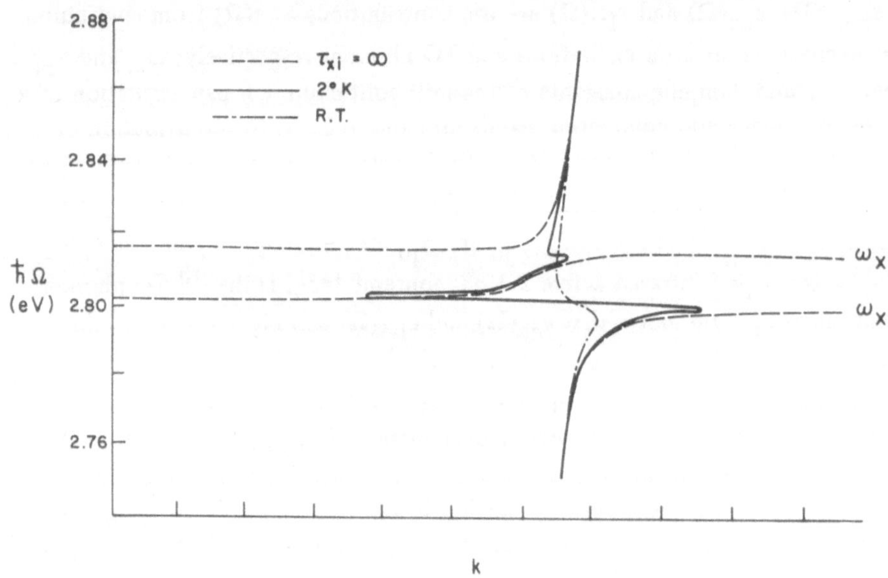

Fig. 10. Polariton dispersion curves for ZnSe based on a two exciton level model. [After S. Ushioda, A. Pinczuk, E. Burstein and D. L. Mills, Proc. 1968 Int. Conf. on Light Scattering Spectra of Solids, G. B. Wright, Ed., p. 347 (Springer-Verlag, N. Y. 1969).]

the admixture of TO phonons at frequencies in the vicinity of the absorption edge is in general negligible, since (except in the case of narrow gap and zero-gap semiconductors) ω_{jo} is very much smaller than ω_g. One can therefore neglect the TO phonon content of the incident and scattered polaritons. The oscillator strengths of the exciton excitations in zincblende and wurtzite type semiconductors are in general quite small and, correspondingly, the exciton contribution, $\epsilon_{ex}(\Omega)$, to the low frequency dielectric constant ϵ_s is quite small. As a consequence, the exciton strengths of polariton modes with frequencies $\Omega(\mathbf{k}) << \omega_{xj}$ are generally very small. On the other hand, the exciton contribution to the dielectric constant can be quite large in the vicinity of $\Omega(\mathbf{k}) = \omega_{xj}$ and the exciton strengths of the polariton can approach unity. The contribution of the continuum e-h pair excitations to the dielectric constant $\epsilon_{cont}(\Omega)$ is appreciable throughout the range of frequencies of interest and consequently the total continuum e-h pair excitation strength (the sum of the strengths of the continuum e-h pair excitations) is appreciable except at frequencies in the vicinity of ω_{xj} (or in the vicinity of ω_{jo}) where the strength of the jth exciton excitation (or that of the jth TO phonon) approaches unity and the strengths of the other excitations approach zero. The latter is a direct consequence of the sum rules on the photon and electric dipole excitation strengths of the polariton modes.[51]

A formulation of the theory of resonance-enhanced Raman scattering in terms of the inelastic scattering of the coupled photon-electric dipole excitation (e.g., polariton) modes by "vibrational" modes has been given by Ovander and co-workers.[24,25] A polariton description has also been used by Thomas and Hopfield[39] to explain the very large cross-sections for light scattering by spin-flip transitions of impurity levels in CdS. In collaboration with D. L. Mills, we have formulated the theory of resonance-enhanced scattering of light by optical phonons in terms of the inelastic scattering of polaritons by phonons via interaction of the phonons with the exciton and continuum e-h pair excitation parts of the polaritons.[18,51] Although the weak interaction of EM radiation with the continuum e-h pair excitation can be treated by perturbation theory, it was also included in the polariton formulation of the Raman scattering process. A polariton theory of resonant-Raman scattering has also been given by Hopfield[52] and by Bendow and Birman.[53]

In the polariton formulation of light scattering, the incident polaritons are scattered by phonons from an initial state $\Omega_i(\mathbf{k}_i,\hat{e}_i)$ to a final state $\Omega_s(\mathbf{k}_s,\hat{e}_s)$ (Fig. 11) via interaction with the exciton and continuum e-h pair excitation parts of the incident and scattered polaritons. The polariton formulation of light scattering by phonons, which is also applicable to the scattering of light by other types of elementary excitations (e.g. plasmons, spin-waves and single particle excitations), leads to a scattering efficiency that is proportional to $[v_p^2(\Omega_s)v_g(\Omega_i)v_g(\Omega_s)]^{-1}$ where $v_p(\Omega_s) =$

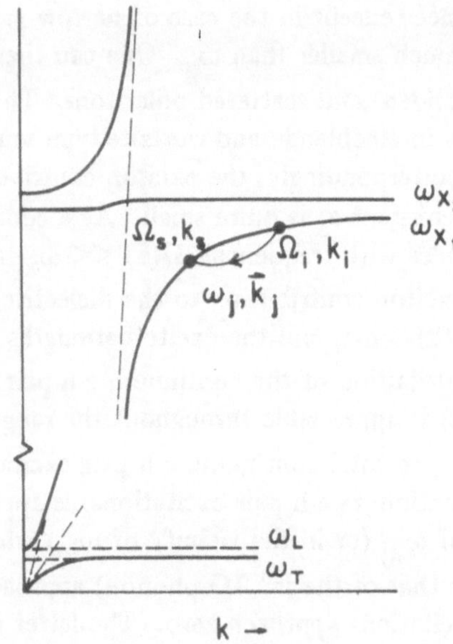

Fig. 11. Schematic diagram showing the scattering of polaritons from an initial state $(\Omega_i, \underset{\sim}{k}_i)$ to a final state $(\Omega_s, \underset{\sim}{k}_s)$.

$(\Omega/k)_{\Omega_s}$ is the phase velocity of the scattered polaritons, and $v_g(\Omega_i) = (\partial\Omega/\partial k)_{\Omega_s}$ and $v_g(\Omega_s) = (\partial\Omega/\partial k)_{\Omega_i}$ are the group velocities of the incident and scattered polariton modes, respectively. Since the phase velocity and the group velocity of polaritons decrease as the frequency approaches a resonance absorption line, this factor leads to a strong enhancement of the scattering efficiency. The frequency variation of the exciton and continuum e-h pair excitation strengths of the polariton modes are additional factors which contribute to the frequency dependence of the scattering efficiency.

MORPHIC EFFECTS

An externally applied force, such as an electric field, which lowers the symmetry of a solid can modify the scattering and absorption of EM radiation by optical phonons. The changes in the properties of the solid which result from the lowering of the symmetry by an external force are called "morphic effects".[54] Morphic effects manifest themselves in the elastic, dielectric and piezoelectric properties of a crystal as new coefficients created by the lowering of the symmetry of the crystal. Thus, electrostriction in centrosymmetric crystals can be regarded as electric field-induced piezoelectricity and as such constitutes a morphic effect. In lattice dynamical phenomena, morphic effects manifest themselves as changes in the selection rules for absorption and scattering of EM radiation by optical phonons and as splittings of the degeneracy of optical phonons. The effects are particularly striking when normally forbidden optical absorption and scattering processes become allowed. The observation of electric field induced absorption by optical phonons in diamond[55] and of electric field induced Raman scattering by optical phonons in $KTaO_3$ and $SrTiO_3$[56-58] are examples of morphic effects in which removal of the center of inversion of a crystal leads to new peaks in the IR absorption and Raman scattering spectra. The surface electric field induced Raman scattering by LO phonons in zinc-blende type semiconductors[32] is an example of morphic effects in which the electric field modifies the selection rules for Raman scattering by optical phonons which are already Raman active.

PHENOMENOLOGICAL THEORY

The phenomenological theory of the morphic effects of externally applied forces on Raman scattering by optical phonons is fairly straightforward. The externally applied force $\mathbf{F}_a$ is treated as a perturbation and the *transition electric susceptibility* for first-order scattering by optical phonons in the presence of the external force $\underset{\sim}{\chi}_j^{(1)}(\mathbf{F}_a)$ is expressed as a Taylor expansion[54] of $\underset{\sim}{\chi}_j^{(1)}$ in powers of $\mathbf{F}_a$ as follows:

$$\underset{\sim}{\chi}_j^{(1)}(\mathbf{F}_a) = \underset{\sim}{\chi}_j^{(1)} + (d\underset{\sim}{\chi}_j^{(1)}/d\mathbf{F}_a)\,\mathbf{F}_a + (d^2\underset{\sim}{\chi}_j^{(1)}/d\mathbf{F}_a^2)\,\mathbf{F}_a^2 + \cdots \tag{5.1}$$

$$= \underset{\sim}{\chi}_j^{(1)} + \underset{\sim}{\chi}_{j\,\mathbf{F}_a}^{(1)} + \underset{\sim}{\chi}_{j\,\mathbf{F}_a^2}^{(1)} + \cdots$$

where $\underset{\sim}{\chi}_{j\,\mathbf{F}_a}^{(1)} = (\partial^2\chi/\partial Q_j\partial \mathbf{F}_a)\,\underset{\sim}{Q}_j\mathbf{F}_a$ represents the change in $\underset{\sim}{\chi}_j^{(1)}$ which is linear in $\mathbf{F}_a$; $\underset{\sim}{\chi}_{j\,\mathbf{F}_a^2}^{(1)} = (\partial^3\chi/\partial Q_j\partial \mathbf{F}_a^2)\,\underset{\sim}{Q}_j\mathbf{F}_a^2$ represents the change in $\underset{\sim}{\chi}_j^{(1)}$ which is quadratic in $\mathbf{F}_a$; and we have omitted the designation that $\underset{\sim}{\chi}_j^{(1)}(\mathbf{F}_a)$ is q-dependent.

In centrosymmetric crystals, the change in $\underset{\sim}{\chi}_j^{(1)}$ induced by an external electric field (i.e. $\mathbf{F}_a = \mathbf{E}_a$) involves a contribution from the electric field induced relative displacements of the atom in the unit cell (provided the crystal has infrared active optical phonons), and a contribution from the electric field induced electronic deformation of the atoms. In crystals lacking a center of inversion, there is an additional

References pp. 77-79

contribution from the electric field induced elastic deformation of the crystal. Thus, in the general case, $\underset{\sim}{\chi}^{(1)}_{jE_a}$, the linear change in $\underset{\sim}{\chi}^{(1)}_j$ induced by an external electric field, is given by

$$\underset{\sim}{\chi}^{(1)}_{j\,Ea} = [\partial\chi_j/\partial E_a]E_a + \sum_k [\partial\chi_j^{(1)}/\partial u_k]u_k(E_a) \qquad (5.2)$$

$$+ [\partial\chi_j^{(1)}/\partial\Sigma(E_a)]\underset{\sim}{\Sigma}(E_a)$$

where $u_k\,(E_a)$ is the electric field induced relative displacement of the atoms in the unit cell associated with a type k infrared active optical phonon; and $\underset{\sim}{\Sigma}(E_a)$ is the electric field induced elastic strain. We note that $u_k(E_a)$ is zero in diamond type crystals (i.e. the optical phonons are infrared inactive) and that $\underset{\sim}{\Sigma}(E_a)$ is zero in centrosymmetric crystals.

$\underset{\sim}{\chi}^{(1)}_{j\,E_a}$ can also be expressed in terms of the atomic displacement Raman tensor, $\underset{\sim}{a}_j = (\partial\chi/\partial u_j)_E$, and the electro-optic Raman tensor, $\underset{\sim}{b} = (\partial\chi/\partial E)_{u_j}$ as follows:

$$\underset{\sim}{\chi}^{(1)}_{j\,E_a} = (d^2\chi/dQ_j dE_a)\,\underset{\sim}{Q}_j E_a + \sum_k\,(d^2\chi/dQ_j du_k)\,\underset{\sim}{Q}_j u_k(E_a) + \qquad (5.3)$$

$$(d^2\chi/dQ_j d\Sigma)\,\underset{\sim}{Q}_j\underset{\sim}{\Sigma}\,(E_a)$$

$$\underset{\sim}{\chi}^{(1)}_{j\,E_a} = \underset{\sim}{a}_{j\,E_a} u_j + \underset{\sim}{b}_{E_a} E_j$$

where

$$\underset{\sim}{a}_{j\,E_a} = (\partial a_j/\partial E_a)_{u_k,\Sigma}\,E_a + \sum_k\,(\partial a_j/\partial u_k)_{E,\Sigma}\,u_k(E_a) + (\partial a_j/\partial\Sigma)_{E,u_k}\,\underset{\sim}{\Sigma}\,(E_a)$$

$$\underset{\sim}{b}_{E_a} = (\partial b/\partial E_a)_{u_k,\Sigma}E_a + \sum\,(\partial a_j/\partial u_k)_{E,\Sigma}u_k(E_a) + (\partial b/\partial\Sigma)_{E,u_k}\,\underset{\sim}{\Sigma}\,(E_a)$$

The phenomenological theory of the morphic effects of an external force F_a on the infrared absorption of EM radiation by optical phonons is formulated in a similar way. The linear electric moment (per unit cell) of the optical phonon $M_j^{(1)}(q)$, which determines the rate of absorption of EM radiation by the optical phonons is expanded in powers of F_a as follows:[54]

$$\underset{\sim}{M}_j^{(1)}(q,F_a) = \underset{\sim}{M}_j^{(1)}(q) + [dM_j^{(1)}(q)/dF_a]\,F_a + [d^2M_j^{(1)}(q)/dF_a^2]\,F_a^2 + \cdots \qquad (5.4)$$

$$= \underset{\sim}{M}_j^{(1)}(q) + \underset{\sim}{M}_{j\,F_a}^{(1)}(q) + \underset{\sim}{M}_{j\,F_a^2}^{(1)}(q) + \cdots$$

where

$$\underset{\sim}{M}_j^{(1)}(q) = [dM/dQ_j(q)]\,\underset{\sim}{Q}_j(q) = [\partial M/\partial u_j(q)]_E\,u_j(q) + [\partial M/\partial E(q)]_{\underset{\sim}{u}_j}\,E_j(q) \qquad (5.5)$$

$$= \underset{\sim}{e}_j^*(q)\,u_j(q) + \underset{\sim}{a}(q)E_j(q)$$

$e_j^*(q) = [\partial M/\partial u_j(q)]_E$ is the q-dependent effective charge (at constant E) of the optical phonons and $(\partial M/\partial E)_{u_j} = \underset{\sim}{a}(q)$ is the q-dependent electric susceptibility per unit cell (at constant u_j).

When the external force corresponds to an electric field (i.e. $F_a = E_a$) the linear change in $M_j^{(1)}(q)$ induced by the electric field is given by

$$\underset{\sim}{M}_{j\,E_a}^{(1)} = \underset{\sim}{e}_{j\,E_a}^*\,u_j + \underset{\sim}{a}_{E_a}E_j \qquad (5.6)$$

where

$$\underset{\sim}{e}_{j\,E_a}^* = (\partial e_j^*/\partial E_a)_{u_k,\Sigma}E_a + \sum_k (\partial e_j^*/\partial u_k)_{E,\Sigma}u_k(E_a) + (\partial e_j^*/\partial \Sigma)_{E,u_k}\underset{\sim}{\Sigma}(E_a)$$

$$\underset{\sim}{a}_{E_a} = (\partial a/\partial E_a)_{u_k,\Sigma}E_a + \sum_k (\partial a/\partial u_k)_{E,\Sigma}u_k(E_a) + (\partial a/\partial \Sigma)_{E,u_k}\underset{\sim}{\Sigma}(E_a)$$

in which we have, for convenience, omitted the designation of q-dependence.

We note that in centrosymmetric crystals $\chi_{j\,F_a}^{(1)}$ and $M_{j\,F_a}^{(1)}$, the parameters which characterize first-order Raman scattering and infrared absorption by type j optical phonons in an external force, are dependent on the parity of the mode.[59] This follows from the fact that the tensor coefficients $(d\chi/dQ_j)$, (dM/dQ_j), $(d^2\chi/dQ_jdF_a)$, (d^2M/dQ_jdF_a), etc. must have even parity in centrosymmetric crystals. This leads to the general result that in the case of odd parity Raman inactive phonons, the atomic displacement Raman tensor in the presence of an external force $\underset{\sim}{a}_{jF_a}$ is zero for all even parity forces such as the elastic strain and even powers of the electric field, but may be non-zero for odd parity forces such as odd powers of the electric field, whereas the effective charge tensor in the presence of an external force $\underset{\sim}{e}^*_{jF_a}$ is zero for odd parity forces and may be non-zero for even parity forces. In the case of even parity (infrared inactive) phonons, $\underset{\sim}{a}_{jF_a}$ is zero for odd parity forces but may be non-zero for even parity forces, and $\underset{\sim}{e}^*_{jF_a}$ is zero for odd parity forces but may be non-zero for even parity. Parity considerations obviously do not apply to crystals lacking a center of inversion.

An alternate procedure can be used to determine the morphic effects of an external force on infrared absorption and Raman scattering by optical phonons.[58] One first establishes the new symmetry of the crystal in the presence of the external force which corresponds to the combined symmetry elements of the force free crystal and of the external force. New symmetries are then established for the optical phonons which conform to the point group symmetry of the "deformed" crystal. The atomic displacement Raman tensor and the effective charge tensor of the optical phonons, in the presence of the external force, correspond to the new symmetries. Although, this approach is a useful one, particularly when the modes are infrared and/or Raman inactive, it has a major drawback in that it does not indicate to what order of the external force the morphic effect occurs.

When the externally applied force is an electric field, the major effect in centrosymmetric crystals is to remove the center of inversion. The electronic states and the optical phonons in the electrically deformed crystals no longer have well defined parity. As a consequence, infrared active, normally Raman inactive phonons may become Raman active; Raman active normally infrared inactive phonons may become infrared active; and phonons which are normally both Raman and infrared inactive, the so called "silent" phonons, may become infrared and/or Raman active. An elastic strain does not remove the center of inversion and therefore cannot induce Raman scattering by odd parity phonons. An elastic strain may, however, induce Raman scattering by Raman inactive phonons in crystals lacking a center of inversion.[59]

ELECTRIC FIELD INDUCED RAMAN SCATTERING BY ODD PARITY OPTICAL PHONONS

An electric field induced first-order Raman scattering by odd parity optical phonons was first observed by Worlock and Fleury in $KTaO_3$ and $SrTiO_3$[56,57,58] paraelectric crystals having the cubic $BaTiO_3$ perovskite structure. Using an ac modulation technique to discriminate against intrinsic higher order scattering, they observed, in the case of $SrTiO_3$, an appreciable first-order scattering by the soft (ferroelectric) F_{1u} symmetry TO phonons at low temperature at fields as low as 200 V/cm. Somewhat higher electric fields were needed to observe scattering by the soft optical phonons in $KTaO_3$. They also observed an appreciable field induced splitting of the degeneracy of the soft optical phonons and of the next higher frequency TO phonons in $SrTiO_3$. Scott *et al*[60] have investigated the electric field induced scattering by polaritons (coupled TO phonon-photon modes) in $SrTiO_3$ and $KTaO_3$, and have obtained, thereby, new and more reliable values for the oscillator strengths and damping constants of the infrared active phonons.

The cubic $BaTiO_3$ (perovskite) structure has three triply degenerate modes of F_{1u} symmetry which are infrared active and one triply degenerate mode of F_{2u} symmetry which is silent. In the presence of an externally applied electric field along [001] = Z the point group symmetry of the perovskite structure is changed from O_h to C_{4v}. As noted by Fleury and Worlock[58] each F_{1u} mode splits into a non-degenerate A_1 mode and a doubly degenerate E mode, and the F_{2u} mode splits into a non-degenerate B mode and a doubly degenerate E mode. The non-zero components of $\underset{\sim}{\chi}^{(1)}_{j\,E_a}$ for the F_{1u} modes are $\chi^{(1)}_{xx,zz} = \chi^{(1)}_{yy,zz}$ and $\chi^{(1)}_{zz,zz}$ for phonons with $\hat{e}_j \parallel \hat{e}_a$, and $\chi^{(1)}_{xz,xz} = \chi^{(1)}_{yz,yz}$ for phonons with $\hat{e}_j \perp \hat{e}_a$. (The first two subscripts designate the components of $\hat{e}_i$ and $\hat{e}_s$, the third subscript designates the component of $\hat{e}_j$ and the fourth subscript designates the components of $\hat{e}_a$, the polarization vector of the applied electric field.) The non-zero components of $\underset{\sim}{\chi}^{(1)}_{j\,E_a}$ for the F_{2u} modes are $\chi^{(1)}_{xx,zz} = -\chi^{(1)}_{yy,zz}$ for the phonons with $\hat{e}_j \parallel \hat{e}_a$, and $\chi^{(1)}_{xz,xz} = \chi^{(1)}_{yz,yz}$ for the phonons with $\hat{e}_j \perp \hat{e}_a$.

Using a scattering configuration in which radiation was incident along [010] = Y and scattered along [100] = X, Fleury and Worlock were able to observe field induced scattering by all of the TO modes in $SrTiO_3$ and by the three lowest $q \simeq 0$ TO modes in $KTaO_3$, but did not observe any field induced scattering by LO phonons in either crystal. They report that the ratio of the field induced scattering intensity of the soft TO phonons with $\hat{e}_j \parallel \hat{e}_a$, to that of the soft phonon with $\hat{e}_j \perp \hat{e}_a$ is 20 in $KTaO_3$ and 7.7 in $SrTiO_3$.

A lattice theory of the electric field induced scattering by the soft TO phonons in cubic $BaTiO_3$ has been developed by Dvorak[61] in which the scattering is attributed to the effects of the electric field induced relative displacements of the atoms. Dvorak relates the scattering efficiency of the optical phonons to the magnitude of the quad-

References pp. 77-79

ratic electro-optic constant which he assumes is due primarily to the contribution from the large electric field induced displacements of the atoms associated with the soft optical phonons. On this basis, he concluded that a field of 10^3V/cm would be sufficient to induce an observable first-order scattering by soft optical phonons in cubic $BaTiO_3$. Dvorak's theory also predicted that the relative scattering intensities of the higher frequency TO phonons relative to that of the soft phonons should go as ω_j^2/ω_{js}^2. The experimental data obtained by Fleury and Worlock do not agree with this relation. Worlock[62] has also developed a lattice theory of the field induced scattering by optical phonons, based on somewhat similar considerations. His prediction of the relative scattering intensities are in somewhat better agreement with experimental data.

ELECTRIC FIELD INDUCED RAMAN SCATTERING BY LO PHONONS IN ZINCBLENDE TYPE SEMICONDUCTORS

The morphic effects of an external electric field manifest themselves, in the case of Raman active phonons, as changes in the polarization selection rules. It is therefore possible, by using scattering geometries in which a zero field scattering does not occur, to observe the field induced scattering by optical phonons which are already Raman active. Thus, Anastassakis *et al*[63] were able to observe the quadratic electric field induced changes in the Raman scattering tensor of the F_{2g} phonons in diamond which normally exhibit scattering only for $\hat{e}_i \perp \hat{e}_s$.

In ZnS type crystals a *backward* scattering by LO phonons does not normally occur for radiation incident along <110> directions. The appearance of LO phonon peaks at frequencies corresponding to unscreened LO phonons in backward scattering from (110) surfaces of ZnS type semiconductors at incident radiation frequencies near the E_1 energy gap was therefore attributed to a resonance enhanced scattering by LO phonons which is induced by electric fields in space charge regions at the surface of the semiconductors.[32] This interpretation was confirmed by the fact that the intensity of the LO phonon peaks could be increased or decreased by externally applied electric fields.

For an electric field along [110] the non-zero components of $\underset{\sim}{\chi}^{(1)}_{j E_a}$ of the F_2 symmetry optical phonons in ZnS type semiconductors, referred to axes X′ = [111], Y′ = [112] and Z′ = [110] appropriate to back scattering from (110) surfaces, are $\chi^{(1)}_{y'y',z'z'}$, $\chi^{(1)}_{x'x',z'z'}$ and $\chi^{(1)}_{y'y',z'z'}$. Thus an electric field induced scattering by the LO phonons should be observable in (x′x′) and (y′y′) (i.e. $\hat{e}_i \parallel \hat{e}_s$) spectra and in x′y′ (i.e. $\hat{e}_i \perp \hat{e}_s$) spectra. Experimentally it is found that the intensities of the peaks in the $\hat{e}_i \parallel \hat{e}_s$ spectra are much stronger than in the $\hat{e}_i \perp \hat{e}_s$ spectra.

In the case of back scattering from (001) surfaces which is normally allowed for LO phonons, but not for TO phonons the non-zero components of $\underset{\sim}{\chi}^{(1)}_{j E_a}$ referred to cubic axes, for E_a along [001] = Z, are $\chi^{(1)}_{xx,zz} = \chi^{(1)}_{yy,zz}$, $\chi^{(1)}_{zz,zz}$ and $\chi^{(1)}_{xz,xz} = \chi^{(1)}_{yz,yz}$. In this scattering configuration a surface electric field induced scattering by LO phonons can be observed in (xx) and (yy) (i.e. $\hat{e}_i \parallel \hat{e}_s$) spectra, while the normally allowed scattering by LO phonons can be observed in (xy) (i.e. $\hat{e}_i \perp \hat{e}_s$) spectra. It is therefore possible to observe both normally allowed and field induced Raman scattering by LO phonons in back scattering from (100) surfaces. We note that the effects of the surface electric field on scattering by TO phonons cannot be observed in any back scattering configuration. Thus back scattering by TO phonons from (100) surfaces does not occur either in the absence or the presence of the surface electric field.

Typical spectra of TO and LO phonons in n-type InSb obtained in back scattering from (110) surfaces are shown in Fig. (12).[40] The intensity of the LO phonon peaks (relative to that of the TO phonon peaks) increases with increasing carrier density becoming larger than that of the TO phonon peak, levels off and then decreases with further increase in carrier density. For all carrier densities, the LO phonon peak appears at the "unscreened" frequency indicating an absence of coupling with plasmons. A plot of the intensity of the LO phonon peak relative to that of the TO phonon peak (which is not affected by the surface electronic field) is given in Fig. (13). Similar results were obtained for InAs and GaSb.[64]

In the case of InSb and GaSb, the absence of any screening of LO phonons at the surface is consistent with the existence of a surface depletion space charge layer in these materials and to the absence of free carriers within the radiation skin depth which is rather small at frequencies near the E_1 gap. In the case of n-type InAs, the absence of screening cannot be attributed to the absence of free carriers within the skin depth since available capacitance and electron tunneling[65,66] data indicates the presence of a narrow accumulation layer at the surface. It has been tentatively suggested[64] that the LO phonons which take part in the surface field induced scattering within the narrow accumulation layer have wave vectors which are appreciably larger than the Fermi-Thomas "screening" wave vector and are therefore uncoupled to the plasmons.

There are other mechanisms which can lead to, otherwise forbidden, scattering by LO phonons from (110) surfaces. These include q-dependent scattering involving either q-dependent terms in the Frohlich interaction matrix elements or q-dependent terms in the energies of electrons and holes in the intermediate state, and impurity induced scattering arising either from changes in symmetry of the medium in the vicinity of the impurity, or from a breakdown of the translational symmetry of the

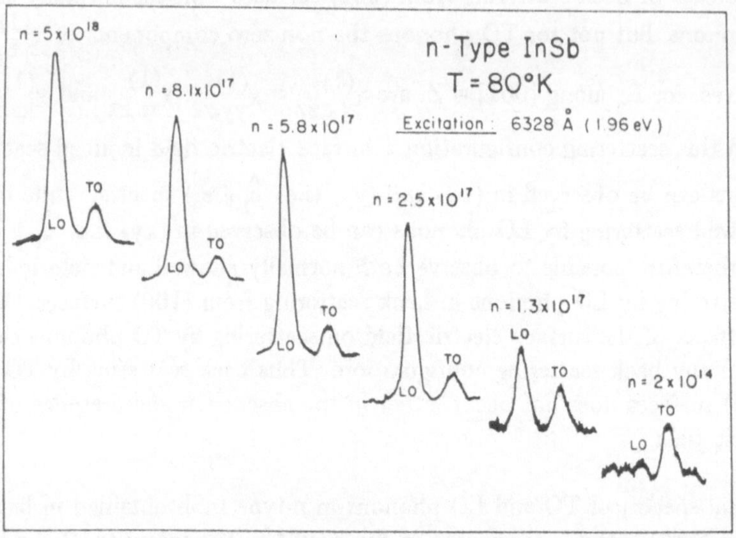

Fig. 12. Recorder traces of back scattering Raman spectra of LO and TO phonons from (110) surfaces of n-InSb at 80°K. [From A. Pinczuk and E. Burstein, Proc. 1970 Int. Conf. on Physics of Semiconductors, p. 727 (U. S. Atomic Energy Commission, Wash., D. C. 1970).]

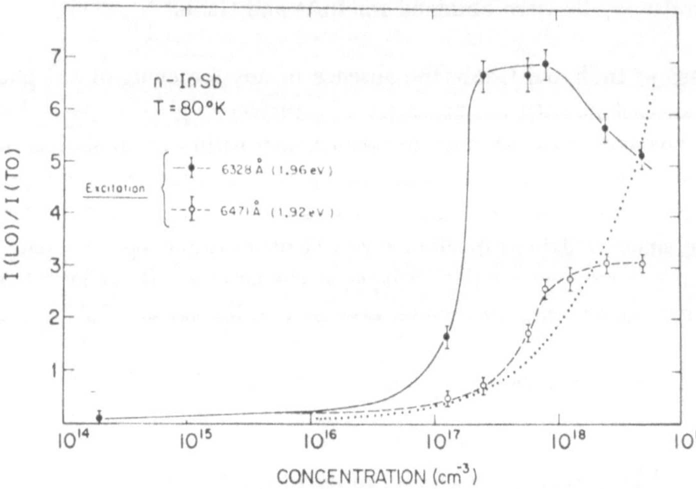

Fig. 13. The ratio I(LO)/I(TO) for back scattering from (110) surface of n-InSb at 80°K plotted as a function of donor density. The dotted curve represents the ratio expected for a field induced scattering tensor which is linear in the electric field. [From A. Pinczuk and E. Burstein, Proc. 1970 Int. Conf. on Physics of Semiconductors, p. 727 (U. S. Atomic Energy Commission, Wash., D. C. 1970).]

medium. Of the two types of mechanisms, only the latter leads to a scattering intensity which increases with increase in impurity (i.e. carrier) density. However, the impurity induced scattering mechanism does not account for the increase and subsequent decrease in the intensity of the LO phonon peaks with carrier density at large carrier densities, nor can it account for the observed effects of the externally applied electric field.

In the case of high carrier density samples (and also in the case of large externally applied electric fields[34]) the intensities of the surface electric field induced LO phonon peaks are considerably greater than the intensities of the allowed TO phonon peaks that appear in the same spectra. This feature together with the observed dependence of the intensity on carrier density are attributed to Franz-Keldysh effects of the electric field on the interband and intraband transitions and to the resultant non-cancellation of the electron and hole intraband Frohlich interaction matrix elements.[41] The net effect of the surface electric field, or equivalently the effect of the bent energy bands at the surface, is to open up a new channel for Raman scattering by LO phonons involving *two band* scattering processes. Cardona[67] and Van Vechten *et al*[68] have used a similar mechanism to account for the observed magnitude of third order non-linear susceptibility coefficients.

As a result of the spatial tilting of the energy bands in the presence of an electric field, the interband transitions ① and ③ (Fig. 2) correspond to tunneling assisted optical transitions (Franz-Keldysh effect) in which $k_{/\!/}$, the component of the k along the directions of the electric field, is not a constant of motion. As a consequence, the electron and hole in the intermediate states are at different points in real space and their contributions to the intraband matrix element of the Frohlich interaction do not cancel.

The intraband Frohlich interaction matrix elements in the presence of an electron field are given by[40]

$$(F^{j}_{\beta\alpha E}) = i(eE_j/qV^{1/2}) \int d^3r \; \varphi^{*}_{\beta E}(r)\,(e^{iq\cdot r_e} - e^{iq\cdot r_h})\,\varphi_{\alpha E}(r) \tag{5.7}$$

where $\phi_{\beta E}(r)$ and $\phi_{\alpha E}(r)$ are the envelope functions of the e-h pair in the intermediate states, which do not have inversion symmetry in the presence of an electric field. (In the absence of a coulomb interaction between the electron and hole in the intermediate states and in a uniform electric field, $\phi_{\beta E}(r)$ and $\phi_{\alpha E}(r)$ correspond to Airy functions). For $\mathbf{q} /\!/ \mathbf{E}_a$ the leading term of $(F^{j}_{\beta\alpha E})$, is q-independent and is given by

$$(F^j_{\beta aE}) \approx i(eE_j/V^{1/2}) \int d^3r\, \varphi^*_{\beta E}(r)\, r_E\, \varphi_{aE}(r) \qquad (5.8)$$

$$= i(eE_j/V^{1/2})(r^a_{\beta aE})$$

where r_E is the component of the relative coordinate of the electron and hole along E_a and $(r^a_{\beta aE})$ is the q-independent electric dipole matrix element for transitions between intermediate states in the presence of an electric field. Since $(F^j_{\beta aE})$ is diagonal in the electronic states, the field induced scattering by LO phonons involving the two band Franz-Keldysh mechanism occurs for $\hat{e}_i \parallel \hat{e}_s$.

It should be noted that the Franz-Keldysh mechanism is also applicable to the electric field induced scattering by odd parity LO phonons in centrosymmetric crystals and is believed to be responsible for the surface electric field induced scattering by LO phonons which has been observed in back scattering from (100) surfaces of p-type samples of PbS and PbTe.[69]

The resonance enhancement of the electric field induced Raman scattering of LO phonons which occurs when ω_i approaches a critical point in the interband density of states will be particularly strong when $(\hbar^2 e^2 E_a^2/2\mu^*_{\parallel})^{1/3} >> \mathcal{E}_G - \hbar\omega_i ; \mathcal{E}_G - \hbar\omega_s$ where $\mu^*_{\parallel}$ is the parallel components of the reduced effective mass tensor of the electron-hole pair. Furthermore, when these two conditions are satisfied, the intensity of the resonance enhanced electric field induced Raman scattering by LO phonons becomes relatively independent of the strength of E_a.

The observed dependence of the intensity of the surface electric field induced scattering by LO phonons in n-InSb on the concentration of carriers in the bulk is attributed to the increase in the electric field in the depletion layer and the associated decrease in width of the depletion layer with increase in donor concentration. At low impurity concentrations, the skin depth is smaller than the width of the depletion layer and the initial increase in the intensity of the LO phonons is due to the increase in electric field. The leveling off in the ratio I(LO)/I(TO) at high concentration is attributed to the fact that at these concentrations the surface field is sufficiently large to satisfy one of the inequalities and the scattering intensity is no longer strongly dependent on the field. The subsequent decrease in I(LO)/I(TO) observed in samples at the highest carrier concentrations in the (6328 Å) spectra is attributed to the decrease in scattering length that occurs when depletion layer barrier width becomes appreciably smaller than the skin depth of the crystal at the incident and scattered radiation frequencies.

CONCLUDING REMARKS

It is now feasible, under resonance enhanced light scattering conditions, to observe "external force" induced scattering (morphic effects) and wave vector dependent scattering (spatial dispersion effects) by Raman active and Raman inactive optical phonons. Moreover one can make effective use of the electric fields in surface space charge regions to observe electric field induced Raman scattering by LO phonons in semiconductors.[64] The latter can, in turn, be used to obtain information about the character of the space charge regions. Wave vector dependent scattering is most readily observed in the vicinity of exciton resonances, and it can be particularly pronounced in the case of resonances associated with impurity-bound excitons.[20]

The ability to observe Raman scattering by normally Raman inactive optical phonons greatly extends the range of materials in which optical phonons can be investigated by first order Raman scattering. An equally important aspect is the fact that external force induced Raman scattering and wave vector dependent Raman scattering, and their dependence on the frequency of the incident and scattered radiation, provide an opportunity for obtaining a deeper understanding of the elementary scattering processes which play a role in the scattering of light by optical phonons.

ACKNOWLEDGMENTS

We wish to acknowledge valuable discussions with E. Ananastasakis, M. V. Klein, A. A. Maradudin, R. M. Martin, D. L. Mills and R. F. Wallis. We also thank R. M. Martin and T. C. Damen for providing us with a copy of their paper prior to publication.

REFERENCES

* *Research supported in part by the U. S. Army Research Office-Durham and Advanced Research Projects Agency.*
1. *Proceedings of the 1968 Int. Conf. on Light Scattering Spectra of Solids, G. B. Wright, Editor (Springer-Verlag, New York, 1969).*
2. *Light is also scattered by fluctuations of the magnetic susceptibility. However the contribution to the scattered intensity from fluctuations in the magnetic susceptibility is very much smaller than that from fluctuation in the electric susceptibility of comparable amplitude, and is therefore, in general, neglected.*
3. *J. M. Worlock and S. P. S. Porto, Phys. Rev. Letters 15, 697(1965).*
4. *R. Loudon, Advances in Physics 13, 423(1964).*
5. *E. Burstein, R. Ito, A. Pinczuk and M. Shand, Proc. 1969 Symposium on the Interactions of Light with Sound Waves, J. Acoust. Soc. (in press).*
6. *In bulk opaque media, the scattering length is equal to $1/(k_i'' + k_s'')$ where k_i'' and k_s'' are*

the imaginary part of the complex wave vectors, $\tilde{k}_i = k_i' + ik_i''$ and $\tilde{k}_s = k_s' + ik_s''$, of the incident and scattered radiation. (D. L. Mills, A. A. Maradudin and E. Burstein, Proc. 1968 Int. Conference on Light Scattering Spectra of Solids, G. B. Wright, Ed., pg. 399 (Springer-Verlag, New York, 1969.)

7. *D. A. Kleinman, Phys. Rev. 126, 1977(1962).*
8. *E. Burstein, A. Pinczuk and R. F. Wallis (to be published).*
9. *R. Loudon, Proc. Roy. Soc., A275, 218(1963).*
10. *S. Ushioda, A. Pinczuk, W. Taylor and E. Burstein, Proc. 1967 Conf. on II-VI Semiconductor Compounds, D. G. Thomas, Ed., p. 1185 (W. A. Benjamin, New York, 1968).*
11. *A. A. Maradudin and S. Ushioda, Phys. Rev. (in press).*
12. *R. Loudon, Proc. 1968 Int. Conf. on Light Scattering Spectra of Solids, G. B. Wright, Ed., p. 25 (Springer-Verlag, New York, 1969).*
13. *E. Burstein, S. Ushioda, A. Pinczuk and J. F. Scott, Proc. 1968 Int. Conf. on Light Scattering Spectra of Solids, G. B. Wright, Ed., p. 43 (Springer-Verlag, New York, 1969).*
14. *A. A. Mooradian and A. McWhorter, Phys. Rev. Letters 19, 849(1967).*
15. *E. Burstein, A. Pinczuk and S. Iwasa, Phys. Rev. 157, 611(1967).*
16. *E. Burstein, E. Anastassakis, D. Portigal and A. A. Maradudin (to be published).*
17. *Y. Toyozawa, Prog. Theor. Phys. Kyoto, 20F, 53(1958).*
18. *D. L. Mills and E. Burstein, Phys. Rev. 188, 1465(1969).*
19. *R. M. Martin and T. C. Damen, Phys. Rev. Letters (in press).*
20. *P. J. Colwell and M. V. Klein, Solid State Communications 8, 2095(1970).*
21. *D. C. Hamilton, Phys. Rev. 188, 1221(1969).*
22. *S. Ushioda, A. Pinczuk, E. Burstein and D. L. Mills, Proc. 1968 Int. Conf. on Light Scattering Spectra of Solids, G. B. Wright, Ed., p. 347 (Springer-Verlag, New York, 1969).*
23. *P. P. Shorygin, Doklady Akad. Nauk SSSR, 87, 201(1952). See review paper by J. Behringer in "Raman Spectroscopy", H. A. Szymanski, Ed., p. 168 (Plenum Press, New York, 1967).*
24. *L. N. Ovander, Optika i Spektroskopia, 4, 555(1958); Soviet Phys.-Solid State, 4, 1081(1962).*
25. *E. M. Verlan and L. N. Ovander, Soviet Phys.-Solid State, 8, 1939(1967).*
26. *R. Loudon, J. Physique, 26, 677(1965).*
27. *R. C. C. Leite and S. P. S. Porto, Phys. Rev. Letters, 17, 10(1966).*
28. *R. C. C. Leite, T. C. Damen and J. F. Scott, Proc. 1968 Int. Conf. on Light Scattering Spectra of Solids, G. B. Wright, Ed., (Springer-Verlag, New York, 1969), p. 359.*
29. *J. F. Scott, R. C. C. Leite and T. C. Damen, Phys. Rev. 188, 1285(1969).*
30. *J. F. Scott, T. C. Damen, W. T. Silfvast, R. C. C. Leite and L. E. Cheesman, Optics Communications, 1, 397(1970).*
31. *J. F. Scott, Phys. Rev. 2B, 1209(1970).*
32. *A. Pinczuk and E. Burstein, Phys. Rev. Letters 21, 1073(1968); Proc. 1968 Int. Conf. on Light Scattering Spectra of Solids, G. B. Wright, Ed., p 429 (Springer-Verlag, New York, 1969).*
33. *R. C. C. Leite and J. F. Scott, Phys. Rev. Letters, 22, 130(1969).*
34. *A. Mooradian, Festerkorper, 9, 74(1969).*
35. *D. L. Stierwalt and A. Nedoluha, Solid State Comm., 8, 309(1970).*
36. *A. Mooradian, Proc. 1968 Int. Conf. on Light Scattering Spectra of Solids, G. B. Wright, Ed., p. 285 (Springer-Verlag, New York, 1969).*
37. *A. Pinczuk, E. Burstein, L. Brillson and E. Anastassakis, Bull. Amer. Phys. Soc., 15, 327 (1970).*
38. *A. Kiel and J. F. Scott, Bull. Amer. Phys. Soc., 13, 1438(1968).*
39. *D. G. Thomas and J. J. Hopfield, Phys. Rev. 175, 1021(1968).*
40. *A. Pinczuk and E. Burstein, Proc. Tenth Int. Conf. on Physics of Semiconductors, p. 727 (U. S. Atomic Energy Comm., Wash., D. C., 1970).*

41. *J. L. Birman and A. K. Ganguly, Phys. Rev. Letters 17, 647(1966); A. K. Ganguly and J. L. Birman, Phys. Rev. 162, 806(1967).*
42. *E. Burstein, S. Ushioda, A. Pinczuk and D. L. Mills, Phys. Rev. Lett., 22, 348(1969).*
43. *J. M. Ralston, R. L. Wadsack and R. K. Chang, Phys. Rev. Lett., 25, 814(1970).*
44. *R. C. C. Leite, J. F. Scott and T. C. Damen, Phys. Rev. Lett., 22, 780(1969).*
45. *M. Klein and S. P. S. Porto, Phys. Rev. Lett., 22, 782(1969).*
46. *T. C. Damen, R. C. C. Leite and J. Shah, Proc. Tenth Int. Conf. on Phys. of Semiconductors, p. 735 (U. S. Atomic Energy Comm., Wash., D. C., 1970).*
47. *R. M. Martin (to be published).*
48. *M. P. Fontana and E. Mulazzi, Phys. Rev. Lett., 25, 1102(1970).*
49. *E. Burstein, Comments on Solid State Physics 1, 202(1968); E. Burstein and D. L. Mills, Comments on Solid State Physics, 2, 93(1969).*
50. *M. Cardona, "Modulation Spectroscopy", Solid State Physics Suppl. 11, H. Ehrenreich, F. Seita and D. Turnbull, Editors, Chapter 2 (Academic Press, New York, 1969).*
51. *E. Burstein and D. L. Mills, Comments on Solid State Physics, 2, 111(1969) and 3, 12(1970).*
52. *J. J. Hopfield, Phys. Rev., 182, 945(1969).*
53. *B. Bendow and J. L. Birman, Phys. Rev., 131, 1678(1970).*
54. *E. Burstein, A. A. Maradudin, E. Anastassakis and A. Pinczuk, Helv. Phys. Acta, Vol. 41. 730(1968).*
55. *E. Anastassakis, S. Iwasa and E. Burstein, Phys. Rev. Letters, 17, 1051(1966).*
56. *P. A. Fleury and J. M. Worlock, Phys. Rev. Lett., 18, 665(1967).*
57. *J. M. Worlock and P. A. Fleury, Phys. Rev. Lett., 19, 1176(1967).*
58. *P. A. Fleury and J. M. Worlock, Phys. Rev. 174, 613(1968).*
59. *E. Anastassakis and E. Burstein, J. P. C. S. (in press).*
60. *J. F. Scott, P. A. Fleury and J. Worlock, Phys. Rev. 177, 1288(1969).*
61. *V. Dvorak, Phys. Rev. 159, 652(1967).*
62. *J. M. Worlock, Proc. 1968 Int. Conf. on Light Scattering Spectra of Solids, G. B. Wright, Ed., p. 411 (Springer-Verlag, New York, 1969).*
63. *E. Anastassakis, A. Filler and E. Burstein, Proc. 1968 Int. Conf. on Light Scattering Spectra of Solids, ibid., p. 421.*
64. *P. Corden, A. Pinczuk and E. Burstein, Proc. Tenth Int. Conf. on Phys. of Semiconductors, p. 739 (U. S. Atomic Energy Comm., Wash., D. C., 1970).*
65. *C. A. Mead and W. G. Spitzer, Phys. Rev. Lett. 10, 471(1963) and Phys. Rev. 134, A713 (1964).*
66. *D. C. Tsui, Solid State Comm., 8, 113(1970).*
67. *M. Cardona, Festkorper Probleme X, 125(1970).*
68. *J. A. Van Vechten, M. Cardona, D. E. Aspnes and R. M. Martin, Proc. Tenth Int. Conf. on Phys. of Semiconductors, p. 82 (U. S. Atomic Energy Comm., Wash., D. C., 1970).*
69. *L. Brillson and E. Burstein, Proc. 1971 Int. Conf. on Light Scattering in Solids (to be published).*

OPTICAL PROPERTIES AND MODEL DENSITY OF STATES*

M. CARDONA and F. H. POLLAK

Physics Department, Brown University, Providence, Rhode Island

The application of simple model densities of states to account for the optical properties of semiconductors and insulators, and their dependence upon various external parameters, is discussed. Special emphasis is placed on the materials of the diamond-zincblende family; several materials with rocksalt structure are also considered. Our model densities of states are obtained by approximating the real energy bands by perfectly parabolic bands extending to infinity, so as to simulate the strongest interband critical points. Critical points with large effective masses (parallel valence and conduction bands) along one or two dimensions give rise to essentially two- and one-dimensional bands.

The simplest model of an insulator, the Penn Model, is an example of a model using one-dimensional bands. The valence and the conduction bands are assumed isotropic with an isotropic gap at the edge of the "spherical" Brillouin zone. This model accounts well for the long-wavelength infrared dielectric constant of zincblende-type materials relating it to the highest maximum in the reflection and absorption spectra (E_2). In order to account for the dispersion in the real part of the complex dielectric constant ϵ_r below the absorption edge, one must add a three-dimensional critical point to represent the lowest direct energy gap E_0. The addition of a spin-orbit-split two-dimensional critical point E_1 and $E_1 + \Delta_1$ enables us to obtain a simple semiquantitative model for the spectral dependence of ϵ_i, the imaginary part of the dielectric constant. This model is compared with experimental results and with the results of detailed computer calculations based on actual band structures. If broadening is included, the Penn model provides a good representation of ϵ_i for amorphous III-V compounds. The modification of these models required by the presence of exciton interaction is discussed.

References pp. 111-112

While the E_0 gap contributes little to the long wavelength ϵ_r, its contribution to the effect of an external perturbation on ϵ_r may be quite sizeable. We examine the application of the model discussed above to interpret the piezoelastic constants of zincblende-type materials, including some effects non-linear in stress which have been observed in the vicinity of E_0. This model can also be used to calculate the temperature dependence of ϵ_r and its dispersion.

The third-order susceptibility (quadratic dependence of ϵ_r on electric field) greatly emphasizes the effect of the E_0 gap. For semiconductors the contribution of E_0 may be even higher than that of the average Penn gap. The third order susceptibility of the alkali halides, however, can be explained simply on the basis of the Penn model. The second order susceptibility (linear electro-optic effect) is also discussed on the basis of simplified models of the optical density of states.

INTRODUCTION

The semiconductors and insulators with diamond, zincblende, and wurtzite (DZW) structure provide excellent grounds to test our knowledge of the optical properties of solids. A large body of information is available concerning the spectral dependence of the linear dielectric constant $\epsilon = \epsilon_r + i\epsilon_i$ for many materials of the family (of the order of 20).[1,2] It is therefore possible to study the systematics of the chemical shifts of observed structure in ϵ from one material to another. Also, the hexagonal wurtzite crystals, with a nearest neighbor structure nearly identical to cubic zincblende, enable us to extend the knowledge gained from the study of the cubic materials to materials of lower symmetry. From the point of view of chemical bonding, the materials of the DZW family, with four valence electrons per atom, are related to the alkali halides (NaCℓ and CsCℓ structures) and to the CaF_2-type materials. It is therefore expected that the detailed understanding of the optical properties of the DZW materials can be carried over, at least in part, to other materials with four valence electrons per atom.[2,3]

An important milestone towards our understanding of the properties of the DZW materials is our present ability to calculate the spectral dependence of ϵ_r and ϵ_i from band structures, obtained either from first principles or generated with the aid of a few adjustable parameters.[4] Such calculations are particularly helpful for the unambiguous identification of observed structure, since it has become clear that assignment to critical points obtained from bands calculated only along high symmetry directions of **k** space is often erroneous. The numerical calculations, however, require lengthy and complex programs and extensive computer time.

The effect of an external parameter (e.g., stress, temperature, electric fields, etc.) on the optical properties (e.g., on ϵ) can, in principle, be also calculated by applying the appropriate perturbation on the band structure, but here the state of the art is, by far, not as advanced as in the case of the unperturbed optical properties. Worthy of special mention are the calculations of the temperature dependence of the optical structure by Walter et al[5] and the formalism recently proposed by Aspnes and Rowe to calculate the Franz-Keldysh effect from the detailed band structure.[6]

In this paper we propose a simple model to explain in a semiquantitative way, the spectral dependence of ϵ_i for the DZW materials. This model is composed of the lowest direct gap (E_0, at $\mathbf{k} = 0$), represented by pairs of three-dimensional parabolic valence and conduction bands extending to infinite $\mathbf{k}$, the E_1 gap, represented by two-dimensional parabolic bands, and the average, nearly isotropic Penn gap[7] which contains most of the oscillator strength and is represented by one-dimensional, parabolic bands. Spin-orbit splitting, usually observed for the E_0 and the E_1 gaps, can be easily included in our model. While this model yields only a gross approximation to ϵ_i, it is expected to approximate well ϵ_r and also the effect of perturbations on ϵ_r below the lowest direct gap E_0.

The one-dimensional Penn bands account well for the zero frequency dielectric constant $\epsilon_r(0)$, without having to invoke the lower energy but weaker E_0 and E_1 gaps. The contribution of the E_0 gap to ϵ_r becomes important as E_0 is approached and is responsible for most of the dispersion in ϵ_r observed below E_0. With regards to the effects of external perturbations on ϵ_r, however, the E_0 contribution becomes more important. The effect of external perturbations can usually be described as a shift in the gaps. For some perturbations (temperature, stress) this shift is sometimes roughly the same for all gaps, while other modulation parameters (electric and magnetic fields) give larger shifts for smaller gaps because of the corresponding smaller effective masses. Therefore, for all perturbations mentioned, the *relative* change in the smaller gaps (E_0, E_1) is larger than that of E_2, and hence the contribution of the smaller gaps to the change in ϵ_r is amplified. Also, the singularity in ϵ_r near E_0 has the form $(\omega - \omega_0)^{1/2}$ (in the absence of excitonic interaction); ϵ_r remains finite near ω_0 (ω is the frequency of the light and ω_0 that of the E_0 gap). The corresponding singularity in the differential effect produced by a change in ω_0 has the form $(\omega - \omega_0)^{-1/2}$ and becomes infinite at E_0. Thus the contribution of the relatively weak E_0 gap to a perturbation in ϵ_r produced by an external agent becomes very large near E_0.

We consider in this paper the contribution of E_0, E_1, and E_2 to the stress, temperature, electric field, and magnetic field induced changes in ϵ_r. Linear and quadratic effects in stress and in electric fields also are considered. Although our detailed model is applicable only to the DWZ materials, a number of results can also be used in other cases. The contributions of the Penn gap ω_g, for instance, can also be used for the alkali halides.[2] For these materials the smaller gaps (E_0)

are also very large (the ratio ω_o/ω_g is close to one), and their contribution to the differential dielectric properties can be neglected except very near E_o. The results for the parabolic gaps of various dimensionalities are quite general and can be applied to any cases in which such gaps are known to be present.

DIELECTRIC CONSTANT AND MODEL DENSITY OF STATES

PENN MODEL: ONE-DIMENSIONAL CRITICAL POINTS

The simplest, although somewhat non-physical model of an insulator is that introduced by Penn.[7] It consists of a free electron gas of valence electrons with an isotropic gap at the Fermi surface. This gap, produced by some Fourier components of the crystal potential, must occur at the edge of the Brillouin zone or rather the Jones zone of the valence electrons. Thus, this Jones zone, which coincides with the Fermi surface, is spherical and hence non-physical. This model was used by Penn to calculate the zero frequency dielectric constant of insulators and its dependence on wave vector. For $\omega = 0$ and $\mathbf{k} = 0$ Penn finds:

$$\epsilon(0) = \epsilon(\omega = 0, \mathbf{k} = 0) = 1 + (\omega_p/\omega_g)^2 F \tag{1}$$

with

$$F = 1 - \omega_g/4\omega_F + 1/3(\omega_g/\omega_F)^2$$

In Eq. (1) ω_p and ω_F are the plasma and Fermi frequencies of the valence electrons, respectively, and ω_g the frequency of the average "Penn" gap. $\epsilon(0)$ is the *electronic* contribution to the dielectric constant, without the reststrahlen contribution. The factor F in Eq. (1) is practically equal to unity, except for the materials with large ω_g for which it can have somewhat smaller values (~ 0.6). We shall assume in our treatment of the DZW materials that $F = 1$. Another factor D, close to unity, was introduced in Eq. (1) by Van Vechten[2] to take into account the polarizability of the core d-electrons in materials of the fourth and lower rows of the periodic table. This factor, never larger than 1.2, will not be considered in this paper since we do not believe the semi-quantitative nature of our considerations warrants its inclusion. For this and other reasons, the expressions derived here may be uncertain to factors of the order of unity.

As mentioned earlier, Eq. (1) is obtained by assuming an isotropic gap ω_g at the edge of the Jones zone and thus one-dimensional bands. If one assumes that the one-dimensional bands extend to infinity and that the matrix elements of the linear momentum $\mathbf{p}$ are constant, one obtains for allowed transitions,

$$\epsilon - 1 = A/\omega^2 [(\omega_g + \omega)^{-\frac{1}{2}} + (\omega_g - \omega)^{-\frac{1}{2}} - 2\omega_g^{-\frac{1}{2}}]. \tag{2}$$

An evaluation of Eq. (2) for $\omega \to 0$ yields the value of A, by comparison with Eq. (1):

$$A = 4/3(\omega_p^2 \ \omega_g^{\frac{1}{2}}) \tag{3}$$

In order to evaluate A, we calculate ω_p from the lattice constant and ω_g from the experimental values of $\epsilon_r(0)$ through Eq. (1).[2] We note that according to Eq. (2) ω_g corresponds to the maximum in $\epsilon_i(\omega)$: the values of ω_g obtained from $\epsilon_r(0)$ agree rather well with the frequency of the highest maximum in the $\epsilon_i(\omega)$ spectrum (E_2) for the DZW materials.[8]

We show in Fig. 1, as a typical example, the $\epsilon_i(\omega)$ spectrum observed for InAs and that calculated numerically from the **k.p** band structure of the material. The agreement between the experimental curve and the full computer calculation is quite good. We have also plotted in Fig. 1 the curve (dotted curve for $\omega > 4.58$ eV):

$$(A/\omega^2)(\omega - \omega_g)^{-\frac{1}{2}}, \tag{4}$$

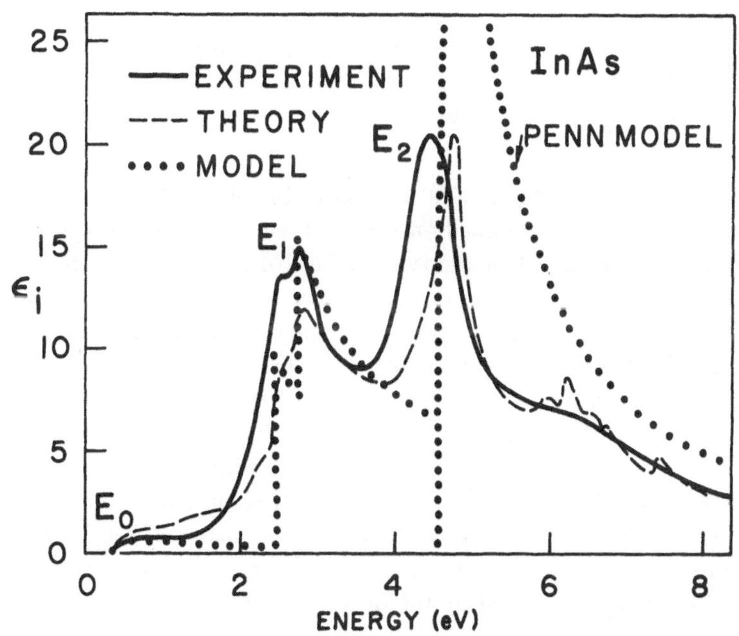

Fig. 1. Experimental ϵ_i spectrum for crystalline InAs [H. R. Philip and H. Ehrenreich, Phys. Rev. *129*, 1550(1963)] compared with the results of computer calculations [C. W. Higginbotham, F. H. Pollak, and M. Cardona, Proceedings of the IX International Conference on Semiconductors, Nauka, Leningrad, 1968, p. 577] and with the predictions of our model.

with A obtained from Eq. (3) for $\omega_o = 14.1$ eV and $\omega_g = 4.58$ eV. This dotted curve represents qualitatively the experimental trend in $\epsilon_i(\omega)$ for $\omega > \omega_g$; its strength is higher than the experimental one, as expected, since part of the oscillator strength, which exists in the Penn model only for $\omega > \omega_g$, has been moved to lower energies in the physical situation.

Equation (2) yields for the dispersion in ϵ_r at long wavelengths:

$$\epsilon_r - 1 = (\omega_p/\omega_g)^2 \; [1 + 35/48(\omega/\omega_g)^2 + \ldots] \tag{5}$$

It is obvious by looking at Fig. 1 that the experimental dispersion in ϵ_r should be stronger than given by Eq. (5), because of the considerable oscillator strength which exists below ω_g. In particular we shall take into account in the following sections the accumulation of such oscillator strength near E_1 and E_o.

It is easy to introduce a phenomenological Lorentzian broadening into Eq. (2); it is done by simply replacing ω by $\omega + i\Gamma$, with Γ the broadening parameter. A separation of the real and imaginary parts of ϵ after this substitution is done yields:

$$\epsilon_i = A\{(\omega^2+\Gamma^2)^{-1} \; (-F[(-\omega_g-\omega)/\Gamma] + F[(\omega-\omega_g)/\Gamma) - $$

$$[\Gamma/\omega(\omega^2+\Gamma^2)] \; (F[(\omega_g+\omega)/\Gamma] + F[(\omega_g-\omega)/\Gamma) - \; 2F[\omega_g/\Gamma]\} \tag{6}$$

with $F[x] = [x^2 + 1]^{-1/2} \, [(x^2 + 1)^{1/2} + x]^{1/2}$.

The function F[x], which corresponds to a Lorentzian-broadened $[x-1]^{-1/2}$ function, has been plotted in Ref. 9. An examination of Fig. 1 suggests that a broadened one-dimensional dielectric constant of the type of Eq. (6) may approximate better the experimental results than the unbroadened version. In particular, such a broadened Penn model may be a good approximation to the ϵ_i of amorphous materials of the DZW family: the destruction of crystalline order has as a result the disappearance of the fine structure in the crystalline ϵ_i. Figure 2 shows the experimental ϵ_i spectrum of GaAs and a fit to it with the function Eq. (6) with $\Gamma = 1.25$ eV and $\omega_g = 3.5$ eV. This fit is good except at low frequencies, where the experimental data suggest a smaller broadening parameter. We note that electroreflectance measurements indicate that all structure observed in crystalline materials (Ge,Se) disappear in their amorphous counterparts with the exception of the lowest direct gap, where, apparently, the broadening introduced by the disorder is also small.[10]

Equation (1) forms the basis of Phillips' definition of ionicity.[2,3] The ω_g gap of a partially ionic material of the DZW family (this can also be extended to the alkali halides) can be written[2]:

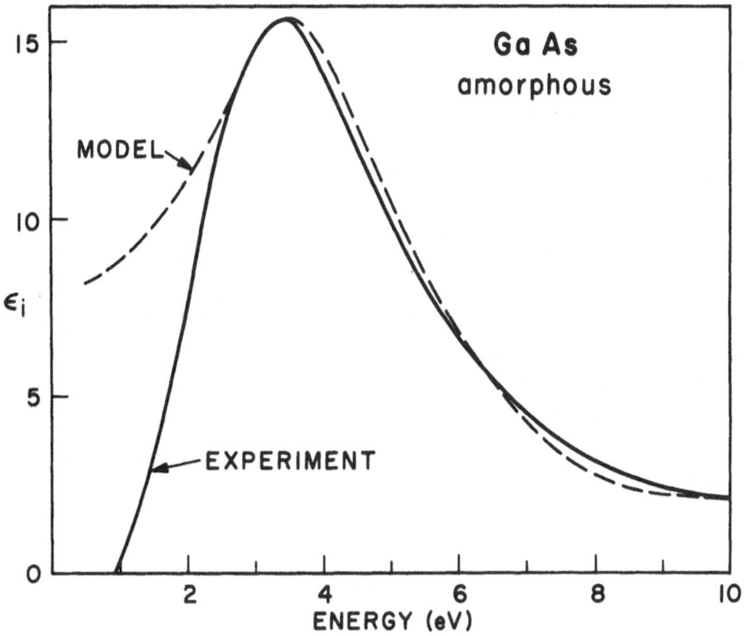

Fig. 2. Imaginary part of the dielectric constant of amorphous GaAs (J. Stuke, private communication).

$$\omega_g = [\omega_h^2 + C^2]^{1/2} \tag{7}$$

where ω_h is the Penn gap of the corresponding isoelectronic group IV material, as obtained from $\epsilon(0)$ through Eq. (1) and C a measure of the ionic (antisymmetric) potential. From the experimental values of $\epsilon(0)$ for an ionic material and for the corresponding isoelectronic group IV material one can determine both ω_g and C. The ionicity index f_i is then defined as:

$$f_i = (C/\omega_g)^2 \tag{8}$$

Van Vechten[2] has pointed out that the covalent gap ω_h is roughly proportional to $a_o^{-2.5}$ ($a_o \equiv$ lattice constant) for all materials of the family. For the zincblende-wurtzite materials, C seems to have a very similar dependence on lattice constant, thus ω_g is also roughly proportional to $a_o^{-2.5}$.

A microscopic justification of the Penn model has been given by Heine and Jones.[11] These authors pointed out that, at least for silicon, the energy gap at the edge of the Jones zone of the valence electrons (fourth Brillouin zone) is constant over a fairly large area which encloses the X point. The gap at the X point is therefore the Penn gap ω_h; it can be calculated rather simply in terms of the standard pseudopotential from factors v(111) and v(220). Heine and Jones found the following approximate expression for the gap at X:

$$\omega_h = 2\left\{v(220) + v^2(111)/(2\pi/a_o)^2\right\} \tag{9}$$

The generalization of Eq. (9) to the zincblende case is [12]:

$$\omega_g = 2\left\{v_s(220) + [v_s^2(111) + v_a^2(111)]/2(\pi/a_o)^2\right\} \tag{10}$$

where the suffixes s and a represent the symmetric and antisymmetric pseudopotential form factors, respectively. We note that Eq. (10), which yields reasonable values of ω_g, at least for the III-V compounds, cannot be brought into the form of Eq. (7). This difficulty could be due to the fact that the gap at the X point, given by Eq. (10), may not be identical to the *average* gap which could still be better represented by Eq. (7).

E_1 GAPS: TWO-DIMENSIONAL CRITICAL POINTS

As shown in Fig. 1, the next prominent structure below E_2 in the optical spectra of the DZW materials are the E_1 peaks (E_1 and its spin-orbit-split mate $E_1 + \Delta_1$). These peaks are due to transitions between the two upper valence bands and the lowest conduction band at points along the [111] direction of **k** space. Along this direction these valence and conduction bands are nearly parallel except very close to **k** = 0. Thus the optical density of states should be well represented by that of a two-dimensional minimum.[13] For the purpose of our discussion, this is equivalent to the results suggested by band calculations: a three-dimensional minimum (M_o) at the edge of the Brillouin zone followed at a slightly higher energy by an M_1-type saddle point inside the zone. The two-dimensional picture, however, has the advantage that it does not require, for its evaluation, the knowledge of the longitudinal mass along the [111] direction: The mass is large and given rather poorly by band calculations.

The contribution of the E_1 gap to the dielectric constant is:[13]

$$\epsilon - 1 = -(B/2\pi)(P^2/\omega^2)\,\ell n[(\omega_1^2 - \omega^2)/\omega_1^2] \tag{11}$$

with

$$B = 4\sqrt{3}\,\pi\mu_\perp/a_o.$$

We are using atomic units ($e = m = \hbar = 1$); $\mu_\perp$ is the reduced transverse mass and P the magnitude of the matrix element of **p**. The contribution of $E_1 + \Delta_1$ is similar to that of E_1, with ω_1 replaced by $\omega_1 + \Delta_1$ and $\mu_\perp$ replaced by the reduced mass for the lower valence band. The three parameters $\mu_\perp$, P, and ω_1 of Eq. (11) are actually related by the **k.p** expression for the effective mass:

$$\mu_\perp(E_1) = [\omega_1 + (\Delta_1/3)]/3P^2$$
$$\mu_\perp(E_1+\Delta_1) = [\omega_1 + 2(\Delta_1/3)]/3P^2 \quad (12)$$

The matrix element P is thus the only parameter that has not been measured directly required to evaluate Eq. (11), since ω_1 and $\omega_1 + \Delta_1$ are simply the energies of the E_1 and $E_1 + \Delta_1$ peaks in the ϵ_i spectrum. This matrix element is obtained from band calculations, but it does not vary much from material to material. If no better values are available, one may simply take an average value $P \simeq 0.6$ for all DZW materials. Equation (11) assumes that the valence and conduction bands are parallel all the way from Γ to L. This is not strictly true as already mentioned: a correction factor smaller than one ($\sim 3/4$) could be introduced in order to take this fact into account. In view of other uncertainties in our derivation (parabolic bands, constant P), which should also introduce factors of the order of unity, we do not include this correction factor in our considerations.

From Eq. (11) we obtain for $\omega > \omega_1$:

$$\epsilon_i = BP^2/2\omega^2 \quad (13)$$

The contribution of E_1 and $E_1 + \Delta_1$ to ϵ_i obtained in this manner for InAs (P = 0.58) is shown in Fig. 1. The experimental E_1, $E_1 + \Delta_1$ structure is approximately reproduced in strength and shape by our two-dimensional model. A similar comparison between experiment and the predictions of Eq. (13) is drawn in Fig. 3 for germanium (P = 0.7) and in Fig. 4 for CdTe (P = 0.52). We notice for CdTe that the asymmetric ϵ_i line shape associated with our model for E_1 and $E_1 + \Delta_1$ does not appear in the experimental curve. This fact is known to be the result of excitonic interaction.[4,14]

E_0 *GAPS: THREE-DIMENSIONAL CRITICAL POINTS*

The materials of the DZW family have a lowest direct gap at **k** = 0. This gap exhibits spin-orbit splitting (E_0, $E_0 + \Delta_0$) and for wurtzite a small crystal field splitting. We shall limit our considerations to the diamond-zincblende family; generalization to wurtzite is possible.

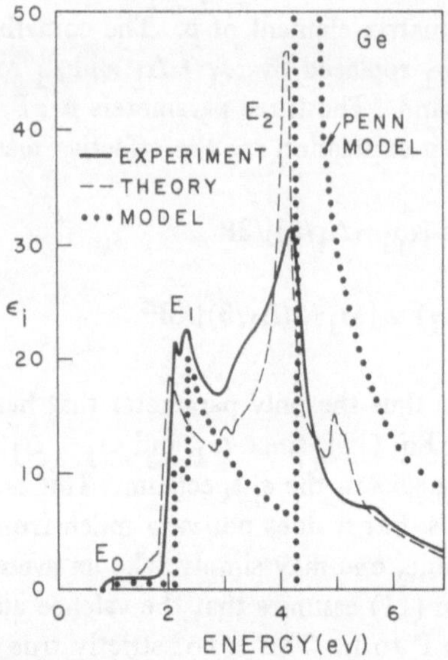

Fig. 3. Experimental ϵ_i spectrum of crystalline germanium, compared with the results of a computer calculation (F. Herman, R. L. Kortum, C. D. Kuglin, and J. L. Shay, in *II-VI Compounds*, edited by D. G. Thomas, Benjamin, New York, 1967). Also, spectrum calculated with our model density of states.

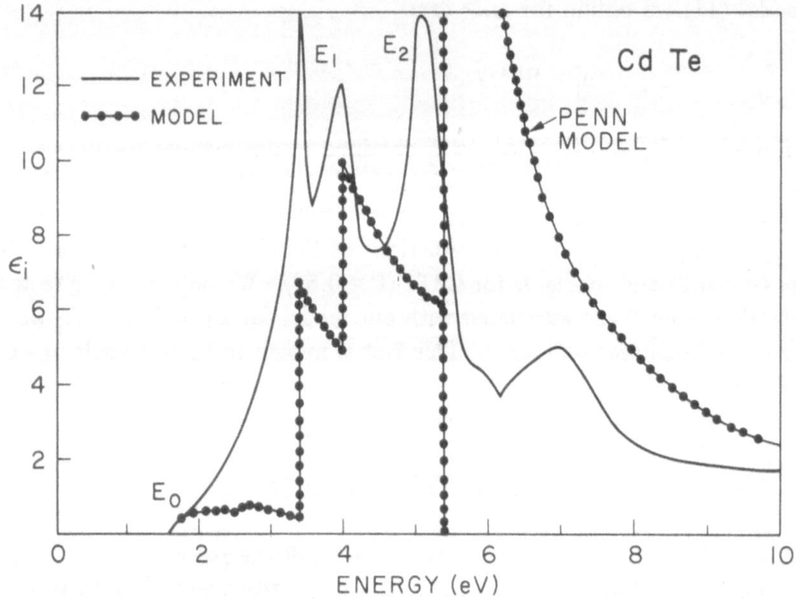

Fig. 4. Experimental ϵ_i spectrum of crystalline CdTe (from Ref. 8) compared with the calculations based on our model density of states.

Under the assumption of a pair (valence-conduction) of parabolic bands with constant matrix elements of **p** the contribution of a three-dimensional M_0 critical point to ϵ is:[15,9]

$$\epsilon - 1 = \tfrac{1}{2}C_0''(\omega_0/\omega)^2\,\omega_0^{-1/2}\,[2\omega_0^{1/2} - (\omega+\omega_0)^{1/2} - (\omega_0-\omega)^{1/2}] \qquad (14)$$

with

$$C_0'' = (4P^2/3)(2\mu^*/\omega_0)^{3/2}\ .$$

The sign of the square roots in Eq. (14) is that which corresponds to a complex plane cut along the $0 \to 1 + i\delta$ ($\delta \to +0$) axis. μ^* is the reduced mass and P an appropriate average matrix element.

When applying Eq. (14) to the E_0 edge of a DZ material, a complication appears in the definition of μ^* because of the valence band degeneracy and its corresponding warping. For sufficiently small gaps, however, ($\omega_0 \lesssim 2$ eV), the valence and conduction bands can be approximately treated with the **k.p** method, neglecting interactions with other bands. The average reduced mass of the E_0 bands thus becomes:[13]

$$\mu^*(E_0) = (3/2P^2)[3/\omega_0 + 1/(\omega_0+\Delta_0)]^{-1} \qquad (15)$$

Equation (14) must be multiplied by 2 in order to take into account the double degeneracy of the E_0 edge. For the non-degenerate $E_0 + \Delta_0$ edge, Eq. (14) remains valid with ω_0 replaced by $\omega_0 + \Delta_0$ and the reduced mass by:

$$\mu^*(E_0+\Delta_0) = (3/4P^2)[1/\omega_0 + 1/(\omega_0+\Delta_0)]^{-1} \qquad (16)$$

The matrix element P for the E_0 gap is usually very close to that used earlier for the E_1 gap. It can be obtained from band calculations and also from experimental cyclotron resonance data. Some uncertainty appears when using Eq. (15) to obtain $\mu^*(E_0)$ at room temperature in the case of small ω_0 materials: although the experimental picture is not yet clear, it seems that Eq. (15) does not yield correct room temperature masses when the room temperature value of ω_0 is used with temperature independent P. The proper temperature behavior of μ^* seems to be represented by Eq. (15) with P temperature independent and ω_0 the zero temperature gap corrected for the effect of thermal expansion only (no electron-phonon interaction). A theoretical investigation of the temperature dependence of $\mu^*(E_0)$ seems possible with the Debye-Waller correction to the pseudopotential form factors[16] but has yet to be performed.

The detailed form of Eq. (14) depends on the assumption of a constant matrix element of P. If, instead, the oscillator strength, i.e., P^2/ω, is assumed constant, the following expression is obtained:[17]

$$\epsilon - 1 = \tfrac{1}{2}C_0'' (\omega_0/\omega)^2 \, \omega_0^{-\frac{1}{2}} [(\omega+\omega_0)^{\frac{1}{2}} - (\omega_0-\omega)^{\frac{1}{2}}]. \qquad (17)$$

Equations (14) and (17) have the same singular behavior near ω_0. Their behavior is also quite similar elsewhere with the exception of a non-dispersive contribution to ϵ_r which corresponds to the large differences in oscillator strengths at larger values of ω between the two models (constant P and constant P^2/ω). We shall use for our discussion Eq. (14); constant P seems to correspond better to the results of band calculations than constant P^2/ω.

Equation (14) applies to any allowed three-dimensional M_0 critical point. Actually the best examples of the predictions of Eq. (14) for the dispersion of ϵ_r have been observed in the lead chalcogenides.[17,18,19] We show in Fig. 5 a fit to the experimental ϵ_r spectrum of PbS near the lowest direct gap obtained with the real part of Eq. (14):

$$\epsilon_r - 1 = \tfrac{1}{2}C_0'' \omega^{-2} [2\omega_0^{\frac{1}{2}} - (\omega+\omega_0)^{\frac{1}{2}} - (\omega_0-\omega)^{\frac{1}{2}}] \; ; \; \omega < \omega_0$$

$$\epsilon_r - 1 = \tfrac{1}{2}C_0'' \omega^{-2} [2\omega_0^{\frac{1}{2}} - (\omega+\omega_0)^{\frac{1}{2}}] \; ; \; \omega > \omega_0 \qquad (18)$$

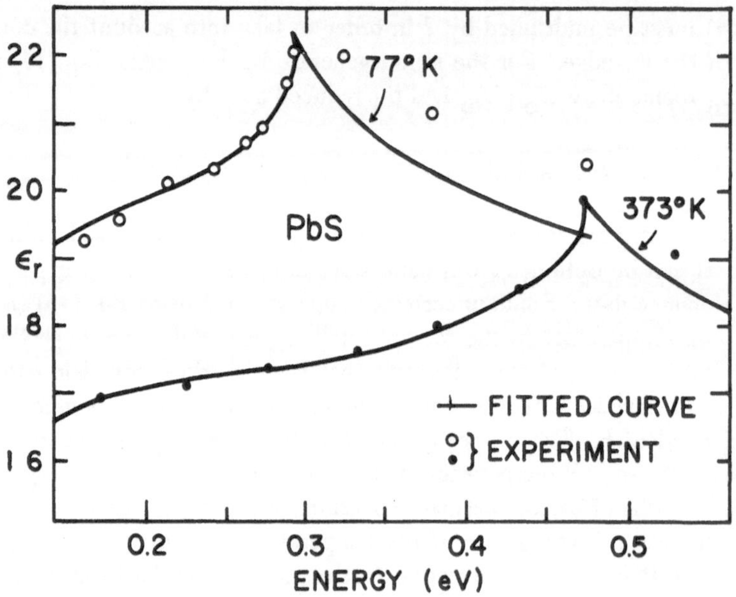

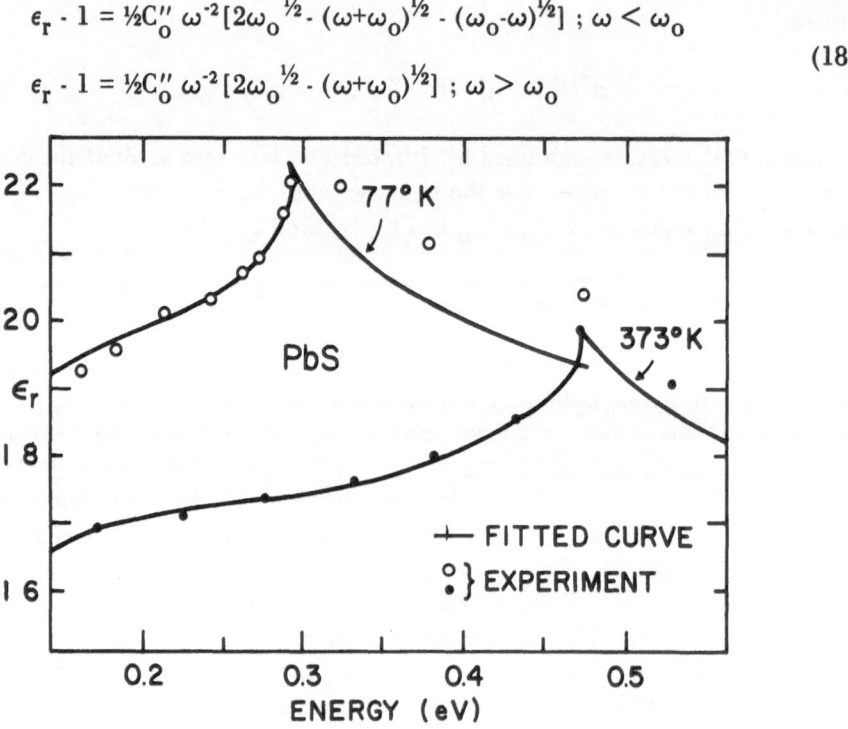

Fig. 5. Fit to the dielectric constant ϵ_r of PbS at two temperatures, obtained with Eq. (14) and a constant dispersionless background. A small free carrier term ($\propto \omega^{-2}$) has also been included. From Ref. 19.

plus a constant term to include the average gap and a small, long wavelength correction for free carrier effects.[19] Fits to the dispersion of ϵ_r below ω_o with Eq. (18) have also been made for Ge and for GaAs.[13] The values of C_o'' obtained are compared with those calculated with Eq. (14) in Table I.

DEPENDENCE OF ϵ_r ON HYDROSTATIC PRESSURE AND ON TEMPERATURE

PENN MODEL: HYDROSTATIC PRESSURE

The effect of hydrostatic pressure on $\epsilon_r(0)$ can be treated with the Penn model [Eq. (1)].[2] The perturbation produces both a change in the plasma frequency ω_p and in the average gap ω_g. The plasma frequency is proportional to $a_o^{-3/2}$. The pressure-induced change in ω_g can be estimated with the prescription given by Van Vechten[2] for the germanium-type materials: $\omega_g \propto a_o^{-2.5}$. An alternative way of estimating the pressure-induced change in ω_g, based on Eq. (9) and the known pressure dependence of the pseudopotential on lattice constant, has been given by Heine and Jones.[11] The results are quite similar to those obtained with the $a_o^{-2.5}$ dependence. From Eq. (1) we find:

$$\begin{aligned} d\ell n\, \epsilon(0)/d\ell n\, a_o &\simeq d\ell n[\epsilon(0)-1]/d\ell n\, a_o = \\ &2[d\ell n\, \omega_p/d\ell n\, a_o - d\ell n\, \omega_g/d\ell n\, a_o] = +2 \end{aligned} \tag{19}$$

The experimental values of $d\ell n\, \epsilon(0)/d\ell n\, a_o$ found in the literature are + 1.8 for Si[20], + 3 [20] and + 4.5 for Ge[21], and + 3.3 for GaAs[22]. A much smaller value of $d\ell n\, \epsilon(0)/d\ell n\, a_o = 0.5$ has been found for ZnS. This result seems to indicate a weaker dependence of the ionic component of the gap C [see Eq. (7)] on lattice constant than $a_o^{-2.5}$; $C \propto a_o^{-1}$ would yield a calculated value of $d\ell n\, \epsilon(0)/d\ell n\, a_o$ close to the observed one. Measurements of the dependence of $\epsilon_r(0)$ on lattice constants for other II-VI and I-VII compounds with zincblende structure are required to elucidate this point.

Vedam and Schmidt[23] have also reported a non-linearity in the pressure dependence of ϵ_r for ZnS at $\omega = 2.1$ eV (measurements were only performed at this photon energy). As we shall see later, the large non-linearity observed cannot be due to the E_o and E_1 gaps since $\omega = 2.1$ eV is not close to ω_o or ω_1. We calculate on the basis of Eq. (1) with $\omega_h \propto a_o^{-2.5}$ and $C \propto a_o^{-1}$ a term quadratic in stress *seven times* smaller than that reported. Therefore the strong non-linearity observed cannot be accounted for; further measurements are needed, in particular of the spectral dependence of the non-linearity.

PENN MODEL: TEMPERATURE DEPENDENCE

The temperature dependence of $\epsilon_r(0)$ can be decomposed into two contributions: the effect of the thermal expansion, which can be obtained from the previous section, and the explicit electron-phonon terms. The effect of the electron-phonon interaction is to change ω_g in Eq. (1), without changing ω_p. The change in ω_g can be estimated with Eq. (10) by multiplying the pseudopotential form factors by appropriate Debye-Waller factors.[12,16] The values of dℓn n/dT (n is the long-wavelength refractive index) calculated by this procedure for group IV and III-V materials are listed in Table II together with experimental values.[12]

E_o GAPS

We consider now the effect on Eq. (14) of a change $\Delta\omega_0$ produced by an external perturbation such as a hydrostatic pressure or a change in temperature. The coefficient C_0'' is not affected much by the stress: Since P^2 is practically stress independent the stress-induced change in μ^* cancels that in ω_0, according to Eq. (15). For the temperature effect we have again the uncertainty in the temperature dependence of μ^* mentioned previously. We shall disregard this uncertainty and assume that the effect on ϵ of a change in temperature is due to the full temperature change in ω_0, with P^2 constant. This assumption is likely to yield an overestimate of the temperature effect on ϵ. From Eq. (18) we obtain by differentiation for $\omega < \omega_g$:

$$\Delta\epsilon_r = \tfrac{1}{4} C_0''^{\,-2}\ [3f(\omega/\omega_0) + g(\omega/\omega_0)](\Delta\omega/\omega_g) \tag{20}$$

where $\Delta\omega_0$ is the change in ω_0 produced by either the pressure or the temperature. The functions f(x) and g(x) are defined as:

$$f(x) = x^{-2}\ [2 - (1+x)^{1/2} - (1-x)^{1/2}]$$
$$g(x) = x^{-2}\ [2 - (1+x)^{-1/2} - (1-x)^{-1/2}] \tag{21}$$

The limit of Eq. (20) for $\omega \to 0$ is zero and thus the three-dimensional critical points yield no contribution to $\Delta\epsilon_r$ at zero frequency: this property is characteristic of the three-dimensional critical points with a reduced mass proportional to the gap and does not hold for other situations (e.g., two-dimensional critical points or three-dimensional critical points off $\mathbf{k} = 0$).

Figure 6 shows the spectral dependence of dℓn n/dT and dℓn n/dp (p = hydrostatic pressure) found experimentally for germanium and the fit to these data with Eq. (20). For $d\omega_0/dT = -4 \times 10^{-4}$ eV $(°C)^{-1}$ and $d\omega_0/dp = 1.3 \times 10^{-5}$ eV bar^{-1}

TABLE I

Values of the coefficient C_0'', which gives the contribution of E_0, to the dielectric constant, for crystalline Ge and GaAs, as calculated with Eq. 14, and also determined from the experimental dispersion, temperature and pressure dependence of the refractive index, and piezobirefringence (PB).

	Calculated	from ϵ_r	from (1/n)(dn/dT)	from (1/n)(dn/dp)	from PB, [100] stress	from PB, [111] stress
Ge	1.9	2.5	2.4	1.9	2.2	1.0
GeAs	1.5	6.6	5.2	4.4	3.1	1.5

TABLE II

Temperature Coefficient of Long Wavelength Refractive Index of Group IV and III-V Semiconductors

	(1/n)(dn/dT) (Theoretical) ($°K^{-1}$)	(1/n)(dn/dT) (Experimental) ($°K^{-1}$)
C	1.1×10^{-5}	0.5×10^{-5}(a)
Si	3.7×10^{-5}	3.9×10^{-5}(b)
		4.8×10^{-5}(c)
Ge	6.1×10^{-5}	6.9×10^{-5}(b)
		1×10^{-4}(d)
GaAs	5.4×10^{-5}	4.5×10^{-5}(e)
GaSb	9.3×10^{-5}	8.2×10^{-5}(e)
InAs	6.7×10^{-5}	--
InSb	15.0×10^{-5}	11.9×10^{-5}(e)
InP	3.6×10^{-5}	2.7×10^{-5}(e)
GaP	2.2×10^{-5}	3.7×10^{-5}(f)
AlAs	4.6×10^{-5}	4.6×10^{-5}(g)
AlP	4.4×10^{-5}	3.6×10^{-5}(g)

[a] *P. T. Narasimhan, Proc. Soc. b 68, 315(1955).*

[b] *M. Cardona, W. Paul, and H. Brooks, J. Phys. Chem. Solids 8, 204(1959).*

[c] *F. Lukes, Czech, J. Phys. B10, 317(1960).*

[d] *F. Lukes, ibid. 743(1960).*

[e] *M. Cardona, Proc. Internat. Conf. Phys. Semiconductors, Prague, 1960 p. 388, Czech. Acad. Sci., Prague and Academic Press, N.Y. 1961.*

[f] *A. N. Pikhtin and D. A. Yaskov, Fiz, Tverd, Tela 9, 145(1967). [English Transl.: Soviet Physics - Solid State 9, 107(1967)].*

[g] *H. G. Grimmeis and B. Monemar, to be published.*

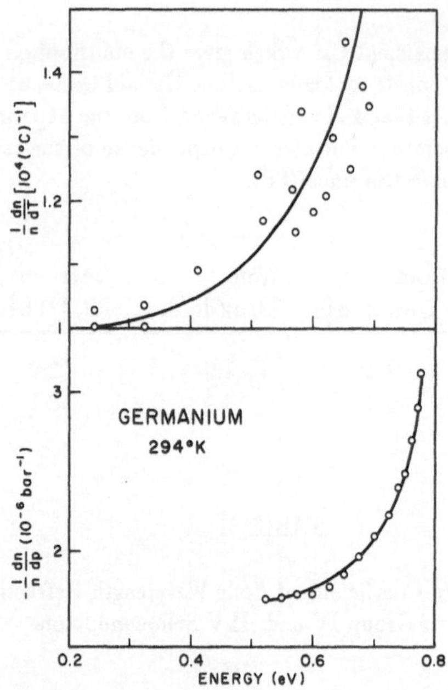

Fig. 6. Spectral dependence of the pressure and temperature coefficients of the refractive index of germanium (from Refs. 21 and 24) and fit with Eq. (20) plus a constant.

the fits yield the values of C''_o listed in Table I. These values are in rather good agreement with the C''_o calculated from the known matrix element P (see Table I).

Figure 7 shows similar results for GaAs. The values of C''_o needed for the fit, with $d\omega_o/dT = -3.85 \times 10^{-4}$ eV $(°C)^{-1}$ and $d\omega_o/dp = 1.2 \times 10^{-5}$ eV bar^{-1}, are also listed in Table I. Most of the C''_o values obtained from fits of experimental data for GaAs are higher than those calculated from the known values of P. This discrepancy is probably due to exciton effects, which make the singularity at E_o stronger.

Equation (20) can also be used to calculate the contribution of E_o to the Faraday rotation[25], with $\Delta\omega = g^*\mu H$ (μ = Bohr magneton, H = applied magnetic and g^* an effective g-factor which represents the splitting of the gap produced by H for right and left circularly polarized light). The corresponding Verdet constant V is:

$$V = \omega(\Delta\epsilon_r)/4H\epsilon_r \tag{22}$$

We show in Fig. 8 a fit with Eq. (20) to the interband Verdet constants observed below the E_o edge of a number of DZ semiconductors.[25]

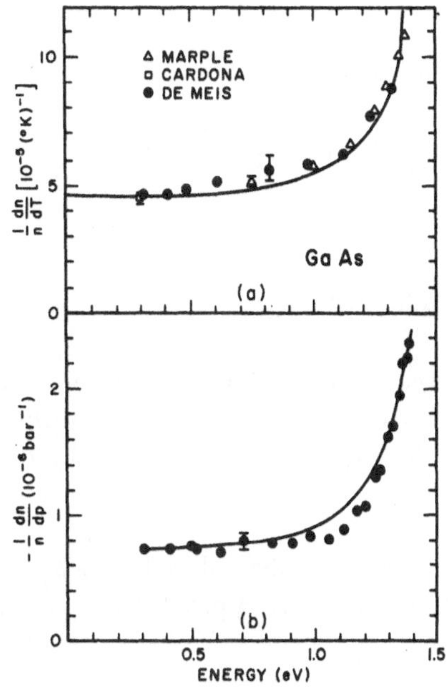

Fig. 7. Spectral dependence of the pressure and temperature coefficients of the refractive index of GaAs (from Ref. 22) and fit with Eq. (20) plus a constant.

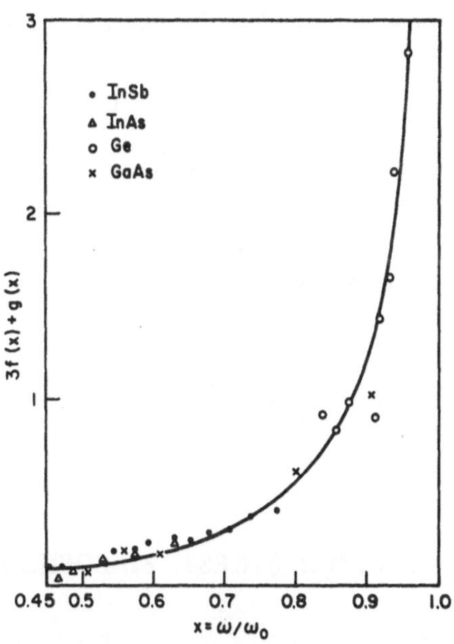

Fig. 8. Fits to the experimentally observed dispersion of the interband Verdet constant of several DZ semiconductors to the function $3f(\omega/\omega_g)+g(\omega/\omega_g)$ of Eq. (20). From Ref. 25.

E_1 GAP

Equation (11) yields for $\omega << \omega_1$:

$$\epsilon_r\text{-}1 = C_1'' \, [1 + (\omega/\omega_1)^2 + \dots] \tag{23}$$

with

$$C_1'' = (2\sqrt{3} \, \mu_\perp P^2)/a_o\omega_1^2$$

Equation (23) should represent correctly the contribution of E_1 to ϵ_r in the region of transparency ($\omega << \omega_o$), since $\omega_o << \omega_1$. The contribution of $E_1 + \Delta_1$ is obtained by replacing in Eq. (23) ω_1 by $\omega_1 + \Delta_1$ and $\mu_\perp$ by the appropriate mass [Eq. (12)].

Taking into account the fact that the reduced mass $\mu_\perp$ is proportional to the gap ω_1, we find for the change in Eq. (23) produced by a pressure or temperature induced change in ω_1:

$$\Delta\epsilon_r = -C_1'' \, (1 + 3\,\omega^2/\omega_1{}^2)\,(\Delta\omega_1/\omega_1) \tag{24}$$

with a similar contribution for the $E_1 + \Delta_1$ gap. It is interesting to note that, contrary to the three-dimensional case, the two-dimensional problem yields a finite contribution to the change in ϵ_r at zero frequency. From the calculated value of C_1'' ($C_1'' = 3.22$[13]) we find for the contribution of E_1 to the pressure coefficient of the refractive index of germanium:

$$d\ell n \, n/dp = -3.6 \times 10^{-7} \, [1 + 3(\omega/\omega_1)^2] \, . \tag{25}$$

The dispersion in Eq. (25) is nearly negligible when compared with that due to the E_o edge [see Fig. (6b)]. The remaining small non-dispersive contribution of Eq. (25) (and that slightly smaller due to the $E_1 + \Delta_1$ gap) can thus be lumped together with the contribution of the average Penn gaps. We note that the three contributions to the hydrostatic pressure and temperature dependence of ϵ_r discussed here have the same sign.

DEPENDENCE OF ϵ_r ON UNIAXIAL STRESS: PIEZOBIREFRINGENCE (PB)

E_o GAP

The pure shear component of a uniaxial stress produces a splitting of the top valence band of a DZ semiconductor (see Fig. 9). This splitting produces a difference

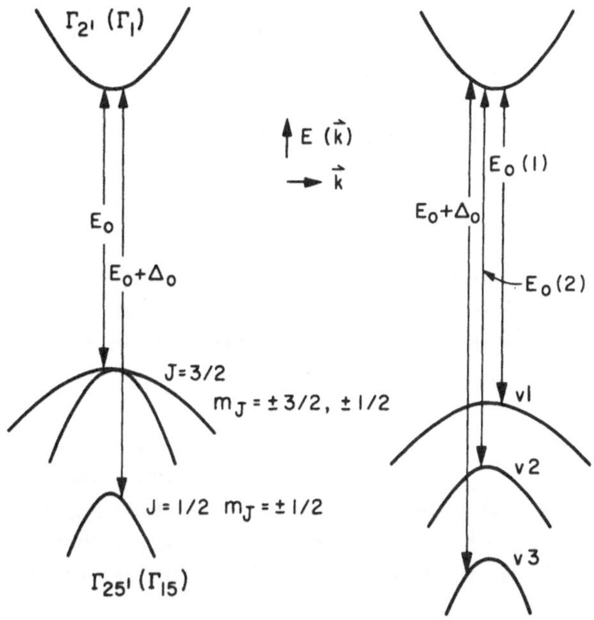

Fig. 9. Upper valence bands and lowest conduction band in DZ crystals in the absence (left) and in the presence (right) of uniaxial stress.

in the refractive indices for light polarized with the electric field parallel and perpendicular to the stress, because of the selection rules associated with the transitions from each of the stress-split bands. There is, at the same time, a stress induced change of the transition probability from bands v_1 and v_3 to the conduction band because of the existence of a non-vanishing matrix element of the stress Hamiltonian between v_1 and v_3.[26] The effect of the band splitting is dominant over that of the change in oscillator strength near E_0: the singular part of the latter effect goes like $(\omega - \omega_0)^{1/2}$ and that of the former like $(\omega - \omega_0)^{-1/2}$. Contrary to the case for hydrostatic pressure, the mass can be assumed to be stress independent: the electron mass, which dominates the reduced mass, is not changed to first order by the splitting of the valence band. The contribution of E_0 and $E_0 + \Delta_0$ to the piezobirefringence for either [001] or [111] stress is:

$$\alpha = [(\epsilon_r)_{\parallel} - (\epsilon_r)_{\perp}]/X = C_0 \{ - g(\omega/\omega_0) + (4\omega_0/\Delta_0)[f(\omega/\omega_0) - (\omega_0/\omega_{0s})^{3/2} f(\omega/\omega_{0s})]\} \quad (26)$$

with $C_0 = 3/8\, C_0'' \,\delta\omega_0/\omega_0$.

The f(x) and g(x) are defined in Eq. (21), $\omega_{os} = \omega_o + \Delta_o$ and $\delta\omega_o$ is the linear splitting of the valence band per unit stress. This splitting can be written in terms of the deformation potentials b and d for [001] and [111] stress respectively:[26]

$$\delta\omega_{001} = 2b(S_{11} - S_{12})$$
$$\delta\omega_{111} = (d/\sqrt{3})S_{44}, \tag{27}$$

where S_{11}, S_{12}, and S_{44} are the elastic compliance constants and X the magnitude of the applied stress. The terms containing g(x) in Eq. (26) represent the contributions of the band splitting while those containing the less dispersive f(x) correspond to stress-induced changes in matrix elements.

Terms in piezobirefringence quadratic in stress have been observed for a number of DZ materials[12,13,27] very near the E_o gap. These terms can be interpreted as due to terms quadratic in $\Delta\omega_o$ in the expansion of the $(\omega - \omega_o)^{1/2}$ dependence of ϵ_r. These terms are proportional to $(\omega - \omega_o)^{-3/2}$ and hence they are very highly dispersive near E_o and may reach observable values. One can fit the observed non-linear birefringence with the expression:

$$(\epsilon_r)_{\parallel} - (\epsilon_r)_{\perp} = \alpha X + \beta X^2 , \tag{28}$$

The coefficient α is given in Eq. (26). The quadratic coefficient β is found to be, by a similar procedure:[12,13]

$$\beta = G[3/4 + 1/4(\delta\omega_H/\delta\omega_o)\, g(\omega/\omega_o) + 1/4(\delta\omega_H/\omega_o)\, h(\omega/\omega_o)] \tag{29}$$

with $h(x) = x^{-2}\,[2-(1-x)^{-3/2} - (1+x)^{-3/2}]$,

$$G = \begin{cases} [4\, C_o b\, (S_{11} - S_{12})]/\omega_o & \text{for [001] stress} \\ (2\, C_o d\, S_{44})/\sqrt{3}\ \omega_o & \text{for [111] stress} \end{cases}$$

In Eq. (29) $\delta\omega_H$ is the shift in ω_o produced by the hydrostatic component of the uniaxial stress per unit stress.

The spectral dependence of α observed for GaSb at room temperature for [001] and [111] stress is shown in Fig. 10, together with a fit using Eq. (26) with C_o as an adjustable parameter plus a constant contribution possibly due to E_1 and to the average gap. The PB reverses sign when approaching the E_o edge, since the contribution of this edge has a sign opposite to that of the non-dispersive contribution of the higher gap. This sign reversal appears for all materials with an E_o gap larger or

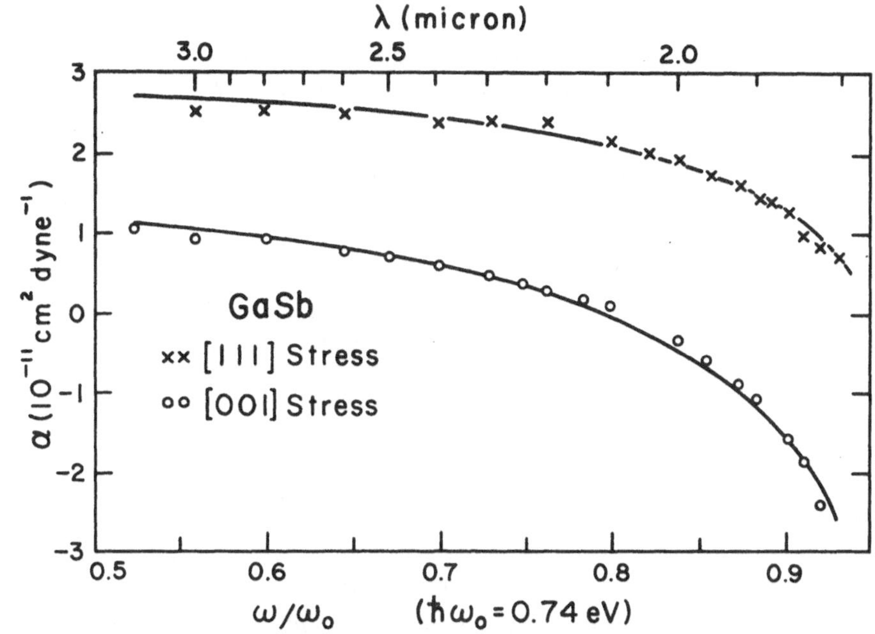

Fig. 10. Linear piezobirefringence coefficients of GaSb as a function of reduced frequency and fit with Eq. (26) plus a constant. From Ref. 12.

equal to that of GaSb (e.g., Ge, GaAs, CdTe, ZnSe). For materials with a smaller E_0 gap (InAs, InSb) the effect of this gap, which varies roughly like ω_0^{-1} when going from one material to another, *dominates* and no sign reversal is found.

The value of C_0 obtained from the fit of $\alpha(\omega)$ can be used to obtain the deformation potentials b and d if the values of P and ω_0 are known. We list in Table III the values of b and d found by this procedure for a number of DZ materials, together with the results of other, somewhat more direct measurements.

Figure 11 shows the spectral dependence of the quadratic PB coefficient β of GaSb for [001] and [111] stress together with the curve obtained with Eq. (27) using for C_0 the values obtained from α. An evaluation of Eq. (29) shows that the main contribution to β comes from the term which contains h(x) and is due to the combined effect of the hydrostatic shift in $\omega_0(\delta\omega_H)$ and the uniaxial splitting $\delta\omega_0$ (this effect can be recognized in Eq. (26) where the increase in the ω_0 gap produced by the hydrostatic component of the compression *decreases* the coefficient C_0).

E_1 *GAP*

The E_1, $E_1 + \Delta_1$ gaps yield also two types of contributions to the linear PB

TABLE III

Deformation potentials b and d obtained with the piezobirefringence technique and with other methods. From Ref. 12.

	Piezobirefringence		Other Methods	
	b (eV)	d (eV)	b (eV)	d (eV)
Ge	-3.0	-2.4	-2.8[a]	-4.95[a]
			-2.6[b]	-4.7[b]
GaAs	-4.1	-6.0	-1.75[a]	-5.55[a]
			-2.0[b]	-6.0[b]
GaSb	-3.0	-5.1	-3.3[a]	-8.35[a]
			-2.0[c]	-4.6[c]
InAs	-1.8	-3.6		
InSb	-1.8	-6.4	-2.0[d]	-4.9[d]
			-2.0±0.15[e]	-5.0±0.5[e]

[a] *A. A. Gavini and M. Cardona, Phys. Rev. 1, 672(1970).*

[b] *F. H. Pollak and M. Cardona, Phys. Rev. 172, 816(1968).*

[c] *C. Benoit a la Guillaume and P. Lavallard, J. Phys. Chem. Solids 31, 411(1970).*

[d] *F. H. Pollak and J. Halpern, Bull. Am. Phys. Soc. 14, 433(1968).*

[e] *C. Benoit a la Guillaume and P. Lavallard, J. Phys. Soc. Japan Suppl. 21, 188(1966).*

below E_o: energy shift and matrix element change. The more dispersive energy-shift contribution is only present for a [111] stress and stems from the intervalley splitting of the bands. The change in matrix elements is produced by the stress-induced mixing of the wavefunctions of the two spin-orbit split valence bands.

The E_1, $E_1 + \Delta_1$ contribution to the linear PB coefficient α below E_o is, for stress along [001]:[13]

$$\alpha = C_1 \left[1 + \tfrac{1}{2}(\omega/\omega_1)^2 - (\omega_1/\omega_{1s})^2 \left(1 + \tfrac{1}{2}(\omega/\omega_{1s})^2\right]\right. \tag{30}$$

and for stress along [111]:

$$\alpha = C_1' \Big\{ 1 + \tfrac{1}{2}(\omega/\omega_1)^2 - (\omega_1/\omega_{1s})^2 \left(1 + \tfrac{1}{2}(\omega/\omega_{1s})^2\right. + (\Delta_1 \mathcal{E}_2/2\sqrt{3}\ \ \omega_1 d)[1 + (\omega/\omega_1)^2 + (\omega_1/\omega_{1s})^3 \left(1 + (\omega/\omega_{1s})^2\right]\Big\} \tag{31}$$

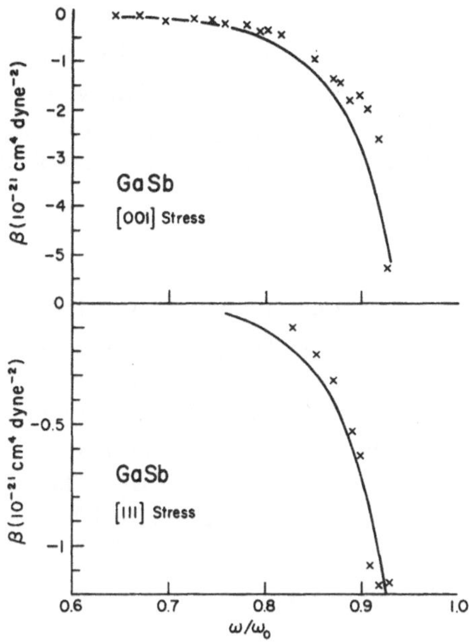

Fig. 11. The quadratic piezobirefringence coefficient β for GaSb at room temperature as a function of reduced frequency and fit with Eq. (29). From Ref. 12.

In Eqs. (30) and (31) $\mathcal{E}_2$ is the shear deformation potential of the bands, $\omega_{1s} = \omega_1 + \Delta_1$ and the constants C_1 and C_1' are defined as

$$C_1 = 2BP^2/\pi \;\; \delta\omega_{001}$$
$$C_1' = 2BP^2/\pi \;\; \delta\omega_{111} \qquad (32)$$

with B defined in Eq. (11) and $\delta\omega$ defined in Eq. (27). We note that the values of b and d are probably close to those used for the E_o edge. Information on b and d specific for the E_1 edge of Ge[28] and GaAs[29] has recently become available; it indicates that b is roughly the same as for the E_o edge while d is smaller.

We notice that the dispersion in Eq. (31) is weak below E_o. In materials with small ω_o (InSb, InAs, GaSb) it can be neglected and the whole contribution of Eq. (31) can be lumped together with the non-dispersive contribution of the average gap discussed below. The interband term in Eq. (31), containing E_2, has a sign opposite to the remaining intraband terms. The intraband term, however, is usually dominant in Eq. (31) so that the sign of the E_1, $E_1 + \Delta_1$ contribution to α is the same as that of the E_o contribution. This is contrary to a recent conclusion by Wardzynski[30] as a result of an algebraic error.

CONTRIBUTION OF THE AVERAGE GAP

We examine now the contribution to the linear PB in the spirit of Eq. (1). The pure shear component of the stress produces no change in the plasma frequency but should cause an anisotropy in the average gap ω_g. By analogy with the case of hydrostatic stress we assume that this anisotropy is also described by:

$$\omega_g \propto a^{-2.5} \propto k_F^{2.5} , \tag{33}$$

where a is the interatomic distance along the direction of the Fermi momentum k_F. This simple reasoning explains the sign reversal in α which occurs when approaching E_0: the contribution of the average gap has a sign opposite to that of E_0 and E_1. Upon compression the lattice constant for the longitudinal direction becomes smaller, the average gap along this direction becomes larger and hence a negative contribution to α results, opposite to the contribution of E_0 and E_1. The following expression for the average gap contribution to the elasto-optic effect has been proposed[4]:

$$\Delta\epsilon_0 = 5\ \mathbf{e}\ \epsilon_0 \tag{34}$$

where ϵ is the dielectric constant tensor and **e** the strain tensor. This expression is the generalization of Eq. (19) to uniaxial stress. One may reduce all components of Eq. (34) by a factor of about one half so as to take into account the fact that the strength of the experimental E_2 peak (see Figs. 1, 3, 4) is roughly one-half of that of the Penn model. The long wavelength PB of a number of DZ materials, after subtraction of the E_0 and E_1 effects is close to one-half of the predictions of Eq. (34).[12]

NON-LINEAR SUSCEPTIBILITIES

SECOND ORDER SUSCEPTIBILITY: FREQUENCY DOUBLING

The second order susceptibility tensor $\chi^{(2)}_{ijk}$, which can be determined in a frequency doubling experiment, has one non-zero component $\chi^{(2)}_{123}$ for a zincblende-type material. This component is zero for germanium. The expression for χ_{123} is somewhat cumbersome[31,32] but consists essentially of sums over the Brillouin zones containing products of three matrix elements of **p** divided by energy denominators. Three bands are necessary to obtain a contribution to $\chi^{(2)}_{123}$; if only two bands are considered the contribution for a given **k** cancels exactly that for -**k**. If spin-orbit interaction is neglected the top valence band and the two lowest conduction bands (Γ_1, Γ_{15}) constitute sets of three bands with a finite contribution to $\chi^{(2)}_{123}$. The spin-orbit interaction splits the valence bands and gives terms linear in **k** for the upper Γ_8 valence bands. These split Γ_8 bands and the lowest conduction band

Γ_1 provide a set of three bands with a finite contribution to $\chi^{(2)}_{123}$. Contrary to statements in the literature[34], this last contribution is negligible compared to the first. It is easy to see that their ratio is of the order of

$$P\omega'_o/P'\omega_o \tag{35}$$

where P is the coefficient of the linear term in **k** for the Γ_8 valence band, P′ the matrix element of **p** between Γ_1 and Γ_{15c} and ω'_o the Γ_{15c} - Γ_{15v} gap. P is smaller than 10^{-2} while P′ is around 0.3 for a III-V compound.[35] Thus the "two optical bands" contribution, involving the Γ_8 and the Γ_1 bands, is usually negligible and the effect must be dominated by the Γ_{15v} - Γ_1 - Γ_{15c} contribution. This contribution can be written, under the assumption of three-dimensional parabolic bands at Γ_{15v}, Γ_1, and Γ_{15c} extending to infinity[32]:

$$\chi^{(2)}_{123} = (iF\mu_{12}\mu_{13})/(\sqrt{2}\,\pi\omega^3) \sum_{i=1}^{6} S_i/(\sqrt{x_i} + \sqrt{x'_i}) \tag{36}$$

with

$$x_i = \mu_{12}\,(\omega_o + n_i\omega)$$

$$x'_i = \mu_{13}\,(\omega'_o + m_i\omega)$$

$$S_i = \begin{matrix} 1, \text{ for } i = 1 \text{ to } 3 \\ -1, \text{ for } i = 4 \text{ to } 6 \end{matrix}$$

$$n_i = -2,\ 1,\ 1,\ -1,\ -1,\ 2$$

$$m_i = -1,\ -1,\ 2,\ -2,\ 1,\ 1.$$

In Eq. (36) $\mu_{k\ell}$ are the reduced masses appropriate to the pair of bands under consideration and F the product of the three matrix elements of **p** (assumed constant) between each pair of bands. It is easy to see that[35]

$$F \simeq i\,P^2\,Q\,V/\omega'_o \tag{37}$$

where P is the matrix element of **p** between Γ_{15v} and Γ_1, Q that between Γ_{15v} and Γ_{15c} and V the antisymmetric coupling between Γ_{15c} and Γ_{15v},[35] practically equal to the constant C defined earlier.

A set of three bands which also gives a finite contribution to $\chi^{(2)}_{123}$ is also associated with the E_1, $E_1 + \Delta_1$ critical points: the Λ_{3v} valence band, the Λ_1 conduction band and the Λ_{3c} second lowest conduction band. The Λ_{3v} - Λ_1 gap can be treated as essentially two dimensional while longitudinal and transverse masses must be considered for the Λ_{3v} - Λ_{3c} gap.

An expression which will not be reproduced here, somewhat more complicated than that of Eq. (36), is obtained.[32] The results of the E_o and E_1 contributions to the real and imaginary parts of $\chi^{(2)}_{123}$ for GaAs is shown in Fig. 12. Figure 13 shows the magnitude $\chi^{(2)}_{123}$, as calculated from the curves of Fig. 1, compared with experimental results by Bloembergen and co-workers.[36] The calculated peak at 0.7 eV is due to the resonance $2\omega \simeq \omega_o$, while that at 1.5 eV, also seen in the experimental points, corresponds to $2\omega \simeq \omega_1$.

Several simple expressions have been proposed for the long wavelength value of the second order susceptibility $\chi^{(2)}_{123}(0)$. Based on the Penn model, Phillips and Van Vechten suggested the expression:[37]

$$\chi^{(2)}_{123}(0) = 0.4\ (a_o C)/(\omega_g^2)\ [\epsilon_r(0) - 1]\ , \tag{38}$$

while Bell proposed:[32]

$$\chi^{(2)}_{123}(0) = 0.24\ (QV/\omega_o^3)\ [\epsilon_r(0) - 1]\ . \tag{39}$$

Both Eqs. (38) and (39) give reasonable agreement with the experimental results (see Table IV).

THIRD ORDER SUSCEPTIBILITY

The third order susceptibility tensor $\chi^{(3)}_{ijk\ell}$ is measured in frequency tripling[38,39] and frequency mixing experiments.[40] The measurements to be discussed here were performed at frequencies (CO_2 laser) low compared with the E_o gap and hence the dispersion in $\chi^{(3)}_{ijk\ell}$ is negligible. Under these conditions there are only two independent components of $\chi^{(3)}_{ijk\ell}$, $\chi^{(3)}_{1111}$ and $\chi^{(3)}_{1122}$; it is convenient, however, to use as independent components $\chi^{(3)}_{1111}$ and the component which corresponds to fields and polarization along [111], $\chi^{(3)}_{\xi\xi\xi\xi}$

$$\chi^{(3)}_{\xi\xi\xi\xi} = 1/3\ [\chi_{1111} + 6\ \chi_{1212}]\ . \tag{40}$$

The third order polarizability for all frequencies tending to zero, $\chi^{(3)}_{ijk\ell}(0)$, is the change in the corresponding component of the polarizability tensor produced by a static field. Other than free carrier effects, two mechanisms are responsible for this change: the field mixing of states within each band (intraband effect) and the

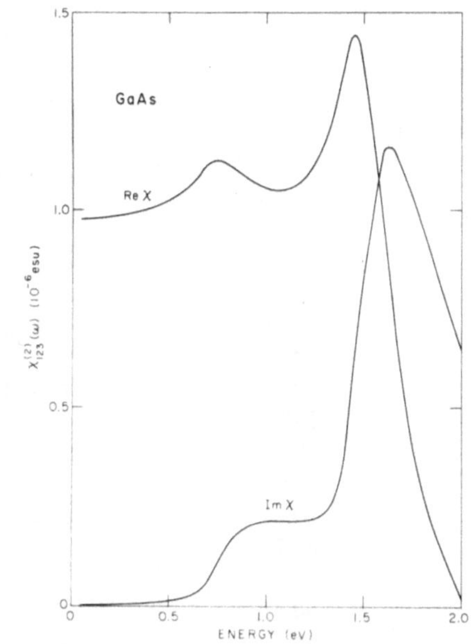

Fig. 12. Real and imaginary parts of $\chi^{(2)}_{123}(\omega)$ for GaAs calculated with a model involving the two lowest conduction bands and the highest band at k = 0 (E_0, E'_0 gaps) and also the corresponding bands along [111] (E_1, E'_1 gaps). From Ref. 32.

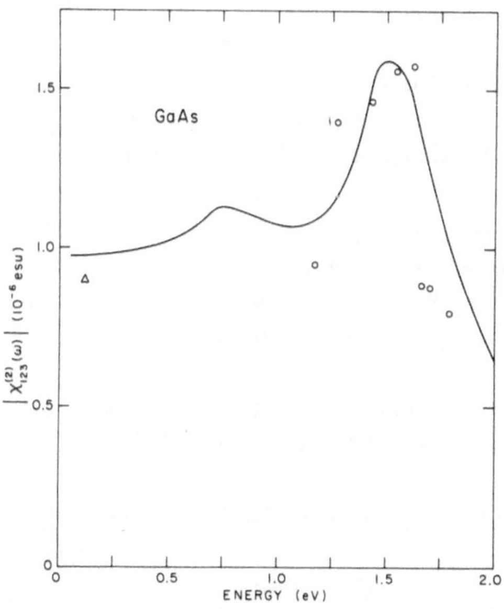

Fig. 13. Magnitude of $\chi^{(2)}_{123}(\omega)$ for GaAs as obtained from Fig. 12. The circles represent experimental data from Ref. 36.

TABLE IV

Experimental long wavelength values of the second order susceptibility of several zincblende-type materials compared with calculations by Phillips and Van Vechten[37] and by Bell[32]. In units of 10^{-8} e.s.u.

	Phillips Van Vechten $\chi^2(0)$	Bell $\chi^2(0)$	Experiment $\chi^2(0)$
GaAs	256	97.7	90 ± 30[a]
GaSb	433	199	302 ± 100[a]
InAs	380	122	200 ± 60[a,b]
InSb	611	271	
GaP	186	70.0	52 ± 17[a]
InP	269	83.7	
AlSb	337	126	
ZnTe	257	50.3	44 ± 16[b]
CdTe	257	51.4	80 ± 30[b]

[a] *J. J. Wynne and N. Bloembergen, Phys. Rev. 188, 1211(1969).*
[b] *C. K. N. Patel, Phys. Rev. Letters 16, 613(1966).*

k-conserving coupling across the gap (interband effect). The intraband coupling results in a decrease of the effective gap at a minimum critical point such as E_o, E_1, and the Penn gap: the band edge is repelled by all other states of the same band. This field-induced decrease in the gap results in an increase in the dielectric constant, and hence in a positive $\chi^{(3)}_{1111}$. The interband effect corresponds to a repulsion between the valence and the conduction bands and hence yields a negative contribution to $\chi^{(3)}_{1111}(0)$. The magnitude of the interband effect is usually three-fifths that of the intraband effect and hence the resulting $\chi^{(3)}_{1111}(0)$ is positive. We note that the intraband effect corresponds to the so-called Franz-Keldysh effect in absorption. Both effects can be obtained from the expression of the polarization in the presence of an electric field E to second order in E:[41]

$$P(\omega,\mathrm{E}) = (Q_o/4\pi^2) \int_{BZ} d^3k \left\{ 2\,\omega_{cv}^{-3}(\mathbf{k},\mathrm{E}) + 40\Omega^3/\omega_{cv}^6 + \ldots \right\} \tag{41}$$

with

$$\Omega^3 = \mathrm{E}^2/8\mu_{||}$$

and

$$Q_o = \pi^{-1}\, |\mathbf{E}\cdot\mathbf{p}|^2$$

In Eqs. (41) ω_{cv} is the energy difference between the conduction and the valence band considering the interband field repulsion and $\mu_{\|}$ is the effective mass along **E**.

The interband effect is obtained from the first summation in curly brackets by replacing ω_{cv} by

$$\omega_{cv}(\mathbf{k},\mathbf{E}) = \omega_{cv}(\mathbf{k}) + 2|\mathbf{E}\cdot\mathbf{P}|^2/\omega_{cv}^3 \tag{42}$$

For parabolic bands one finds indeed that the interband effect on χ has negative sign and a magnitude three-fifths that of the intraband effect (second summand in curly brackets).

The contributions of the average gap, the E_o gap and the E_1 gap to the intraband $\chi^{(3)}_{1111}(0)$ are[41]:

$$\begin{aligned} &\text{Penn:} && 0.63\ \omega_p^{4/3}/\omega_g^4\,[\epsilon_r(0) - 1] \\ &E_o: && 0.06(P^2\mu^{1/2}/\omega_o^{9/2})[1 + 1.85\,(\omega/\omega_o)^2 + \ldots] \\ &E_1: && 0.52\ P^2/a_o\omega_1^{5} \end{aligned} \tag{43}$$

In Eqs. (43) we have neglected spin-orbit interaction. It can be easily taken into account by replacing ω_1 by $\omega_1 + \Delta_1$ and ω_o by $\omega_o + \Delta_o/3$ (provided $\Delta_o << \omega_o$). In principle the results of Eq. (43) must be multiplied by 2/5 in order to take into account the interband effect. A calculation of the average gap (Penn) case including band non-parabolicity[41] yields instead of the coefficient 0.63 of Eq. (43) 0.4, with a negligible interband contribution. We use this result in our tabulation (Tables V and VI). For the semiconductors (Table VI) a factor of ½ has been introduced in the ω_g contribution for the reasons discussed earlier. This factor has not been used for the alkali halides (Table V) since for these materials the E_o and E_1 contribution has not been considered explicitly. We note that the E_o contribution becomes large (dominant) for materials with a small ω_o: Eqs. (43) give contributions inversely proportionally to a large power of the gap ($\omega_o^{9/2}$ for E_o). For the alkali halides $\omega_o \simeq \omega_g$ and hence $\chi^{(3)}_{1111}(0)$ can be treated exclusively with the Penn model. The results obtained with the first of Eqs. (43) and the parameters listed in Ref. 2 for these materials and for MgO are presented in Table V together with experimental data.[39,40] The agreement between theory and experiment is good, especially if one considers that Eqs. (43) are uncertain to factors of the order of unity. The experimental bands from one material to the other are well described by the calculated values. Table IV presents experimental values of $\chi^{(3)}_{1111}(0)$ for a number of semiconductors, together with the calculated contributions of the E_o, E_1, and Penn gap

TABLE V

Experimental values of the third order susceptibility $\chi^{(3)}_{1111}(0)$ for MgO and several alkali halides. Also, values predicted with the Penn model (Eq. 39). In units of 10^{-14} e.s.u.

$\chi^{(3)}_{1111}(0)$	MgO	LiF	NaF	NaCℓ	KCℓ	KBr
Calculated	2.1	0.14	0.18	2.3	4.2	4.6
Experiment[a]	2.4	0.3	0.25	1.2	1.2	2.8
Experiment[b]				1.7	1.9	3.0

[a]*See Ref. 39*

[b]*See Ref. 40*

TABLE VI

Contributions to $\chi^{(3)}$ from the three regions discussed in the text. All numbers except ratios are in units of 10^{-10} e.s.u. The numbers in parentheses in the last line are experimental values. From Ref. 41.

Crystal	Si	GaAs	Ge	InAs	PbS	InSb
E_o	0	0.034	0.33	3.6	28	43
E_1, <100>	0.012	0.022	0.10	0.04	- -	0
E_1, <111>	0.016	0.029	0.13	0.05	- -	0
E_G	0.039	0.022	0.08	0.1	- -	0
Sum, <100>	0.051	0.078	0.51	3.7	28	43
Ratio to Ge, (theory)	0.10	0.15	1	7.3	55	84
Ratio to Ge, (Experiment)	0.06[a]	0.12[a]	1(+1.0)[a]	- -	- -	- -

[a]*Ref. 38.*

(only the E_0 contribution is given for PbS since the E_1 and Penn model may not apply to it). The non-linear susceptibilities are now several orders of magnitude larger than for the alkali halides. The experimental results are also well described (within factors of the order of unity) by the sum of the E_0, the E_1, and the Penn gap contributions. It is interesting to point out that anisotropy $[\chi^{(3)}_{1111}(0) \neq \chi^{(3)}_{\xi\xi\xi\xi}(0)]$ is observed in the semiconductors. The E_0 and ω_g contributions are isotropic and the anisotropy calculated for the E_1 contribution is insufficient to explain the experimental results. It has been pointed out[41] that the observed anisotropy can be explained by introducing internal field corrections. Such corrections are beyond the scope of this paper.

REFERENCES

* *Supported by the National Science Foundation and the Army Research Office, Durham.*

1. *J. C. Phillips, Solid State Physics, Vol. 18, Academic Press, New York 1966, p. 1.*
2. *J. A. Van Vechten, Phys. Rev. 182, 891(1969).*
3. *J. C. Phillips, Covalent Bonding in Crystals, Molecules, and Polymers, University of Chicago Press, 1969.*
4. *M. Cardona, J. Res. of the NBS, 74A, 253(1970).*
5. *J. P. Walter, R. R. L. Zucca, M. L. Cohen, and Y. R. Shen, Phys. Rev. Letters 14, 102(1970).*
6. *D. E. Aspnes and J. E. Rowe, Solid State Communications, 8, 1145 (1970).*
7. *D. Penn, Phys. Rev. 128, 2093(1962).*
8. *M. Cardona, J. Appl. Phys. 36, 2181(1965).*
9. *M. Cardona, Modulation Spectroscopy, Academic Press, New York, 1969.*
10. *H. Piller, B. O. Seraphin, K. Markel, and J. E. Fischer, Phys. Rev. Letters 23, 775(1969). G. Weiser and J. Stuke, Phys. Status Solids 35, 747(1969).*
11. *V. Heine and R. O. Jones, J. Phys. C. 2, 719(1969).*
12. *P. Y. Yu and M. Cardona, Phys. Rev., 2, 3193 (1970).*
13. *C. W. Higginbotham, M. Cardona, and F. H. Pollak, Phys. Rev. 184, 871(1969).*
14. *E. O. Kane, Phys. Rev. 180, 852(1969).*
15. *L. I. Korovin, Soviet Phys., Solid State 1, 1202(1959).*
16. *C. Keffer, T. M. Hayes, and A. Bienenstock, Phys. Rev. Letters 21, 1676(1968).*
17. *F. Stern, Phys. Rev. 133, A1653(1964).*
18. *J. N. Zemel, J. D. Jensen, and R. B. Schoolar, Phys. Rev. 140A, 330(1965).*
19. *M. Cardona, in Solid State Physics, Nuclear Physics, and Particle Physics, edited by I. Saavedra, Benjamin, New York, 1968.*
20. *M. Cardona, W. Paul, and H. Brooks, J. Phys. Chem. Solids 8, 204(1959).*
21. *N. J. Trappenniers, R. Vetter, and H. A. R. de Bruin, Physica 45, 619(1970).*
22. *W. M. de Meis, Technical Report No. HP-15, Gordon McKay Lab, Harvard University, 1965.*
23. *K. Vedam and E. D. D. Schmidt, Phys. Rev., 150, 766(1966).*
24. *F. Lukes, Czech. J. Phys. B10, 743(1960).*
25. *L. M. Roth, Phys. Rev. 133, A542(1964).*
26. *F. H. Pollak and M. Cardona, Phys. Rev. 172, 816(1968).*
27. *A. Feldman and D. Horowitz, Jr., J. Appl. Phys. 39, 5597(1968).*
28. *D. D. Sell and E. O. Kane, Phys. Rev. 185, 1103(1969).*
29. *J. E. Wells, Ph.D. Thesis, University of Illinois, 1969.*

30. *W. Wardzynski, J. Phys. C. 3, 1251(1970).*
31. *P. N. Butcher and T. P. McLean, Proc. Phys. Soc. 81, 219(1963); 83, 579(1964).*
32. *M. I. Bell, Proceedings of the Conference on Electronic Density of States. Gaithersburg, Maryland, 1969, in press.*
33. *A. G. Aronov and G. E. Pikus, Soviet Phys., Solid State 10, 648(1968).*
34. *V. S. Bagaev, T. Ya. Belousova, Yu. N. Berozashvili, and D. S. Lordkipanidze, Soviet Phys.-Semiconductors 3, 141 8(1970).*
35. *M. Cardona, J. Phys. Chem. Solids 24, 1543(1963).*
36. *R. K. Chang, J. Ducuing, and N. Bloembergen, Phys. Rev. Letters 15, 415(1965).*
37. *J. C. Phillips and J. A. Van Vechten, Phys. Rev. 183, 709(1969).*
38. *J. J. Wynne, Phys. Rev. 178, 1295(1969).*
    J. J. Wynne and G. D. Boyd, Appl. Phys. Letters 12, 191(1968).
39. *C. C. Wang and E. L. Baardsen, Phys. Rev. 185, 1079(1969); 1B, 2827(1970).*
40. *P. D. Maker and R. W. Terhune, Phys. Rev. 137, A801(1965).*
41. *J. A. Van Vechten, M. Cardona, D. E. Aspnes and R. M. Martin, Proceedings of the 10th International Conference on Semiconductors, Cambridge, Mass., 1970, p. 82.*

LIGHT MODULATION BY REFLECTION

L. T. KLAUDER, JR., and J. G. GAY

Research Laboratories, General Motors Corporation, Warren, Michigan

ABSTRACT

The possibility of modulating a light beam by reflecting it from the surface of a crystal whose surface dielectric constant is modulated by an electric field is investigated theoretically. Since the dielectric constant may depend on distance in from the surface of the crystal it was necessary to develop a computer program which calculates the reflectivity of a medium with a non-uniform dielectric constant. To get an idea of the effect of spatial variations in the dielectric constant, the program is first used to calculate the reflectivity of media with simple ramps in their dielectric constant. The results show that a dielectric constant whose ramp extends at least one-third of a wavelength of light into the medium looks homogeneous to the light.

To estimate the effectivensss of a reflection modulator, the surface dielectric constant of CdS near a band edge is simulated both with and without a perturbing electric field and the reflectivity program is used to compute both the perturbed and unperturbed reflectivity. The best results, for a field of 10^6 volt/cm, are that a contrast ratio (ratio of reflectivities) of 17 can be obtained at an efficiency (largest reflectivity) of 1.7 percent.

The usual way to modulate a light beam with a crystal is to use it to alter the intensity or polarization of the beam during transmission. We investigate theoretically the possibility of modulating a beam by reflecting it from the surface of a crystal whose optical properties are modulated by a surface electric field.

It is known from electroreflectance theory and experiments[1] that one can make large changes in the dielectric constant in the vicinity of the exciton energy in semiconductors like CdS by applying an electric field to the surface. The nature of the

References pp. 128-129

change is illustrated in Fig. 1. The solid curve represents the imaginary part of the dielectric constant, ϵ_2, or equivalently the absorption coefficient, of a semiconductor when there is no field. The sharp peak is due to an exciton bound state and the absorption at higher energy is the band edge absorption. When a field is applied the bound exciton state turns into a resonant state with an intrinsic width. If the field is high enough (about 10^6 volt/cm in CdS) the exciton becomes very broad and the band edge acquires a tail so that the absorption coefficient goes over into the dashed curve. The decrease in ϵ_2 at the exciton peak depends on temperature but at 80°K is of the order of a factor of 10. We will consider the possibility of using this change to modulate a reflected laser light beam by altering the reflectivity minimum at the Brewster angle for a beam polarized with its magnetic vector transverse to the plane of incidence (T.M. polarization). When T.M. polarized light is incident at the Brewster angle as indicated in Fig. 2, the reflected ray is perpendicular to the refracted ray. There is no reflection because the polarization of the medium is along the direction of the reflected ray. Any changes in the optical properties of the medium will spoil this condition and result in the appearance of a reflected beam.

The Brewster angle phenomenon is an idealization valid only at the interface between two uniform and non-absorbing media[2]. Since a real crystal surface may not be sharp on the scale of a wavelength of light the Brewster angle will be prespoiled to some extent. Also the perturbing electric field will be attenuated with

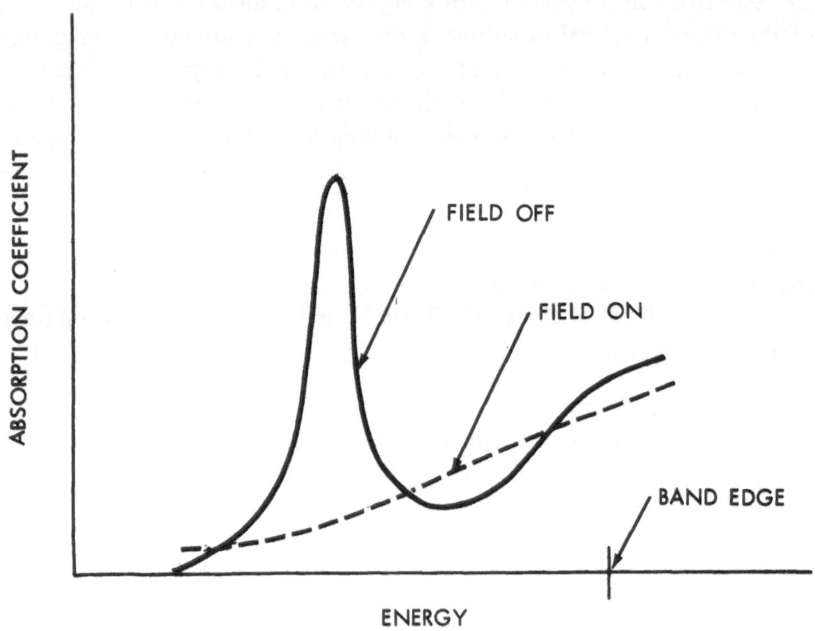

Fig. 1. Schematic of the change in the absorption coefficient of a crystal with application of an electric field.

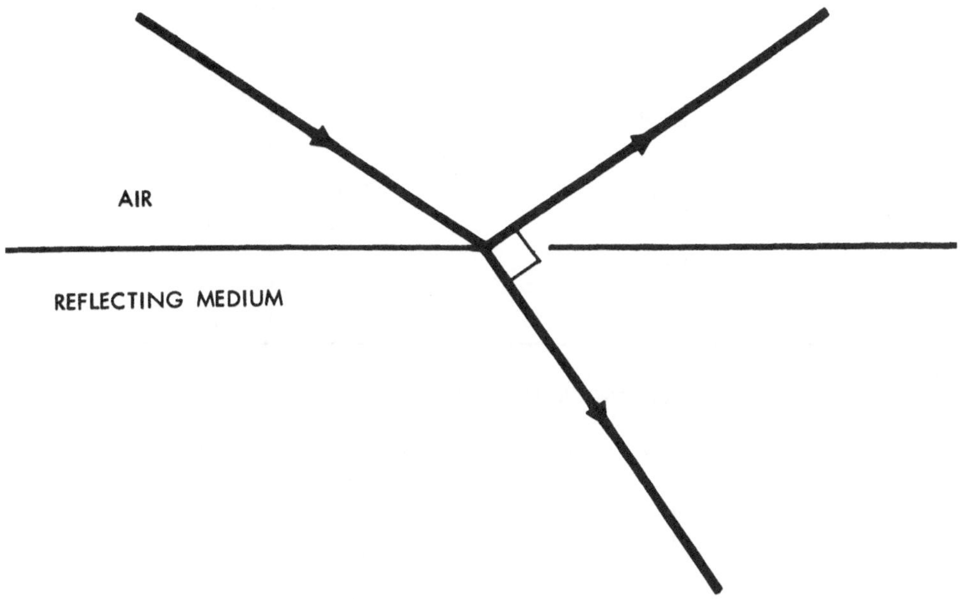

Fig. 2. Geometry of the Brewster angle. The polarization of the reflecting medium is parallel to the reflected ray preventing the emission of a reflected wave.

distance into the crystal depending on the conductivity or level or doping of the crystal. These facts mean that dielectric constant will vary with distance into the crystal as illustrated schematically in Fig. 3. Consequently, in order to estimate the effectiveness of the postulated reflection modulator, we must be able to calculate the reflectivity coefficient of a crystal with spatially dependent dielectric constant.

Aspnes and Frova[3] have developed a perturbation technique based on the WKB approximation which does this provided the part of the dielectric constant which varies is small relative to the uniform or bulk dielectric constant. Since we anticipate situations in which this requirement is violated, we developed a computer program which computes the reflectivity by numerically integrating the wave equation. This is done as follows[4]. We know that in the bulk where ϵ is uniform the solution is just a right going plane wave whose direction is determined by Snell's Law, i.e., by the angle of incidence and the bulk dielectric constant. One can then take this as a starting solution at the right boundary of the surface layer and numerically integrate the wave equation to the surface. This surface solution is then matched to an incident and reflected wave in the air to obtain the reflection coefficient.

There are several methods which can be used to obtain a numerical solution of the wave equation in this context. We elected to use an approach which is

References pp. 128-129

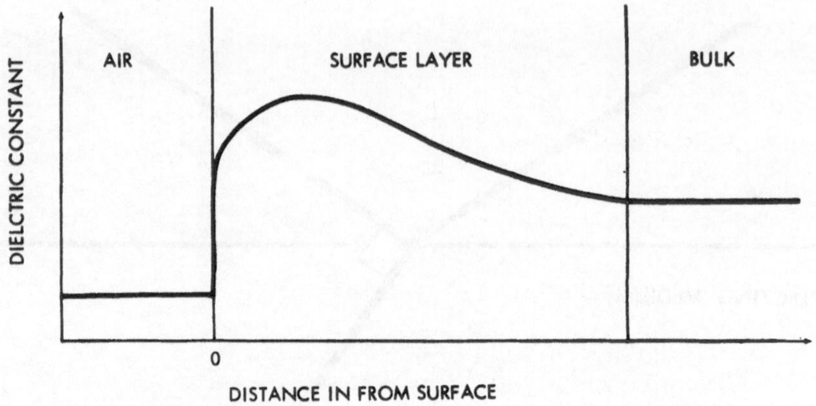

Fig. 3. Schematic of the variation of the dielectric constant with distance in from the crystal surface.

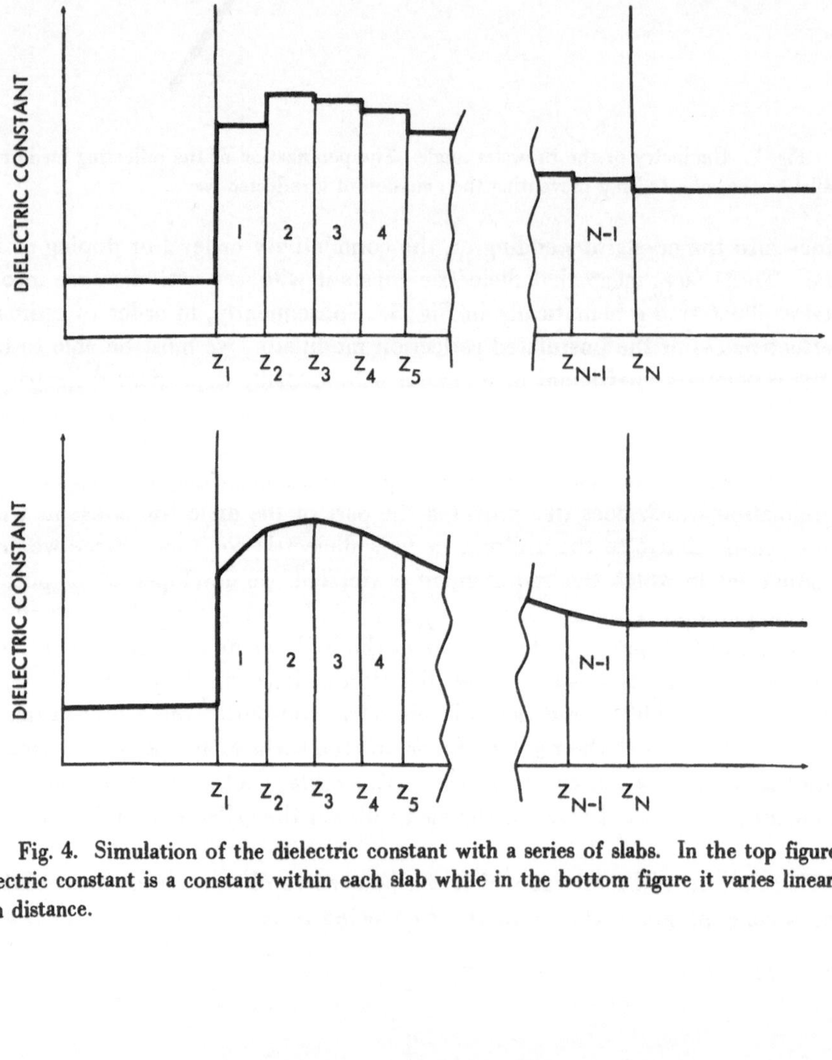

Fig. 4. Simulation of the dielectric constant with a series of slabs. In the top figure the dielectric constant is a constant within each slab while in the bottom figure it varies linearly with distance.

commonly taken in the formal theory of stratified media[4]. We divide the surface layer into a number of thinner slabs and treat the dielectric constant within each slab as being independent of position. This is illustrated at the top of Fig. 4. The solutions of the wave equation in each slab are then plane waves, and the solution of the wave equation for the substitute profile can be written exactly in terms of a product of complex two-by-two matrices, there being one such "characteristic" matrix for each thin slab. There is a standard theorem[5] on the existence of solutions of ordinary differential equations which indicates that the solution for the substitute profile converges to that for the original profile as the number of thin slabs tends to infinity, the thickness of each slab tending to zero. (The number of slabs required for our purposes turned out to be of the order of 50.) This method of solution was chosen on the basis of the ease with which it can be coded rather than on the basis of efficiency.

The procedure is illustrated in more detail by the following matrix equation

$$[a_o, b_o] = [1, 0]\, F_{n,\ell}\, M_{N-1}\, M_{N-2} \cdots M_2 M_1\, F_{o,r} \tag{1}$$

The vector [1,0] on the right is a vector of solution coefficients whose components specify the relative amounts of right-going and left-going plane wave in the solution. The fact that it is [1,0] indicates that the solution at the right boundary of the surface layer is a right-going plane wave. $F_{N,\ell}$ is a 2x2 matrix which converts the vector of solution coefficients into a vector whose components are the tangential fields E and H. M_{N-1} is a complex 2x2 matrix which propagates the [E,H] vector through the (N-1)*st* slab. M_{N-1} propagates the vector of field components through the (N-2)*nd* slab and so on until the surface is reached. At the surface $F_{o,r}^{-1}$ converts the [E,H] vector into the vector $[a_o, b_o]$ of coefficients for the solution in the air. The reflectivity is then just $|b_o/a_o|^2$.

We may observe that for the case of T.E. polarization (electric vector perpendicular to the plane of incidence) there are several forms of simple analytic variation of the dielectric constant with distance for which the wave equation has well known functions as its solutions. Thus, for example, if the substitute dielectric profile is chosen to consist of slabs in each of which the variation is linear with distance, as illustrated at the bottom of Fig. 4, the solutions are Airy functions. The treatment which is obtained in this case is completely analogous to that described earlier. If the given profile were well approximated by a small number of linear slabs, one might expect this latter approach to be more efficient. The reason that we do not pursue this approach is that we are primarily interested in the case of T.M. polarization for which case the wave equation has known solutions only when the dielectric constant is independent of position.

Before discussing the results of applying the reflectivity program to a dielectric constant profile simulating a real reflection modulator, we wish to show some results for dielectric constants with simple linear ramps as illustrated in Fig. 5. These ramp dielectric constants are not intended to represent a real material. Our aim in studying them was to see if we could get a rule-of-thumb as to how far into the medium a perturbation has to extend before it begins to look as though it were present throughout the medium. However, the results can be used to illustrate some general features of reflection modulators.

Note that when B = 0 the dielectric constant becomes homogeneous and equal to ϵ and when $B = \infty$ it again becomes homogeneous and equal to $\epsilon + \Delta\epsilon$.

Figure 6 is a computer drawn plot of R as a function of ramp length for normal incidence and $\epsilon = 5$, i.e., no absorption. The light used had a wavelength of approximately 5000 Å in air and thus approximately $\lambda = 2200$ Å in the medium. These curves show two features which are characteristic of both polarizations for all angles of incidence we have tried. These are that R changes from its B = 0 value to its $B = \infty$ value by the time $B = 1/3\ \lambda$, and that R subsequently oscillates about its $B = \infty$ value. Thus we seem to have our rule-of-thumb but have not tested it for other ramp shapes.

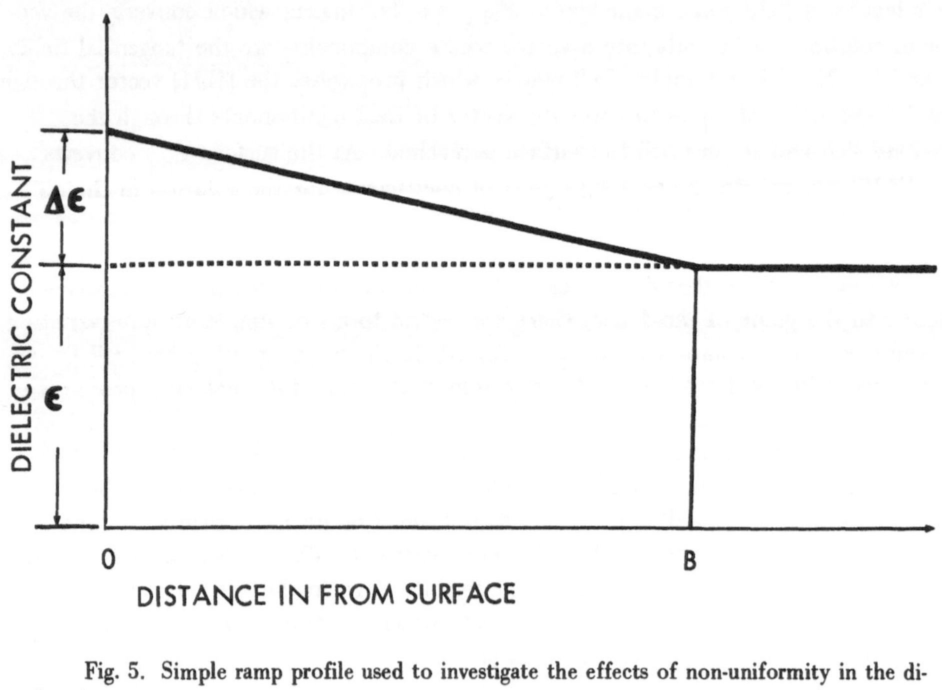

Fig. 5. Simple ramp profile used to investigate the effects of non-uniformity in the dielectric constant.

If we could cause the ramp to appear and disappear we could use this as a modulator. For example, if we used a ramp of height 0.5 and length 800 Å we could get a contrast ratio of 16.6/14.6 approximately 1.1. However the efficiency, which is given by the larger of the two reflectivities, is only about 0.17. Figure 7 shows the same curves except that the imaginary part of ϵ is made 1. This corresponds to an absorption length of approximately 1800 Å. The only significant change is that the oscillations about the $B = \infty$ value subside more rapidly.

Figure 8 shows the TM reflectivity at the Brewster angle for $\epsilon = 5$. Note that the curves for $\Delta\epsilon = \pm 0.1$ have merged and that the reflectivity is 0 at $B = 0$. Thus if this arrangement were considered as a modulator the contrast ratio would be infinite. However, the reflectivity with the $\Delta\epsilon = 0.5$ ramp turned on is only 0.0006 which indicates rather poor efficiency.

When we add an imaginary part of 1 to ϵ we see, in Fig. 9, the effect of prespoiling the Brewster phenomenon, the $B = 0$ reflectivity has risen to 0.0015. Thus, while the perturbed reflectivity has risen by a factor of 5 to 0.0030, the contrast ratio has dropped to 2. This is typical of the behavior found with more realistic models for a modulating medium: One is always faced with a trade-off between efficiency and contrast ratio.

With these examples in mind let us estimate the sorts of contrast ratios and efficiencies we can obtain by modulating the surface dielectric constant of a real crystal. To do this we have to construct a reasonable model of the surface dielectric constant of the crystal.

We choose to work with CdS and to modulate it with a depleting field. In CdS the effective mass approximation is valid and gives for the imaginary part of the dielectric constant near a band edge[6]

$$\epsilon_2(E) = (2\pi e\hbar/m)^2\ |P_{cv}|^2$$

$$\times\ \sum_n [F_n/(E_{cv}-E_n)^2]\ \delta[E-(E_{cv}-E_n)] + F(E)\ S(E)/(E_{cv}+E)^2 \qquad (2)$$

where the energy E is measured from the top of the valence band. E_{cv} is the gap energy, and P_{cv} the conduction to valence band momentum matrix element. The first term gives the contribution of the bound exciton states. The E_n are binding energies and the F_n are enhancement factors which take account of the increase in absorption due to the electron-hole Coulomb attraction. The second term gives the continuum enhancement. S(E) is the joint density of states and F(E) the continuum enhancement. To take account of lifetime effects Eq. (2) has to be folded with a Lorentzian

References pp. 128-129

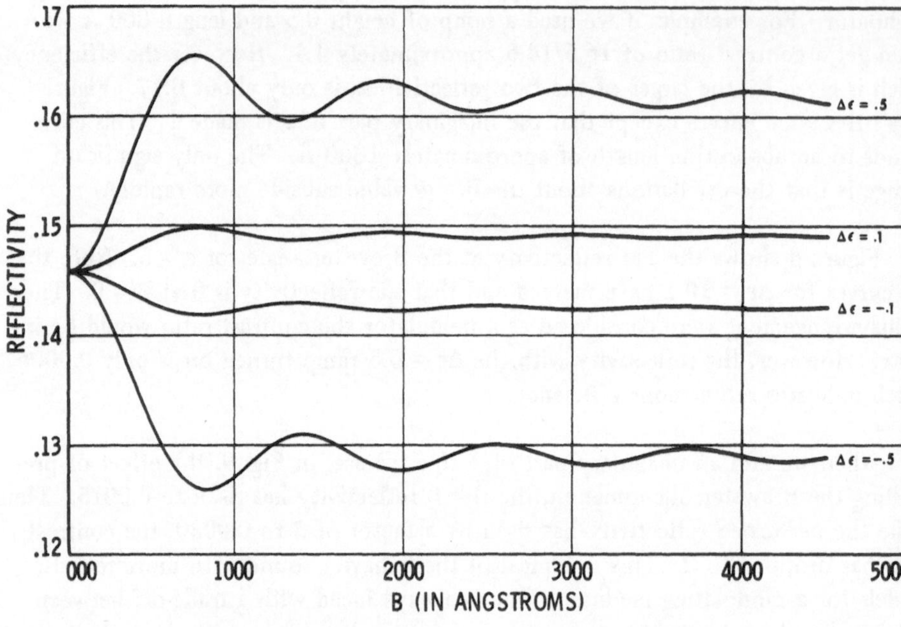

Fig. 6. Reflectivity as a function of ramp length for normal incidence and $\epsilon = 5$.

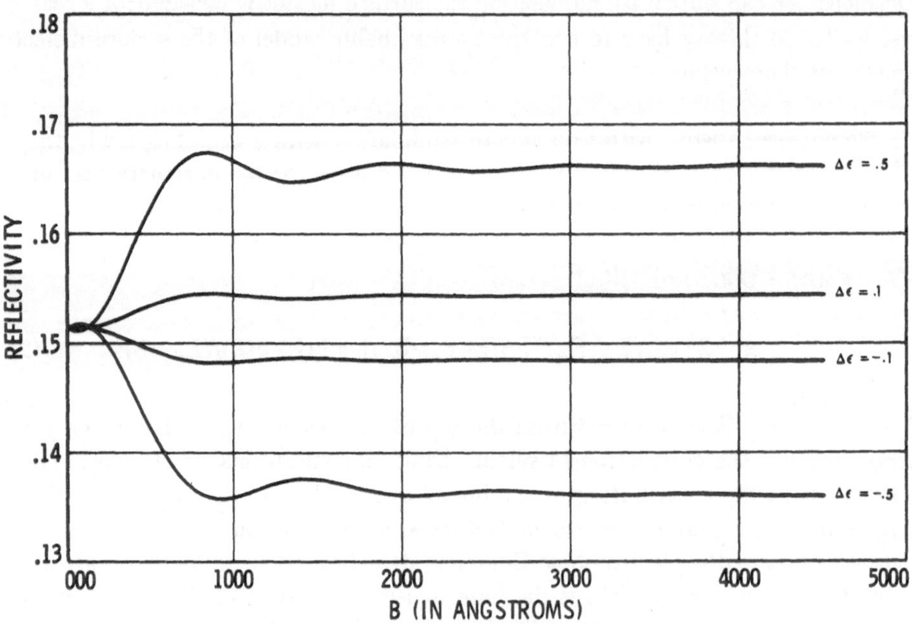

Fig. 7. Reflectivity as a function of ramp length for normal incidence and $\epsilon = 5 + i$.

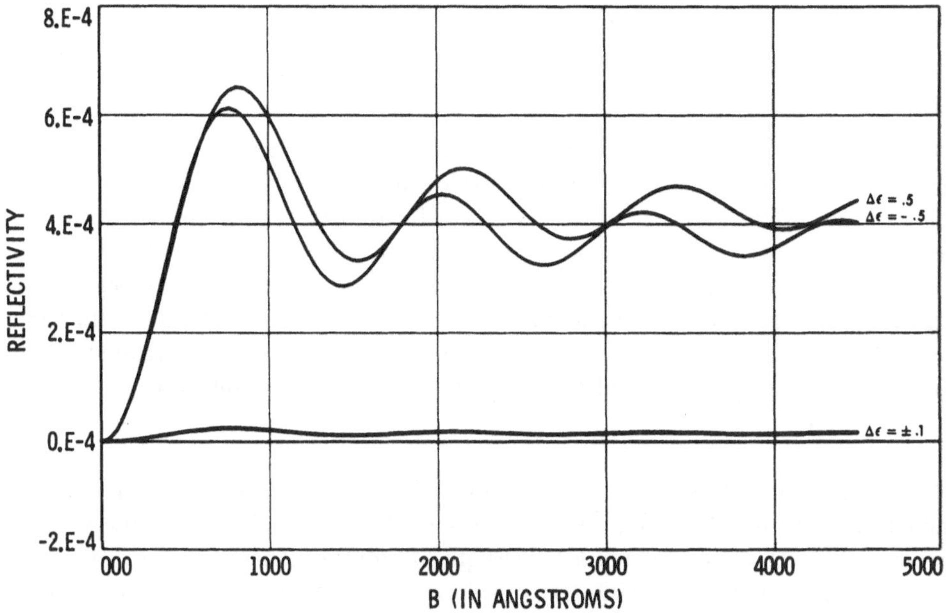

Fig. 8. Reflectivity as a function of ramp length at the Brewster angle for $\epsilon = 5$.

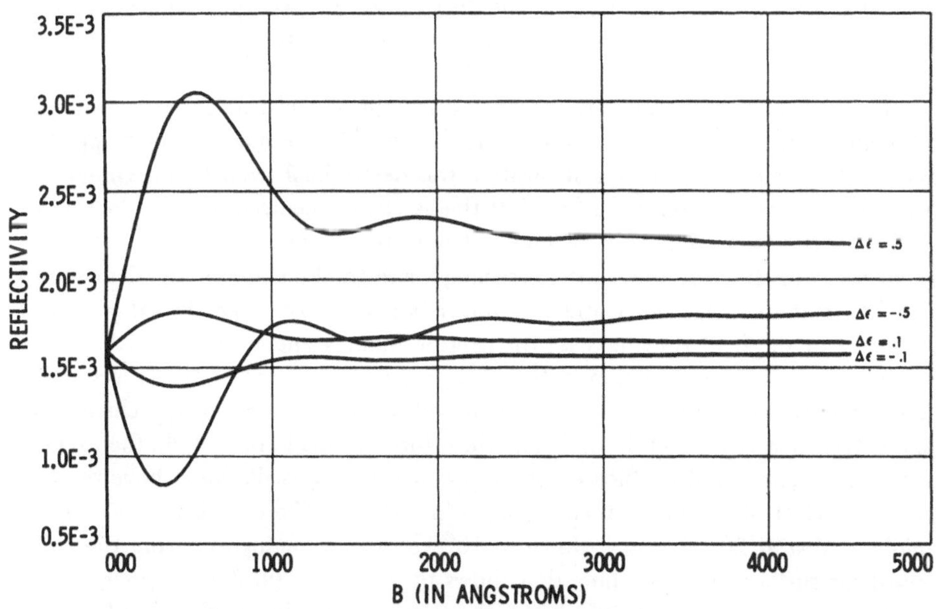

Fig. 9. Reflectivity as a function of ramp length for $\epsilon = 5 + i$ and the angle of incidence equal to the Brewster angle for $\epsilon = 5$.

$$L(E) = (\Gamma/\pi)\,[\Gamma^2 + (E\text{-}E')^2]^{-1} \qquad (3)$$

where Γ is a half width.

Away from the surface the various parameters in the expression for ϵ_2 can be determined from effective mass theory or from experimental data. However, the effect of the surface is largely unknown except for the effect of intrinsic surface fields which we assume are too low to be significant (10^4 volt/cm or less in CdS). We *assume* that the surface affects only the bound exciton states and progressively diminishes their binding energy and enhancement factor from 150 Å out to the surface.[7] This can be qualitatively justified by noting that the surface represents a potential barrier to electrons and holes and thus places a constraint on the exciton wave function.

The dielectric constant as a function of energy relative to the band gap energy and distance in from the surface is shown in Fig. 10. Zero meV corresponds to approximately 2.6 eV or a vacuum wavelength of 5000 Å. The crystal is at 80°K so that away from the surface the exciton shows up as a sharp absorption peak at -30 meV with a half width of 3 meV[8]. The absorption beginning just below 0 meV is the continuum or band edge contribution. The crystal is doped to a donor density of 6 x 10^{18} with a carrier concentration of 5 x 10^{16} so all excited bound states are screened out of existence. The decrease in the exciton enhancement and binding energy as the surface is approached shows up clearly.

We now apply as a modulation a depleting field of 10^6 volt/cm at the surface. The field is rapidly attenuated in the crystal because it is terminated on stripped donors. The field as a function of position was determined from Fermi-Thomas theory which gives not only the field but the carrier concentration as a function of position. Thus we were able to include not only the direct effect of the field on ϵ_2 but also the effect of free carrier screening[9] and the blocking or Burstein shift effect[10]. However, the direct effect of the field is far larger than the other two. The effect of the field is to increase the exciton binding energy, broaden the exciton and put a tail on the continuum states. The magnitudes of these effects were estimated from the work of Blossey[11]. A half width and a binding energy change ΔE were determined as a function of field. No effort was made to include the asymmetric line shape found by Blossey or the Franz-Keldysh oscillations above the band gap. The effect of the field on ϵ_2 is shown in Fig. 11. The field is attenuated rapidly so that at 200 Å in it is down to 10^4 volt/cm. Coming out from the bulk toward the surface, the first thing that shows up at about 200 Å is the increase in the binding energy of the exciton which, however, is quickly overshadowed by the onset of broadening which smears the exciton and band edge into a more or less uniform absorption from 120 Å to the surface.

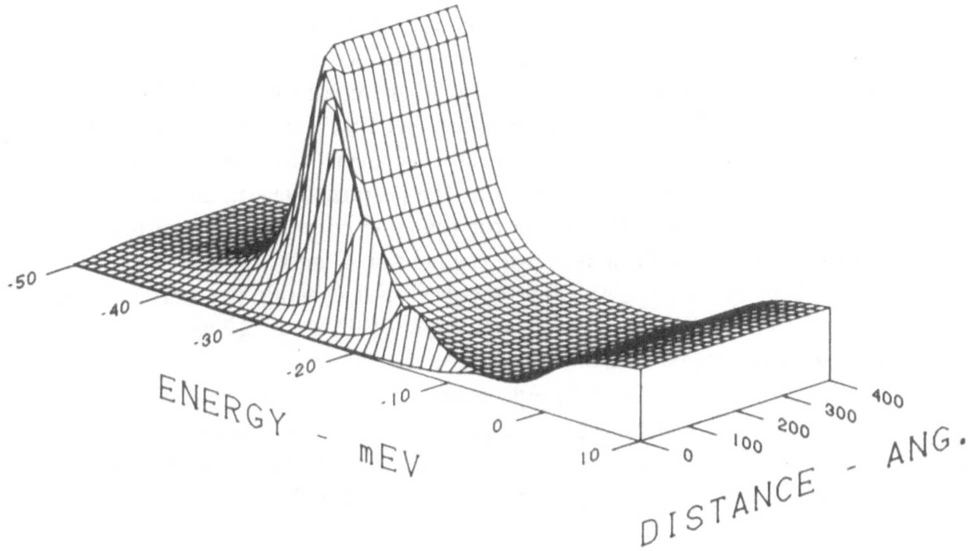

Fig. 10. Imaginary part of the dielectric constant of a doped CdS crystal without field applied as a function of energy and distance in from the surface of the crystal.

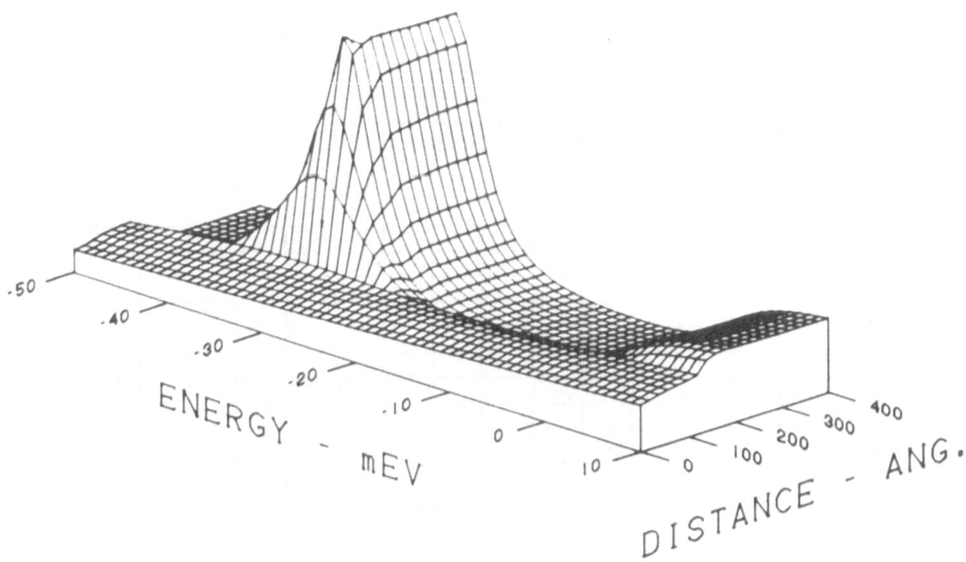

Fig. 11. Imaginary part of the dielectric constant of a doped CdS crystal with field applied as a function of energy and distance in from the surface of the crystal.

References pp. 128-129

Thus the perturbation is basically this smearing of the absorption in the first 150 Å. To compute the reflectivity changes caused by the perturbation, the unperturbed and perturbed ϵ_2's (along with the corresponding real parts) were fed to the reflectivity program which computed an unperturbed reflectivity R and a perturbed reflectivity R′. Figure 12 shows a plot of the contrast ratio, R′/R, versus energy for T.M. polarization at 69.1° incidence. At first glance this looks good with a contrast ratio of almost 14 at the peak at -3 meV. The difficulty is that the efficiency at the peak is only 0.0016, that is, the perturbed crystal reflects only 0.0016 of the incident light. The reason for the good contrast ratio and poor efficiency is that 69° is near a Brewster angle for light at -3 meV. We can trade off contrast for efficiency by going to other energies. For example if we are content with a contrast ratio of 2 we can get an efficiency of 0.024 at -31 meV. Curves like this for different angles of incidence form a surface which is shown in Fig. 13. The fin near 0 eV is the region of largest contrast ratio. In the figure it is truncated at 10 but the contrast ratio goes up to 43 at 68°. The efficiency is only 0.0007, however. Contrast ratios of the order of 2 and efficiencies of the order of 0.03 can be found at the beginning of the fold near -30 meV.

The modulations achieved with this doped crystal were due to the direct effect of the field. We can maximize this effect by going to an undoped crystal, i.e., one

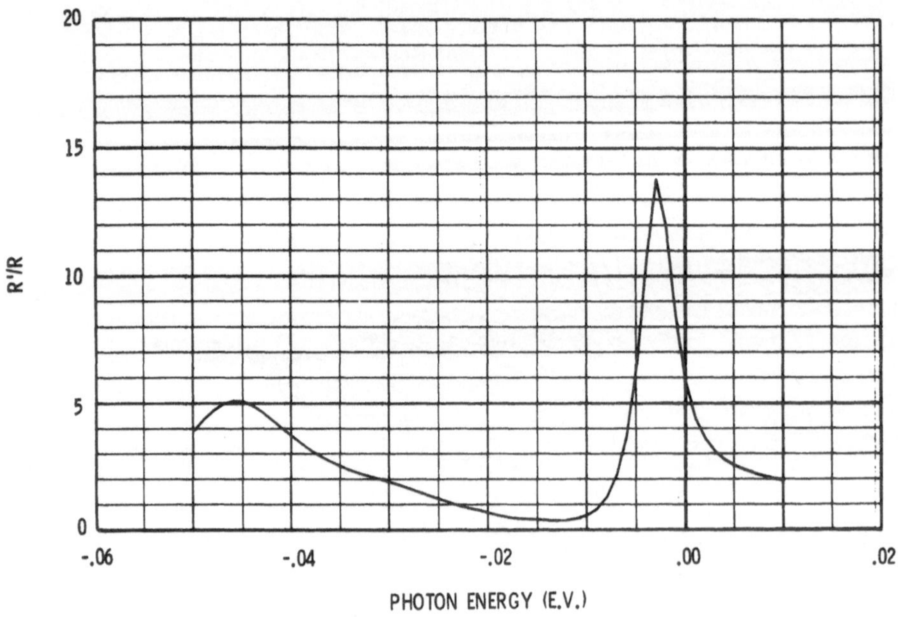

Fig. 12. Contrast ratio R′/R obtained with the doped CdS crystal at an angle of incidence of 69.1°.

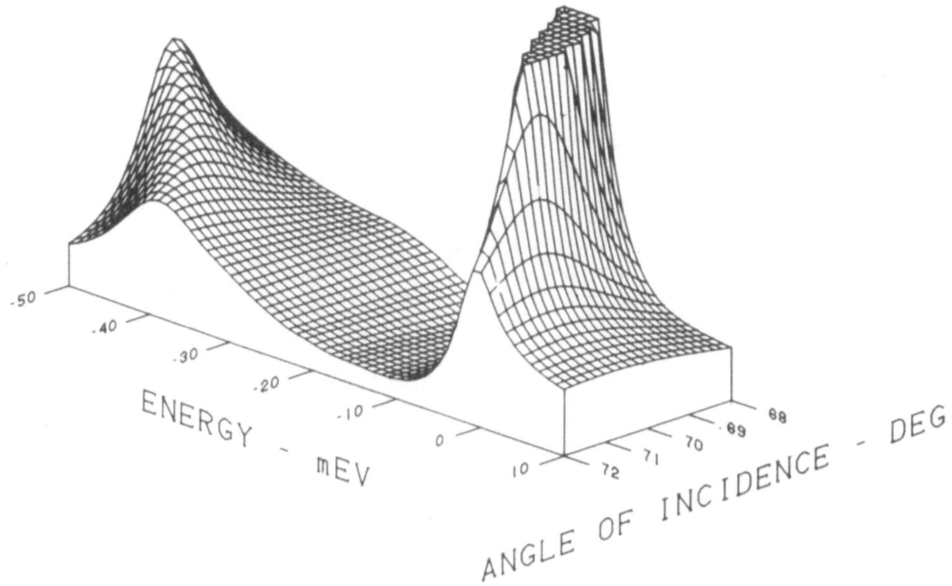

Fig. 13. Contrast ratio R′/R of the doped crystal for a range of angles of incidence. The peak is truncated at 10.

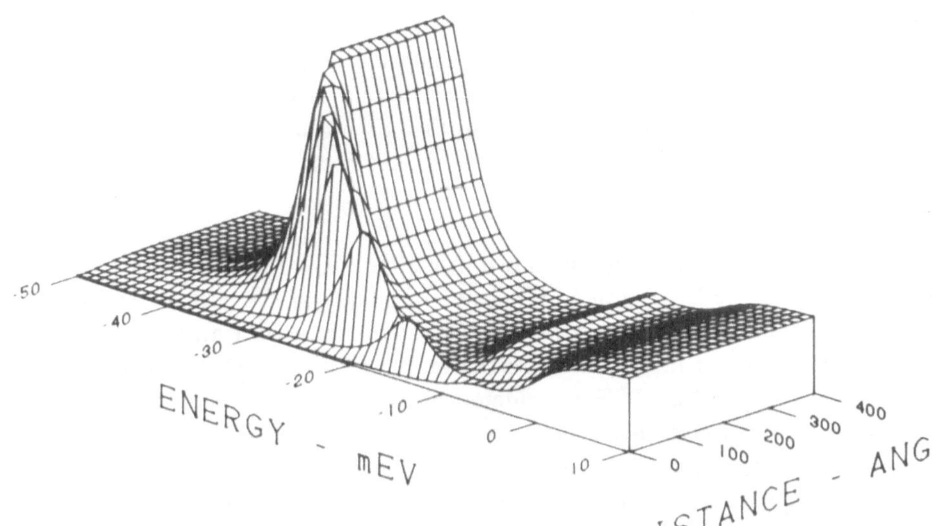

Fig. 14. Imaginary part of the dielectric constant of an undoped CdS crystal without field applied as a function of energy and distance in from the surface of the crystal.

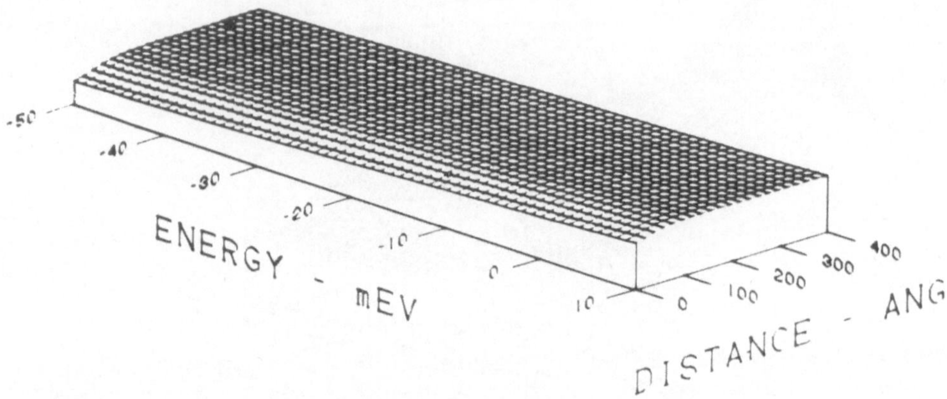

Fig. 15. Imaginary part of the dielectric constant of an undoped CdS crystal with field applied as a function of energy and distance in from the surface of the crystal.

with a donor density of 10^{16} or less. The dielectric constant shown in Fig. 14 is for an undoped crystal before the field is applied. Surface effects are the same as before and the only significant change is the appearance of an excited exciton state at -7 meV. Now when the field is applied the structure of the dielectric constant becomes completely smeared out as shown in Fig. 15. In fact, since the attenuation length of the field is 10,000 Å or greater, one should imagine that this flattened ϵ extends indefinitely into the crystal.

A plot of R'/R versus energy is shown in Fig. 16. The results are not too good. The best contrast ratio is 12 with an efficiency of 0.0012 at the peak at -16 meV and 67.6° incidence. Efficiencies never exceed about 0.01 even with contrast ratios of 2 to 3. On the other hand, if we interchange R and R' and take the contrast ratio to be R/R' rather than R'/R we obtain a somewhat better result which is shown in Fig. 17. The maximum contrast ratio occurs on the ridge at about 67° incidence and is 17 with an efficiency of 0.017. By moving along the ridge toward the front we improve the efficiency at the expense of the contrast ratio until near 72° we get efficiencies of 0.030 and contrast ratios of 3.

Table I shows a summary of the results obtained for the two crystals. These values are only estimates since they depend on our surface dielectric constant which

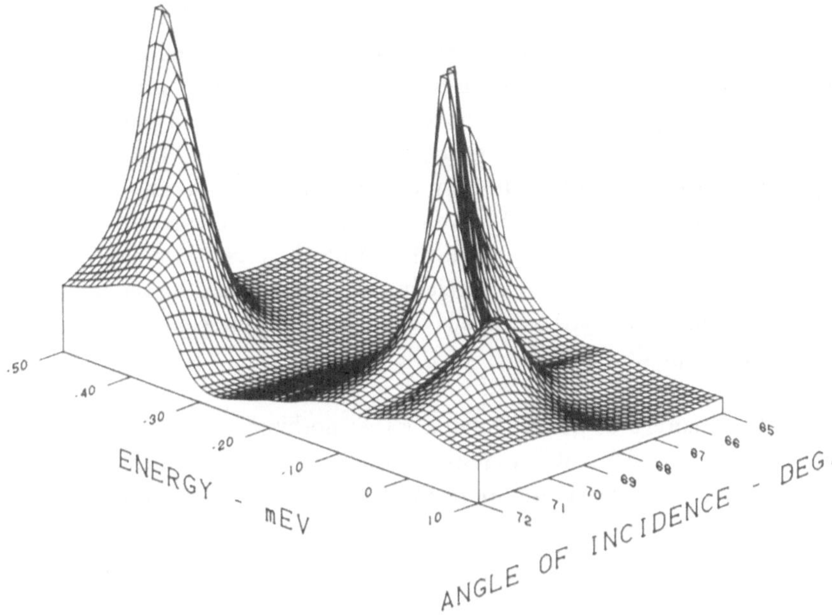

Fig. 16. Contrast ratio R′/R of the undoped crystal for a range of angles of incidence. The peak is truncated at 10.

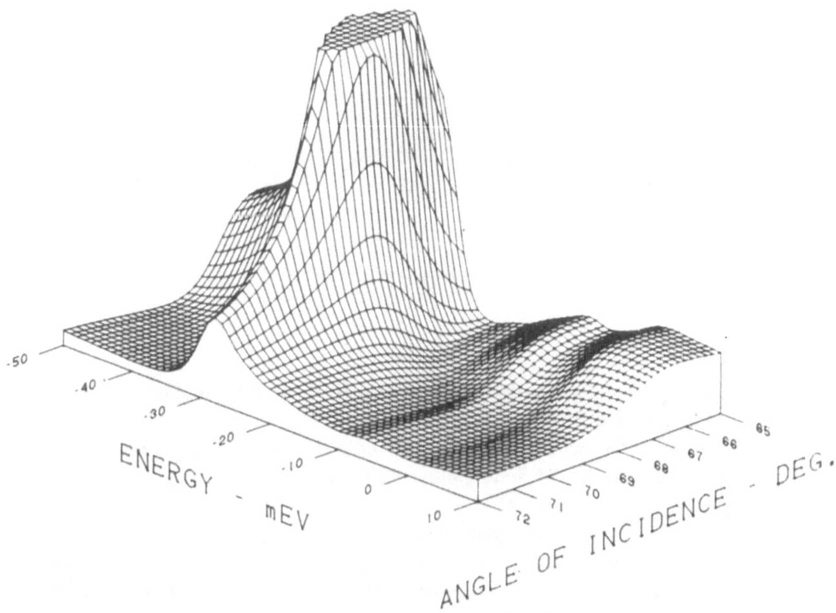

Fig. 17. Contrast ratio R/R′ of the undoped crystal for a range of angles of incidence. The peak is truncated at 10.

References pp. 128-129

is suspect in a number of ways. First, the effect of the surface was to some extent guessed. Second, the fields we assumed are high. An internal field of 10^6 volt/cm requires an external field of 10^7 volt/cm because of the dielectric properties of CdS. We repeated the calculation for the undoped crystal with an internal field of 10^5 volt/cm and found essentially no change in the R/R′ results. However one cannot go much lower than 10^5 volts/cm internal because the exciton broadening starts in abruptly between 10^4 and 10^5 volt/cm[11]. Third, we assumed no intrinsic surface field. This may not be true. Fourth, in CdS there are three narrowly spaced band edges--each with an exciton. A specially cut crystal would be required to isolate one of the bands as we have assumed. We believe the interference of the three band edges would degrade the effect.

A possible advantage of the modulator is that nothing in the crystal has to move for it to operate. All that is required is the presence of the field. Thus its frequency response would be limited only by the frequency with which the field could be applied.

TABLE I

Summary of Performance of Simulated Modulators

	Contrast Ratio	Efficiency
Doped crystal - maximum contrast ratio	43	0.0007
Doped crystal - efficiency of 3%	2	0.03
Undoped crystal - maximum contrast ratio	17	0.017
Undoped crystal - efficiency of 3%	3	0.03

REFERENCES

1. *B. O. Seraphin, Semiconductors and Semi-Metals (R. K. Willardson and A. C. Beer, Eds., Academic Press, New York, 1969), Vol. 4.*
2. *The effect of modulating with an electric field near the Brewster angle has been investigated by Fischer and Seraphin (Solid State Comm. 5, 973(1967) in the approximation of a homogeneous crystal dielectric constant.*
3. *D. E. Aspnes and A. Frova, Solid State Comm. 7, 155(1969).*
4. *M. Born and E. Wolf, Principles of Optics (Pergamon Press, London, 1970) Chapter 1.*
5. *E. A. Coddington and N. Levinson, Theory of Ordinary Differential Equations (McGraw-Hill Book Company, Inc., 1955), Chapter 1.*
6. *R. J.Elliott, Polarons and Excitons (C. G. Kuper and G. D. Whitfield, Eds., Oliver and Boyd, Ltd., London, 1963), p. 269.*

7. *M. F. Degan and M. D. Glinchuk, Soviet Physics - Solid State 5, 2377(1964).*
8. *The various parameters required in Eqs. (2) and (3) to specify the bulk dielectric constant may be obtained from D. G. Thomas and J. J. Hopfield, Phys. Rev. 116, 573(1959).*
9. *W. A. Albers, Jr., Phys. Rev. Letters 23, 410(1969).*
10. *J. G. Gay and L. T. Klauder, Jr., Phys. Rev. 172, 811(1968).*
11. *D. F. Blossey, Ph.D. Thesis, University of Illinois (1969).*

THE ACOUSTOOPTIC INTERACTION

R. W. DIXON

Bell Telephone Laboratories, Incorporated, Murray Hill, New Jersey

ABSTRACT

A short review of the physical concepts which allow optical beam intensity, direction, phase, etc., to be controlled by interaction with an ultrasonic wave is given. This includes brief consideration of the recent elegant additions to the theory of the photoelastic interaction by Nelson and Lax. The distinction between Bragg and Raman-Nath diffraction is made and acoustooptic diffraction in optically anisotropic media is mentioned in order to illustrate the importance of phase matching. The Bragg diffraction regime in which the diffracted optical energy is confined near one angular direction is shown to be of use in optical beam probing experiments. Investigation of such subjects as ultrasonic attenuation, acoustic nonlinear phenomena, acoustic diffraction, acoustoelectric effects, and the photoelastic properties of materials is possible using this technique.

Considerable evolution of acoustically driven devices which perform some optical control function has occurred recently. Part of this evolution has been due to the discovery of better acoustooptic materials, several of which are mentioned. Well understood devices include optical modulators, switches, sequential scanners, and random access deflectors. Consideration of the way in which acoustic waves interact with multiwavelength optical beams has led to the invention of a new class of devices which can act as multiwavelength modulators and tunable optical filters. Intracavity acoustooptic laser modulators (e.g. for creating short optical pulses in conjunction with mode locked lasers or for use as variable duration laser output couplers) have also become important.

References pp. 148-149

INTRODUCTION

This article is intended as a brief review of a few of the areas of acoustooptics which the author feels are of current interest. The selection is intended to emphasize the major evolutionary steps in the field which have occurred in the past several years. These include, for example, generalizations of the theory of the photoelastic interaction, considerations of multiple optical wavelength acoustooptic interactions, and applications of acoustooptic devices to such areas as fast pulse modulators and extraction of optical power from lasers. Because of space limitations a thorough general review is not intended, although it is hoped that enough literature references are included that the reader can easily explore in greater depth any aspects of the subject which he may find of interest.

NORMAL BRAGG DIFFRACTION

Figure 1 shows the basic geometry with which we shall be concerned. An ultrasonic beam is generated by a plane transducer of width L. Frequencies of interest range to at least a few GHz, but the region from 100-500 MHz is emphasized. The acoustic wavelength Λ is considered to be small compared with L so that the acoustic energy is confined to a collimated beam with a small well defined diffraction angle. An optical beam intersects the ultrasonic beam in the medium and a portion of the incident light is diffracted into other directions. Conservation

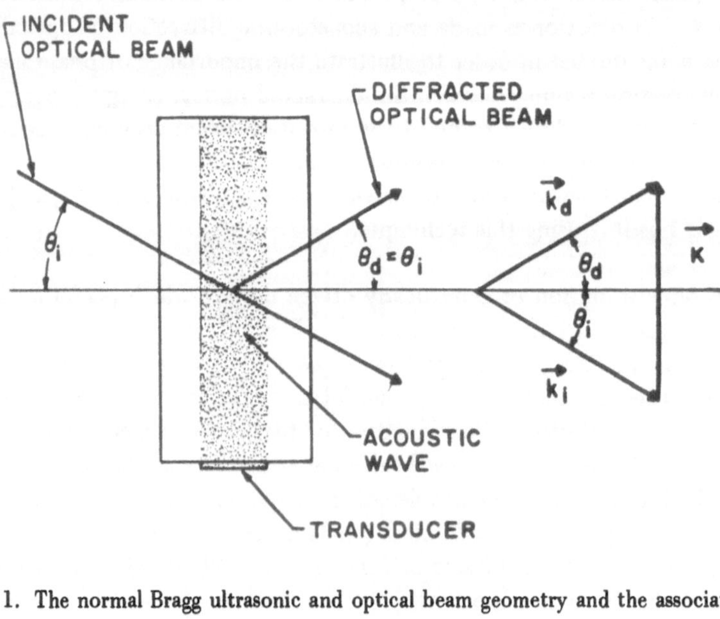

Fig. 1. The normal Bragg ultrasonic and optical beam geometry and the associated wave vector diagram.

of energy and wave vector allow the deduction of the scattering geometry. In this paper restriction is made to interactions which occur in the Bragg diffraction regime, wherein the diffracted light is confined near a single angular direction. The Bragg angle in optically isotropic media is defined by,

$$\sin \theta_B = \tfrac{1}{2}(|\mathbf{K}_a|/|\mathbf{k}_o|), \tag{1}$$

where $\mathbf{K}_a$ and $\mathbf{k}_o$ are the acoustic and optic k-vectors respectively. Much of the older literature of light scattering by injected sound is concerned with lower ultrasonic frequencies, with the Raman-Nath form of diffraction being the limiting case for normal incidence. Physically, the qualitative definition of the Bragg regime is easily seen, since the requirement that the diffracted light be confined near one angular direction is equivalent to stating that the acoustic diffraction angle, in the plane of the acoustooptic interaction, should be less than the angle through which the light is diffracted. Wave-vector conservation diagrams such as that one shown in Fig. 1 are often useful in thinking about concepts of this type.[3] Thus

$$\Delta\theta_{\text{acoustic}} < 2\theta_{\text{Bragg}} \tag{2a}$$

means that for Bragg scattering one requires

$$\Lambda/L < |\mathbf{K}_a|/|\mathbf{k}_o|; \text{ or } \lambda_o L/n\Lambda^2 > 1, \tag{2b}$$

where λ_o is the free space optical wavelength, n is the refractive index and Λ is the acoustic wavelength. The interested reader is referred to the paper by Klein and Cook,[2] and to the references given there, for a discussion of the distinctions between Raman-Nath, Bragg, and intermediate diffraction processes.

PHOTOELASTICITY

Historically, the acoustooptic diffraction process has been viewed as one in which a propagating strain wave modulates the refractive index according to[4]

$$\Delta B_{ij} = p_{ijmn} S_{mn} \tag{3}$$

where $B_{ij} = \epsilon_{ij}^{-1}$ is the optical indicatrix, p is the photoelastic tensor, and S is the strain tensor. With this definition the photoelastic tensor was symmetric on exchange of the first two indices and of the last two indices.

One of the most interesting recent occurrences has been a rethinking of the theory of the photoelastic interaction. By formulating a rotationally invariant classical theory of the optical-elastic interaction, Nelson and Lax[5] have shown that the displacement gradients should be considered as the fundamental elastic variables in a theory of photoelasticity. This overturns the long held view that the strain was the proper independent variable in Eq. (3). The essential point is that changes in the optical indicatrix can arise not only from strains, but also from rotations (i.e., antisymmetric as well as symmetric combinations of displacement gradients can scatter light). Thus Eq. (3) should be generalized to the form

$$\Delta B_{ij} = p_{ijmn}S_{mn} + p'_{ijmn}R_{mn}, \tag{4}$$

where S_{mn} and R_{mn} are the strain and rotation defined in terms of elastic displacements by

$$\begin{aligned} S_{mn} &\equiv \tfrac{1}{2}(\partial u_m/\partial x_n + \partial u_n/\partial u_m) \\ R_{mn} &\equiv \tfrac{1}{2}(\partial u_m/\partial x_n - \partial u_n/\partial x_m) \end{aligned} \tag{5}$$

and p and p′ are the symmetric and antisymmetric photoelastic coefficients respectively.

Equation (4) can also be written

$$\Delta B_{ij} = \bar{p}_{ijmn}(\partial u_m/\partial x_n) \tag{6}$$

where, however, $\bar{p}$ is not symmetric on exchange of (mn). $\bar{p}$ reduces to the standard Pockel's definition when rotations are absent. The second term in Eq. (4) will be nonzero in crystal classes which permit optical anisotropy. The existence of the second term in Eq. (4) destroys the traditionally assumed symmetry on exchange of the final photoelastic indices. The distinction between Eq. (3) and (4) becomes important for shear strains propagating in birefringent crystals. There, in theory, rotations can play as significant a role as strains in the photoelasticity of strongly birefringent crystals in the presence of shear distortions.

Nelson and Lazay[6] have in fact recently made the first observation of light scattering from rotations, by working with thermally generated sound in rutile. They conclude that at 5145 Å, $\bar{p}_{2332} \simeq -0.0255$ while $\bar{p}_{2323} \simeq +0.0009$. These coefficients had previously been thought, on the basis of symmetry, to be equal! In terms of the symmetric and antisymmetric p's defined in Eq. (4), this means $p_{2323} \cong -0.0123$, but $p'_{2323} \simeq +0.0132$.

One should also be aware that an indirect photoelastic effect, analogous to the indirect electrooptic effect, can exist in piezoelectric crystals. The indirect effect was first derived by Chapelle and Taurel[7] but has been ignored until rediscovered in recent times.[8,5] It is not a tensor and does not have the symmetry of the direct effect.

Thus the foundations of the theory of photoelasticity have been generalized and are considerably more complicated than was previously realized. While more experimentation needs to be done in order to assess the quantitative importance of the new ideas, for many practical purposes it is clear that the symmetry generalization will be relatively unimportant, because one rarely deals with photoelastic components in strongly birefringent crystals and even less often with shear strains in strongly birefringent materials. Nevertheless care should be exercised. The older formulations could be in serious quantitative error in strongly piezoelectric and optically anisotropic crystals. In fundamental studies of such things as the changes in photoelasticity encountered in phase transitions, in ferroelectrics for example, it will be necessary to be extremely careful. Formulation of the theories of higher order effects, such as acoustically induced optical harmonic generation will generally also require consideration of rotations.[9]

ANISOTROPIC BRAGG DIFFRACTION

Generalizations in the historic statement of the Bragg condition, Eq. (1) have also been made. The normal Bragg condition is simply derived from energy and wave-vector conservation. As with the photoelastic interaction, however, it has been instructive to ask about the influence of optical anisotropy, in this case on the phase matching conditions, and it was found[10,11] that the ordinary Bragg conditions require generalization whenever the refractive indices of the incident and diffracted optical waves are different. Equations (7a,b) show the form which the Bragg conditions take in this case. The subscripts i and d

$$\sin \theta_i = (\lambda_o/2n_i\Lambda)\,[1 + (\Lambda^2/\lambda_o^2)(n_i^2 - n_d^2)] \tag{7a}$$

$$\sin \theta_d = (\lambda_o/2n_d\Lambda)\,[1 - (\Lambda^2/\lambda_o^2)(n_i^2 - n_d^2)] \tag{7b}$$

denote incident and diffracted respectively, and θ_i $[\theta_d]$ is the angle between the incident [diffracted] optical **k**-vector and the normal to the acoustic **K**-vector. In the normal Bragg diffraction process $\theta_i = \theta_d = \theta_{Bragg}$. Figure (2) shows the experimental confirmation of these ideas for shear waves propagating in the x-crystallographic direction in crystal quartz. The solid lines are theoretical and the open circles are experimental. As can be seen there exists a frequency f_{min} in cases of anisotropic Bragg diffraction below which light cannot be diffracted by the ultrasonic beam. Phase matching is impossible. At the limiting frequency f_{min}, the diffraction geometry becomes collinear and has been studied at some length because of the ease with which the interaction length and the coherence length of the diffraction process can be varied.[10] This collinear geometry has also been utilized for producing acoustically driven optical filters, of which more will be said later.[12,13] The value of $f_{min} = (v/\lambda_o)\,|n_i - n_d|$ is often in a very important range for acousto-

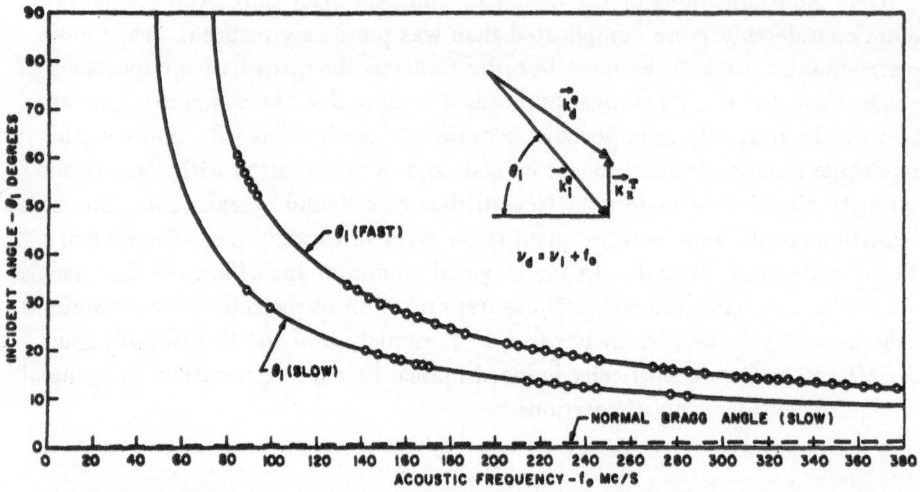

Fig. 2. Comparison of theory with experiment for the anisotropic Bragg diffraction process from x directed shear waves in crystal quartz. Compare the normal Bragg angle (in the absence of optical anisotropy).

optic experiments, for example the values at 0.63μ for shear waves in crystal quartz are 48 and 75 MHz respectively. For x-directed shear waves in rutile f_{min} = 2.4 GHz. Such effects are even important in modifying the normal scattering geometry and in increasing the number of lines in the observed spectra in thermal Brillouin scattering experiments.[11,14]

OPTICAL BEAM PROBING

Having commented on the new formal foundations of photoelasticity theory and on the subtleties of phase matching, let us return to the discussion of Bragg diffraction. One of the most important uses of Bragg diffraction in the past five years has been as a tool for understanding the properties of elastic waves. The technique has been called optical beam probing.[15] The first uses of this technique were for the measurement of ultrasonic velocity and loss; and of the angular energy distribution in a sound column. That is, the incident optical angle with respect to the direction of the ultrasonic beam may be changed, and the angular properties of the sound column investigated; and the distance from the transducer to the optical beam may be varied, and the decrease in ultrasonic energy with increasing distance studied. Ultrasonic decay information may also be obtained from the angular distribution of the scattered light when a well-collimated incident beam is used. Measurement of ultrasonic attenuation in liquid helium in the range a = 475 cm^{-1} was an early success of the technique.[16] The advantage which the optical beam probing technique has of being able to follow the ultrasonic wave point by point inside the (transparent) diffracting medium is very useful, and an important advantage of the technique as compared with purely ultrasonic methods.

A somewhat more recent area in which optical beam probing techniques have proven useful is in the study of the diffraction and focusing of ultrasonic waves. For example Cohen and Gordon,[17] using the configuration shown in Fig. (3), determined ultrasonic spatial intensity distributions by moving a narrow optical beam through the acoustic beam. Figure (4) shows the intensity distribution before and after focusing of a 325 MHz ultrasonic wave in fused silica, and Fig. (5) shows the ultrasonic wave traversing the focal plane of the curved reflecting surface shown in Fig. (3). These distributions agreed with those calculated by a diffraction integral over the transducer area. Similar calculations could be performed for the angular distribution of the ultrasonic energy. Figure (6) shows a typical experimental result for the focusing curved surface shown in Fig. (3), taken by varying the angle between the probing optical beam and the ultrasonic beam about the Bragg angle. As can be seen the technique yields beautiful definition and detail of the ultrasonic diffraction processes.

To make the point even more emphatic, the theoretical and experimental diffraction patterns for diffraction of an ultrasonic wave generated by a plane trans-

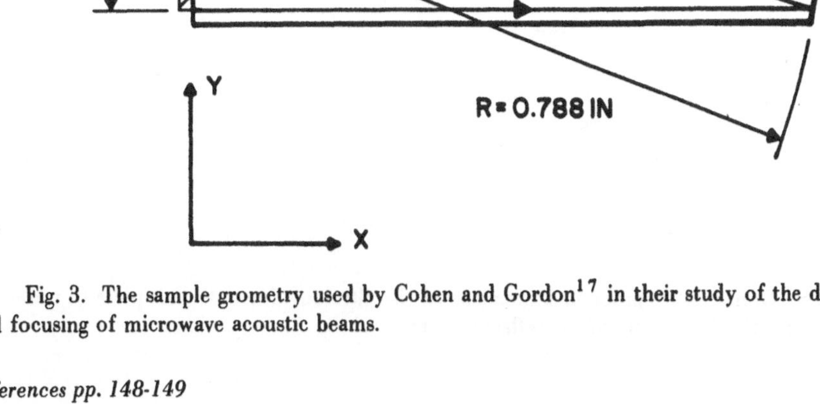

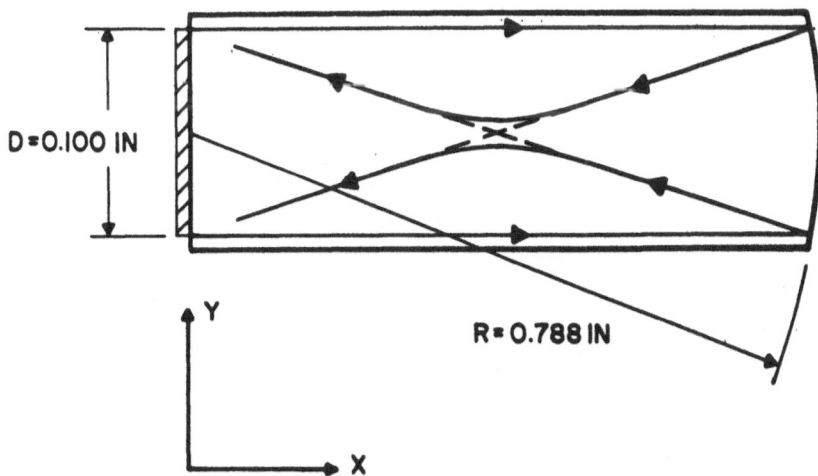

Fig. 3. The sample grometry used by Cohen and Gordon[17] in their study of the diffraction and focusing of microwave acoustic beams.

References pp. 148-149

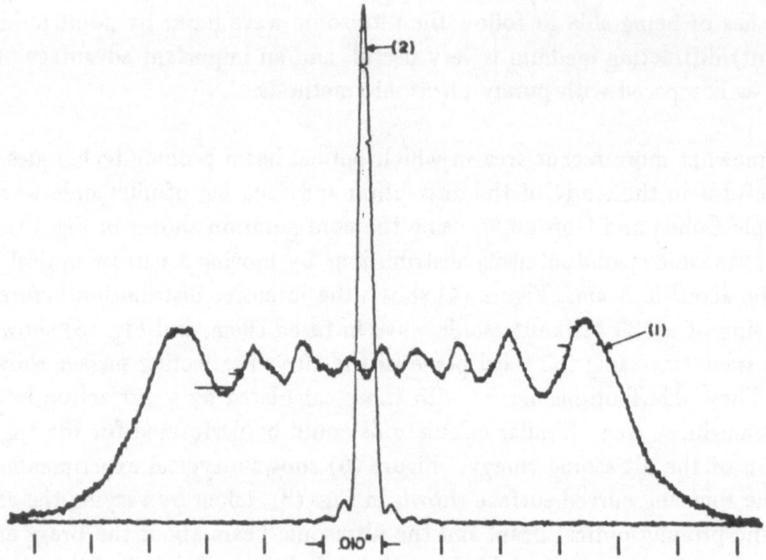

Fig. 4. The ultrasonic intensity of a 325 MHz ultrasonic beam before and after focusing by the cylindrical surface shown in Fig. 3. The intensity scale of the focused beam has been reduced five times.

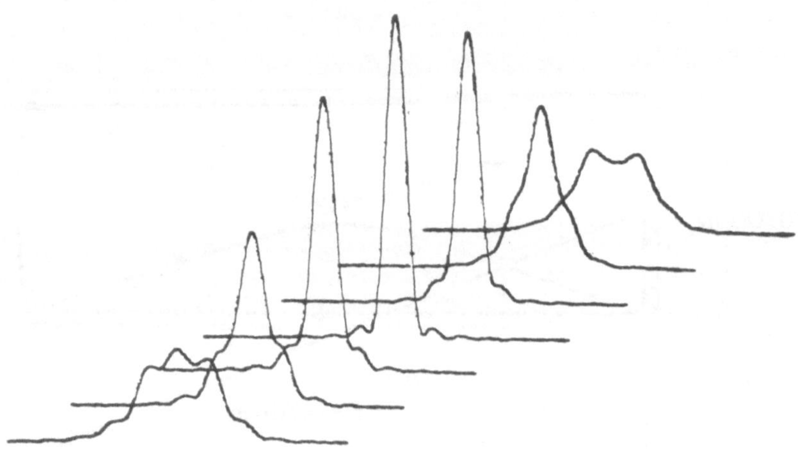

Fig. 5. Optical beam probe measurements of the ultrasonic intensity as a pulse of ultrasound traverses the focal plane of the reflecting surface of Fig. 3.

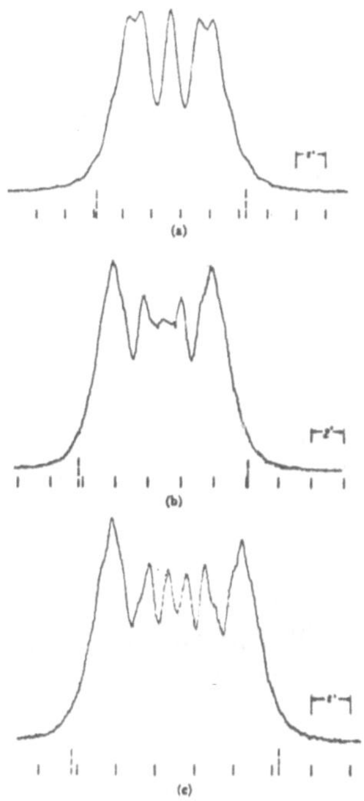

Fig. 6. The angular ultrasonic intensity distribution generated at the ultrasonic focus at a) 200 MHz, b) 325 MHz, and c) 425 MHz. Favorable comparison with standard optical diffraction patterns can be made. See e.g., F. A. Jenkins and H. E. White, *Fundamentals of Optics*, 3rd ed., McGraw-Hill, New York, 1957, p. 374.

ducer on silicon are shown in Fig. (7).[18] In this case knowledge of the characteristics of ultrasonic propagation in ultrasonically anisotropic media must be understood in order to calculate the Fresnel integrals. Ultrasonic diffraction in crystals, even near relatively high symmetry pure mode directions, is typically very different from diffraction in isotropic materials due chiefly to deviations between the directions of energy flow and wave-vector. Investigation of such differences is very difficult using purely ultrasonic techniques.

Another of the areas in which recent progress has been made using the optical beam probing ideas, is in nonlinear acoustics. Several years ago the author performed an experiment in which two oppositely directed transverse acoustic waves were mixed in a piece of fused quartz to product a longitudinal elastic wave at the same frequency.[19] Energy and wave-vector conservation were used to show the existence of a phase matching condition which for plane waves is

$$\omega_1/\omega_2 = (1 + v_T/v_L)/(1 - v_T/v_L) \tag{8}$$

References 148-149

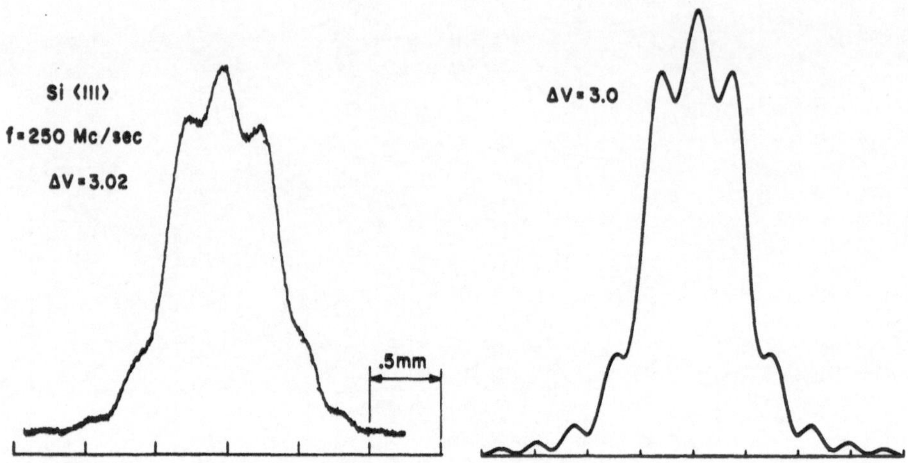

Fig. 7. Comparison of theory with experiment for the intensity distribution of ultrasound generated in a plane transducer in the acoustically anisotropic material silicon.[18]

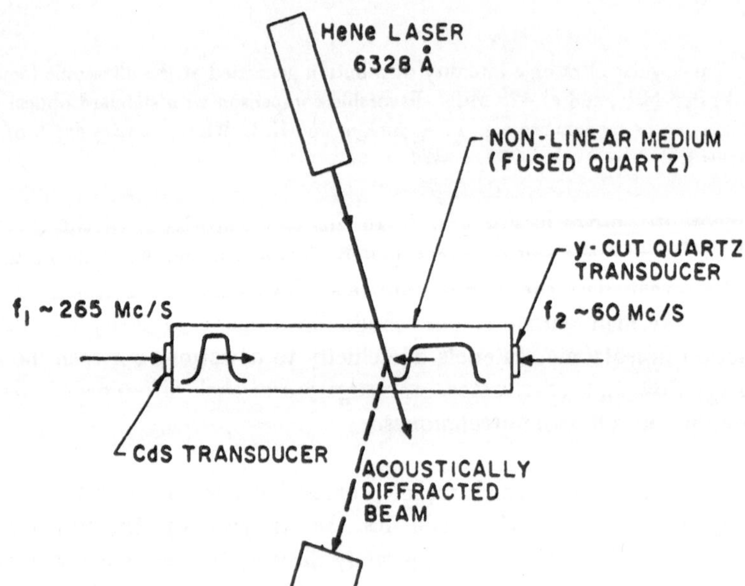

Fig. 8. Sample geometry for the study of the nonlinear interaction of two transverse ultrasonic waves to produce a phase-matched longitudinal wave.

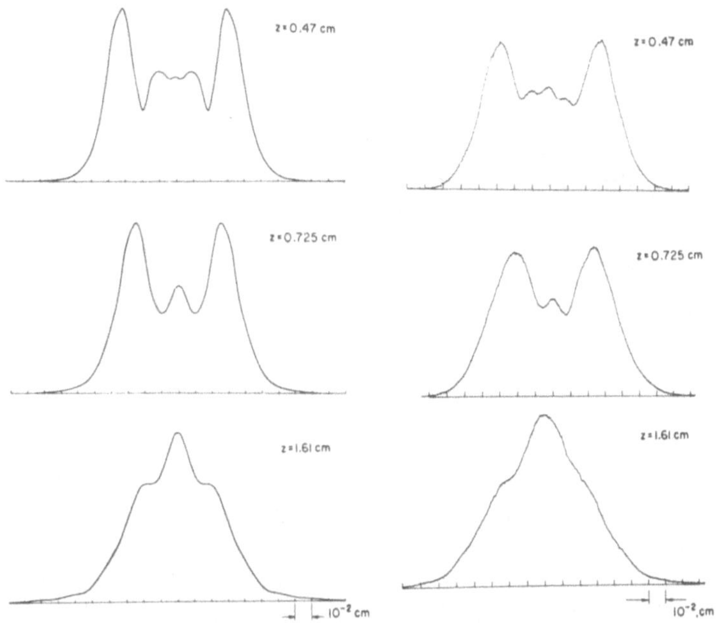

Fig. 9. Comparison of the theoretical and experimental intensity profiles of a second harmonic acoustic wave at several distances from the plane transducer. The nonlinear medium was crystal quartz and the ultrasonic beam parallel to the z-axis.

where ω_1 and ω_2 are the frequencies of the oppositely directed transverse ultrasonic waves, while v_T and v_L are the transverse and longitudinal ultrasonic velocities respectively. The experimental geometry is shown in Fig. (8). When the frequency of one of the waves deviated from that for perfect phase matching the intensity of the generated longitudinal sum frequency wave at $\omega_3 = \omega_2 + \omega_1$, which was being generated showed the typical $[(\sin x)/x]^2$ behavior which one associates with such phase matching. The experiment showed very directly the power of optical techniques for measuring nonlinear acoustic phenomena.

More recently Cohen[20] has studied elastic second harmonic generation using Bragg diffraction techniques. In addition to observing the growth of the elastic second harmonic intensity, it was also possible to see the diffraction of second harmonic ultrasonic waves which are created by bulk nonlinearities. Figure (9) shows the theoretical diffraction patterns predicted for the second harmonic elastic wave generated by a longitudinal ultrasonic wave produced in a plane transducer on a z-cut crystal quartz bar. The experimentally obtained intensity distributions are also presented for comparison. While some differences (due mostly to transducer nonuniformities) are obvious, the general agreement of theory and experiment is excellent. From these measurements of harmonic beam intensity, components of the third order elastic tensor may be measured,[20] although the large number of independent third order elastic tensor components would make a comprehensive study of this interaction very tedious.

There are other areas related to the optical beam probing concept which could be mentioned. Examples include the subject of optical reconstruction of acoustic holograms by means of surface wave sampling[21] and studies of surface acoustic waves, including nonlinear effects among these waves.[22]

APPLICATIONS

Considerable interest in the acoustooptic interaction has occurred because this interaction can be used to perform various optical beam control functions.[3] Among these are optical modulation, switching, random access or sequential deflection, etc. Most of the early developments have been adequately reviewed.[1,3,23] Only three areas, which the author feels are of particular current interest, will be considered here. These are (1) the search for new materials for acoustooptic devices, (2) improvements in the application of acoustooptic devices for fast modulators, and (3) the application of acoustooptic devices to multioptical wavelength interactions.

In most acoustooptic applications one is interested in conserving electrical driving power, e.g. in achieving the largest modulation possible per unit of incident electrical power. The figures of merit which should be used to characterize acousto-optic devices were discussed several years ago,[24,25] as were the values of these figures of merit for several potentially useful materials.[25] Pinnow has recently made a further analysis of the various considerations pertinent to the selection of acoustooptic materials.[26] The technique which has been found most often useful in measuring photoelastic parameters is the comparative technique of Dixon and Cohen[27] which measures light scattering from a standard material (often fused silica) and compares it with the light scattered from the sample under test. Table 1 shows a short selection of materials which are currently being used for acoustooptic devices. These include fused silica, iodic acid, gallium phosphide, and lead molybdate. In a typical application, using a single plane transducer a $PbMo0_4$ deflector operated with a bandwidth of 80 MHz (90 to 170 MHz) and used one watt electrical power to deflect more than 50% of the incident optical energy (5145 Å argon laser output).[28] Lead molybdate is about 25 times more efficient as a modulating medium than is fused silica. A description of the details of device design is clearly

TABLE 1

FIGURES OF MERIT $M_2 = n^6p^2/\rho v^3$ FOR SEVERAL IMPORTANT ACOUSTOOPTIC MATERIALS

Material	$\lambda(\mu)$	M_2 (relative to fused silica)	Useful Transmission Range (μ)
Fused silica	0.63	1	0.2-2.5
$LiNbO_3$	0.63	4.6[a]	0.5-4.5
HIO_3	0.63	55[b]	0.4-1.3
$PbMoO_4$	0.63	23.7[c]	0.4-3.9
GaP	0.63	29.7[a]	0.6-10.0
As_2S_3 glass	1.15	230[a]	0.6-11
Ge	10.6	540[d]	2-15
Te	10.6	4400[e]	4.5-20
TeO_2	0.63	23[f]	0.35-5

REFERENCES IN TABLE 1

a. R. W. Dixon, J. Appl. Phys. 38(1967) 5149.
b. D. A. Pinnow and R. W. Dixon, Appl. Phys. Letters 13(1968) 156.
c. D. A. Pinnow, L. G. Van Uitert, A. W. Warner, and W. A. Bonner, Appl. Phys. Letters 15(1969) 83.
d. R. L. Abrams and D. A. Pinnow, J. Appl. Phys. (to be published).
e. R. W. Dixon and A. N. Chester, Appl. Phys. Letters 9(1966) 190.
f. N. Uchida and Y. Ohmachi, J. Appl. Phys. 40(1969) 4692.

beyond the scope intended for this review, but it is possible to gain some qualitative appreciation for the state of the art by realizing that 100 resolvable spot, 1μsec random access deflectors are now possible,[29] as are real time sequential television deflectors using crystals instead of liquids.[30]

Fast pulse modulators are complementary devices to the scanners just mentioned. Maydan[31,32] made very careful studies of this application and it is worth considering some of their characteristics. Figure (10) shows the geometry. In these devices the optical beam is typically focused to a relatively small dimension (to decrease the acoustic transit time across the optical beam - which in turn determines the modulator's risetime). Risetimes of about five nanoseconds are now common while 2-3 nanoseconds are possible with difficulty. Figure (11) shows an application in which the modulator is used as an intracavity laser modulator. Such a system can be used to extract mode-locked laser pulses or more generally as a variable output laser coupler. By careful design such a device produces average

References 148-149

pulsed output powers which are equal to the cw laser power available with optimal transmission mirrors and no modulator present in the cavity. Figure (12) shows light pulses being extracted from a He-Ne laser (6328 Å) by such a device at about a 10^6 sec^{-1} rate. In this case each pulse has about 25 nsec duration. The cavity power level as a function of time is also shown. When the pulse repetition rate is

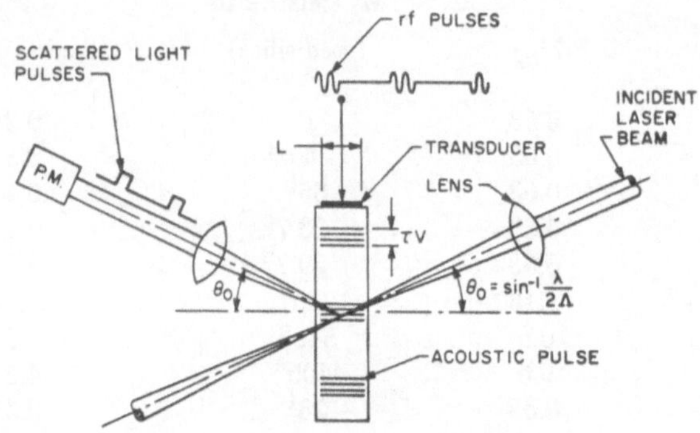

Fig. 10. Diffraction geometry for a pulsed acoustooptic modulator.

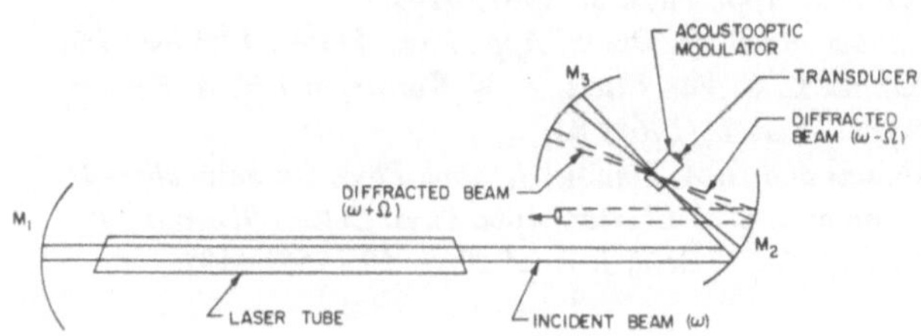

Fig. 11. Application of an acoustooptic device as an intracavity laser modulator. The modulator is placed at the center of curvature of the mirror M_2.

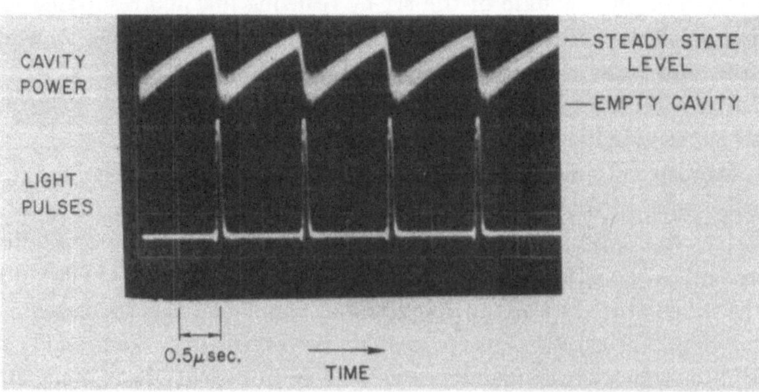

Fig. 12. Light pulses being extracted by an intracavity acoustooptic modulator from a He-Ne laser. The optical power level inside the laser cavity is also shown for comparison.

also shown. When the pulse repetition rate is reduced much below 10^6 sec^{-1} the average pulsed power falls off because of the He-Ne build-up characteristics. For these low repetition rates, however, the peak output power is still measured to be as much as 100 times the maximum cw power available from the laser.

In a similar application in the Nd:YAlG laser, an intracavity Brewster cut fused silica acoustooptic modular was used to couple optical power from the laser.[33] Q-switching was possible for repetition rates below about 30 kHz, while cavity dumping without Q-switching occurred for repetition frequencies between 125 kHz and several MHz. Again the average power reached the maximum cw power capability of the laser used (2W). Such devices are, in the author's opinion, significantly superior to other available devices. The driving powers are generally comparable to high-efficiency electrooptic modulators, but the acoustooptic devices are generally simpler and much less critical to adjust. Extinction ratios are very high and no temperature stabilization is required. The high optical quality of the modulating media available allow the modulator to be used inside the laser cavity if desired.

MULTIWAVELENGTH ACOUSTOOPTICS

Thus far we have not utilized the dependence of the Bragg relationships [Eqs. (1) and (7)] on optical wavelength. The most obvious multiwavelength acoustooptic device would utilize approximately the same configuration as shown in Fig. (1), operate in the normal Bragg regime, and be useful, for example, as a multiwavelength laser modulator for a laser which is capable of operating simultaneously at more than one wavelength. For example, the author has independently modulated the 5145 and 4800 Å transitions of an argon laser using acoustic frequencies near 250 MHz and an acoustic beamwidth of 1 cm.[34] As shown by the k-vector diagram in Fig. (13) the multiple wavelength incident beams typically have the same direction as the multiple wavelength modulated output beams. Alignment of other components is therefore straightforward.

Since the same Bragg angle is used for all wavelengths, the Bragg condition gives the relationship between the optical wavelengths and the ultrasonic frequencies required to modulate each. Neglecting dispersion the relationship is simply $\lambda_i f_i$ = constant. The smallest wavelength separation $\Delta\lambda$ between wavelengths which can be independently modulated in this simple device is

$$\Delta\lambda \cong (2nv^2 \cos\theta)/(Lf^2) = (2n\Lambda^2 \cos\theta)/L \tag{9}$$

where Λ is the acoustic wavelength and L is the beamwidth over which the acoustooptic interaction occurs. Optical multiplexing is another obvious application of such

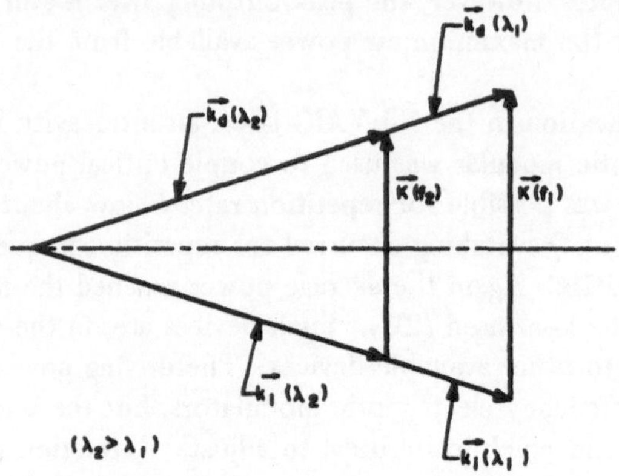

Fig. 13. A k-vector diagram showing two incident optical plane waves of different wavelengths (λ_1 and λ_2) being diffracted in the normal Bragg process by two plane ultrasonic waves of different frequency (f_1 and f_2).

a device. Acoustooptic devices can also, in theory, be used to perform the multi-wavelength optical processing functions associated with electronically tunable optical filters, electronically tunable optical spectrometers, etc.[34]

Harris and co-workers[12,13] have, for example, fabricated an electronically tunable acoustooptic filter using $LiNbO_3$. They have used the collinear acousto-optic interaction in which co-directional acoustic and optical beams can be phase matched near the acoustic frequency

$$f_{min} = (\Delta n v)/\lambda_o \qquad (10)$$

using the anisotropic Bragg diffraction regime.[10,11] The geometry is schematically indicated in Fig. (14). Note that the diffracted (output) wave and the incident (input) wave have different polarizations. The optical propagation direction is chosen so that this implies different wavevector magnitudes as well. The first filter had a tuning range from 7000 to 5500 Å for an ultrasonic frequency range of 750 to 1050 MHz and used $LiNbO_3$ as the modulating material. A more recent design using $CaMoO_4$ can be tuned from 6700 to 5100 Å by varying the acoustic driving frequency between 40 and 58 MHz.[35]

The time interval required to tune the filter from one optical passband to another is determined by the transit time of the acoustic wave over the interaction pathlength of the device. A typical value is 10μsec.

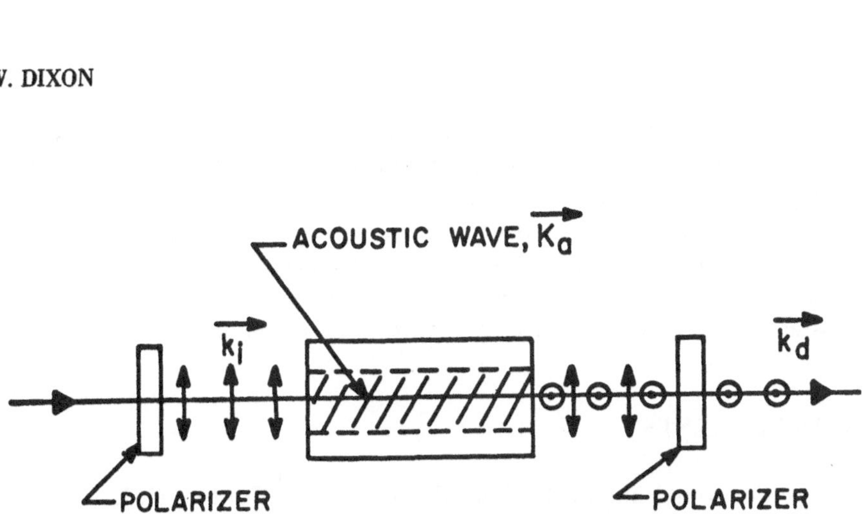

Fig. 14. Schematic diagram of an acoustooptic filter operating in the collinear anisotropic Bragg diffraction mode.[10,11,12,13]

The bandwidth of an acoustooptic filter is obviously important and can be approximately obtained directly from wavevector conservation arguments. Let the collinear interaction have a wavevector mismatch Δk, where

$$\Delta k \equiv K_a - (k_i - k_d). \tag{11}$$

K_a is the acoustic k-vector; k_i and k_d are the incident and diffracted optical k-vectors. On resonance $\Delta k = 0$ and $K_a = nk_o$, or $f_{min} = \Delta nv/\lambda_o$. This fixes the relationship between optical and acoustic wavelengths. Maximum transmission occurs at this point. When the optical wavevector is changed to k_o', keeping the acoustic frequency fixed, Δk becomes nonzero. For a typical diffraction process one (nominally) expects a $[(\sin \Delta kL)/(\Delta kL]^2$ dependence on phase, where L is the interaction path length. Thus the approximate bandwidth will be given by $\Delta kL = \pi$, or neglecting dispersion, $\Delta k = \Delta nk_o - \Delta n'k_o' \simeq 2\pi\Delta n\Delta v$. Thus $\Delta v = (2L\Delta n)^{-1}$ is the wavenumber bandwidth. For a more detailed bandwidth analysis the reader is referred to the literature.[13] In the $LiNbO_3$ device reported by Harris, et al., the passband at a given ultrasonic frequency was about 2 Å, while for the $CaMoO_4$ device the passband was about 8 Å.

SUMMARY AND CONCLUSION

A few selected subjects illustrating basic features of the acoustooptic interaction have been discussed. The generalization of the theory of photoelasticity to include interactions in optically anisotropic media is particularly intriguing because it changes the long held generally accepted understanding of the basic symmetry relationships appropriate for this interaction. Optical beam probing was mentioned because it has proven to be so fruitful a technique for investigating various physical phenomena involving elastic waves. Finally a few acoustooptic device applications were mentioned because of their importance in performing optical beam control functions and because this is a dynamic growth area at the present time.

ACKNOWLEDGMENTS

The author would like to thank M. G. Cohen, E. I. Gordon, M. Lax, D. Maydan, and D. F. Nelson for important contributions and conversations during the preparation of this manuscript.

REFERENCES

1. *General references include: E. G. Coker and L. N. G. Filon, A Treatise on Photo-Elasticity, Cambridge University Press, London, 1931; C. F. Quate, C. D. W. Wilkinson and D. K. Winslow, Proc. IEEE, 53(1965) 1604.*
2. *W. R. Klein and B. D. Cook, IEEE Trans. on Sonics and Ultrasonics, SU-14(1967) 123.*
3. *E. I. Gordon, Appl. Optics, 5(1966) 1629.*
4. *F. Pockels, Ann. d. Physik, 37(1889) 144, 269, 372; and 39(1890) 440.*
5. *D. F. Nelson and M. Lax, Phys. Rev. Letters, 24(1970) 379; and D. F. Nelson and M. Lax, Phys. Rev. B3(1971) 2778.*
6. *D. F. Nelson and P. D. Lazay, Phys. Rev. Letters 17, (1970) 1187.*
7. *J. Chapelle and L. Taurel, C. R. Acad. Sci., 240(1955) 743.*
8. *G. A. Coquin, 1969 St. Louis Ultrasonics Symposium.*
9. *G. D. Boyd, F. R. Nash, and D. F. Nelson, Phys. Rev. Letters 24(1970) 1298.*
10. *R. W. Dixon, IEEE J. Quantum Electronics, QE-3(1967) 85.*
11. *R. W. Dixon, Proc. Symp. Modern Optics, Polytechnic Press, Brooklyn, New York 1967, p. 265.*
12. *S. E. Harris, S. T. K. Nieh, and D. K. Winslow, Appl. Phys. Letters 15(1969) 325.*
13. *S. E. Harris and R. W. Wallace, J. Opt. Soc. Amer. 59(1969) 744.*
14. *V. Chandrasekharan, Proc. Ind. Acad. Sci. A33(1951) 183.*
15. *M. G. Cohen and E. I. Gordon, Bell System Tech. Journal 44(1965) 693.*
16. *E. I. Gordon and M. G. Cohen, Phys. Rev. 153(1967) 201, and M. A. Woolf, P. M. Platzman, and M. G. Cohen, Phys. Rev. Letters 17(1966) 294.*
17. *M. G. Cohen and E. I. Gordon, J. Appl. Phys. 38(1967) 2340.*
18. *M. G. Cohen, J. Appl. Phys. 38(1967) 3821.*
19. *R. W. Dixon, Appl. Phys. Letters 11(1967) 340.*
20. *M. G. Cohen, IEEE J. Quantum Electronics WE-6(1970) 25.*
21. *A. Korpel and P. Desmares, J. Acous. Soc. Amer. 45(1969) 881; A. Korpel and L. W. Kessler, Proc. Third Int. Symp. on Acoustic Holograph, New Port Beach, Calif., 1970.*
22. *R. Adler, A. Korpel, and P. Desmares, IEEE Trans. on Sonics and Ultrasonics SU-15(1968) 157; E. G. Lean and C. C. Tseng, J. Appl. Phys. 41(1970) 3912.*

23. *R. W. Dixon, IEEE Trans. on Electron Devices ED-17(1970) 229.*
24. *E. I. Gordon, IEEE J. Quantum Electronics QE-2(1966) 104.*
25. *R. W. Dixon, J. Appl. Phys. 38(1967) 5149.*
26. *D. A. Pinnow, IEEE J. Quantum Electronics QE-6(1970) 223.*
27. *R. W. Dixon and M. G. Cohen, Appl. Phys. Letters 8(1966) 205.*
28. *D. A. Pinnow, L. G. Van Uitert, A. W. Warner, and W. A. Bonner, Appl. Phys. Letters 15(1969) 83.*
29. *G. A. Coquin, J. P. Griffin, and L. K. Anderson, IEEE Trans. on Sonics and Ultrasonics SU-17(1970) 34.*
30. *A. Korpel, R. Adler, P. Desmanes, and W. Watson, Proc. IEEE 54(1966) 1429.*
31. *D. Maydan, IEEE J. Quantum Electronics QE-6(1970) 15.*
32. *D. Maydan, J. Appl. Phys. 41(1970) 1552.*
33. *D. Maydan and R. B. Chesler, J. of Appl. Phys. 42, (1971) 1031.*
34. *R. W. Dixon, unpublished.*
35. *S. E. Harris, S. T. K. Nieh, and R. S. Feigelson, Appl. Phys. Letters 17(1970) 223.*

LIGHT SCATTERING AND BIREFRINGENCE IN $BaTiO_3$ CERAMICS

W. A. ALBERS, JR. and M. KAPLIT

Physics Department, Research Laboratories, General Motors Corporation, Warren, Michigan

INTRODUCTION

In the last few years Land and his coworkers at the Sandia Corporation Laboratories have demonstrated that electrically-controlled birefringence and light scattering can be realized in lead-zirconate-titanate (PZT) ferroelectric ceramics.[1-7] The electrically controlled birefringence is related to the change induced by an electric field in the statistically-averaged birefringence of the crystallites comprising the ceramic. The electrically controlled light scattering has been interpreted by Nettleton in terms of domain wall displacement.[8] By appropriate choice of modified PZT ceramics and device configuration, it has been possible to devise electrically controlled light shutters, spectral filters, optical memories, light modulators, variable contrast black-and-white displays and multicolor displays.[1-4] Subsequently other laboratories have investigated various aspects of image storage and display based on the electrically-controlled birefringence in modified PZT ferroelectric ceramics.[9-12] Meanwhile Heartling at Sandia was able to improve the optical transparency of PZT by the addition of various modifiers, culminating in his preparation of the first optically transparent ferroelectric ceramic, lead lanthanum zirconate titanate (PLZT).[13-16] The development of the transparent PLZT may well enable the electrically controlled birefringence in these ferroelectric ceramics to realize practical application.

The purpose of the work reported in the present paper is to characterize another ferroelectric ceramic, barium titanate ($BaTiO_3$), as a possible electrically controlled light scattering and birefringence material. The goal is to relate the electrooptic properties of the single crystal material and ceramic processing variables such as grain size, density and sintering temperature to the electrooptic properties of the ceramic material. The partial realization of this goal is reported, but in addition, we

References pp. 166-167

report the observation of two electrooptic effects which, to the knowledge of the authors, have not been previously reported. The first of these is an electrically induced change in the light intensity scattered through large angles by a translucent $BaTiO_3$ ceramic. The other is a mode of electrically controlled birefringence in a longitudinal electrooptic configuration which results when the ceramic is poled in the longitudinal direction (as contrasted to the usual requirement of transverse poling). Both of these effects are discussed in some detail.

The paper is presented in three parts. We first discuss the results of our investigations of light scattering in $BaTiO_3$ ceramic slabs, indicating how the scattered light characteristics enable one to separate the scattered light from the unscattered transmitted light, thus facilitating the independent observations of electrooptic effects of scattered light and transmitted light birefringence. The second part describes the electrooptic effect associated with the electric field induced change in birefringence of the unscattered transmitted light. The longitudinal electrooptic coefficient of $BaTiO_3$ ceramic is determined and compared to the similar value of PZT. Finally, the last part is devoted to a discussion of the relationship of single crystal $BaTiO_3$ electrooptic properties to those observed in the ceramic. These correlations suggest how the electrooptic properties of ferroelectric ceramics may be improved.

LIGHT SCATTERING IN FERROELECTRIC CERAMICS

UNPOLARIZED FERROELECTRIC CERAMICS

For the purposes of this discussion a ferroelectric ceramic can be considered as a cohesive aggregate of relatively small (about one to ten micron) single crystal grains. The grains are assumed to be randomly oriented in the ceramic prior to any application of external stimuli such as an electric field or a stress or induced strain. Ferroelectric ceramics are prepared from pure fine powder material in either of two ways. One is to form a pellet of the powder by pressing at room temperature and then heating this pellet to a temperature of about 200°C to 400°C below the melting point of the material.[17] The sintering that occurs yields a ceramic which can subsequently be cut and shaped into desired geometries. The other is to simultaneously heat and compress a compacted pellet of the powder. This so-called hot pressing technique for preparing ceramics usually yields higher densities consistent with smaller grain sizes than does the simpler press and then heat procedure.[17]

The processing of a ceramic involves cutting, lapping, and polishing operations which can induce substantial strain in the material. For this reason all samples were carefully annealed after completion of shaping and surface preparation operations. The annealing was carried out at temperatures ranging from 500°C to 800°C in an air or oxygen atmosphere. The annealing step was assumed effective in relieving any possible induced anisotropies related to the processing.

The relative transparency of $BaTiO_3$ slabs of varying density is illustrated in the photograph of Fig. 1. These three slabs are all about ten mils thick but exhibit densities ranging from 95% to 99% of theoretical going from left to right in the figure. The increasing transparency with increasing density is quite apparent. The poor transparency of the lower density samples is associated with their relatively high porosity. The pores scatter incident light and render the ceramic translucent. It is clear that the porosity must be minimized in order to employ ceramics as optical image-processing elements.[15,18,19]

The optical properties of the unpolarized ceramics can be demonstrated in another way. In Fig. 2 is plotted the angular distribution of light intensity resulting from illumination of three ferroelectric ($BaTiO_3$) ceramic slabs of densities comparable to those shown in Fig. 1. The light source was a helium-neon laser, the beam from which was normally incident on a large area face of the ceramic slab. The intensity of light at a given angle relative to the incident light direction was determined by a combination aperture and photomultiplier which allowed an angular resolution of ± ½ degree. The data are consistent with the photograph of Fig. 1; the large-angle scattering is relatively intense for the low density material and virtually all the transmitted light is scattered. As the density increases, an unscattered forward transmitted component of light appears and the large-angle scattered light is reduced in intensity. The large angle scattered light is characterized by an angular intensity dependence of $\cos^2\theta$, where θ is the scattering angle relative to the incident light direction. This has been found to be true for all sample densities studied. Also, the scattered light was observed to be completely depolarized, while the unscattered transmitted light exhibited the polarization of the incident light.

The $\cos^2\theta$ dependence of the scattered light intensity is consistent with the theoretical predictions of light scattering theory.[20,21,22] Scattering theory for particles of dielectric constant ϵ imbedded in a dielectric medium, in the form of a slab, of dielectric constant ϵ_0 yields a $\cos^2\theta$ dependence if the wavelength of the light is large compared to the dimensions of the scattering particles. Further, the theory indicates that the scattered light is depolarized if multiple scattering and/or deviations of scattering centers from sphericity exist.[21] The size and concentration of pores in the ferroelectric ceramics are such as to satisfy the various assumptions of the theory. We thus conclude that the pores of the ceramic material scatter light isotropically in quantitative agreement with simple theory.

The isotropic and depolarized nature of the scattered light provides a means of separating the scattered and the unscattered light so that the transmitted component representing transparency can be studied directly. Since the ceramic materials prepared in the laboratory exhibit varying densities and thus a variety of scattered light intensities, the ability to separate the scattered and unscattered components becomes quite important when investigating the influence of electric fields and the attendant electrooptic effects.

References pp. 166-167

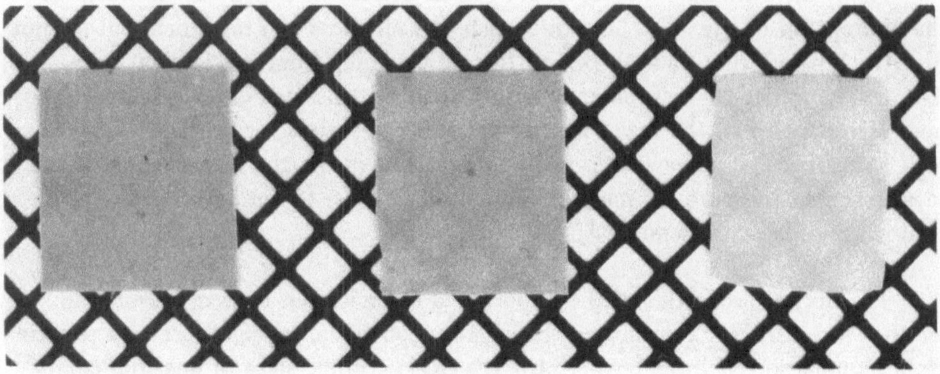

a b c

Fig. 1. Illustration of relative transparency exhibited by $BaTiO_3$ ceramic slabs of the same thickness but varying density; (a) 95.0%, (b) 97.6%, and (c) 99.1% of theoretical bulk density.

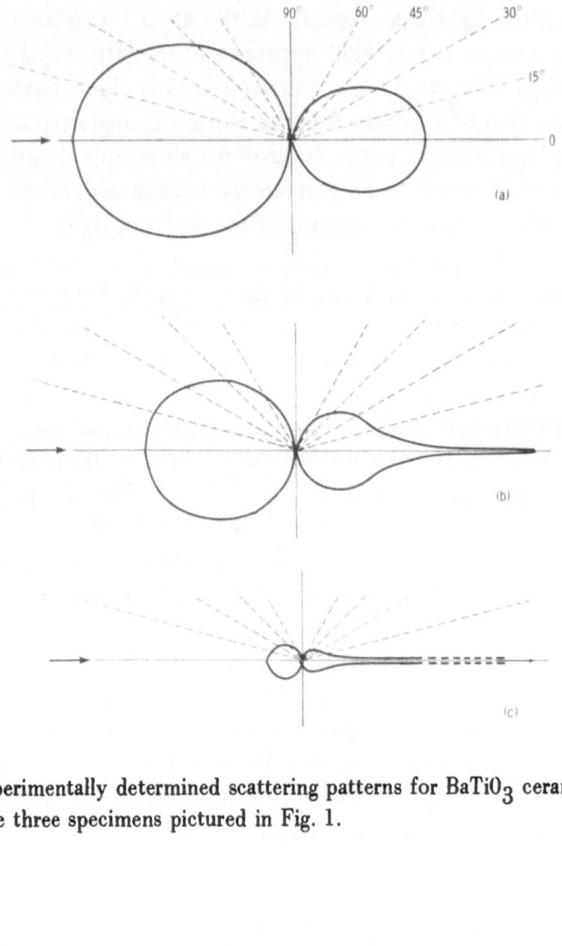

Fig. 2. Experimentally determined scattering patterns for $BaTiO_3$ ceramic slabs of densities comparable to the three specimens pictured in Fig. 1.

POLARIZED CERAMICS

The application of an electric field polarizes (poles) a ferroelectric ceramic, and it becomes uniaxially birefringent with its optical axis presumably parallel to the direction of the poling field.[23] As a consequence of the spontaneous polarization unique to ferroelectric materials, a hysteresis in the polarization, P, as a function of electric field, E, develops as illustrated schematically in Fig. 3. Having poled a ferroelectric material it requires a coercive field E_c to reduce the polarization to zero. Any degree of remanent polarization P_r between the extremes $\pm P_s$ of the saturation polarization can be induced by suitable adjustment of the applied electric field. The refractive index along the optical axis is proportional to the polarization and thus exhibits the same type of hysteresis and electric field dependence. The optical birefringence therefore can be electrically controlled. It is the electrically controlled birefringence of ferroelectric ceramics which offers attractive electrooptic device applications as has been so adequately demonstrated by the Sandia group.

We have investigated the influence of poling on the isotropically scattered component of average density ferroelectric ceramics. Figure 4 illustrates the typical reduction in the scattered light intensity upon poling a 0.010 inch thick $BaTiO_3$ slab in a direction normal to the large area surfaces. The ferroelectric disc was poled at room temperature with a field of approximately $3E_c$ (see Fig. 3) applied for several hours. The scattered light was observed to continue to exhibit the $\cos^2\theta$ dependence characteristic of the unpoled material and was completely depolarized. This is readily understood in terms of the simple scattering theory already discussed, which predicts that the scattered light intensity is a function of the difference in the squares of the refractive indices of the scattering centers, n, and the slab medium, n_o, in the form $(n^2-n_o^2)^2/n_o^2$.[20,21,22] Thus the change in refractive index upon application of the field is reflected in a change in the intensity of the scattered light.

Figure 5 shows the variation of scattered light intensity in a fixed direction as the applied electric field is varied so as to traverse a complete electrical hysteresis loop. The variation in intensity reflects the variation in the index of refraction, the "butterfly" loop being typical of many other polarization-dependent properties of ferroelectric ceramics. It is thus concluded that the effect of electric fields on the scattered component of light transmitted through a ferroelectric slab is consistent with what we understand about such materials within the framework of simple scattering theory.

Two final remarks about the scattered component are appropriate before we go into a discussion of the electrooptics of the unscattered transmitted component. First, we have observed that the influence of the electric field on the large angle scattered light intensity is reduced when this scattered light intensity is smaller. This is understandable in terms of the multiple scattering concept since for higher densities

References pp. 166-167

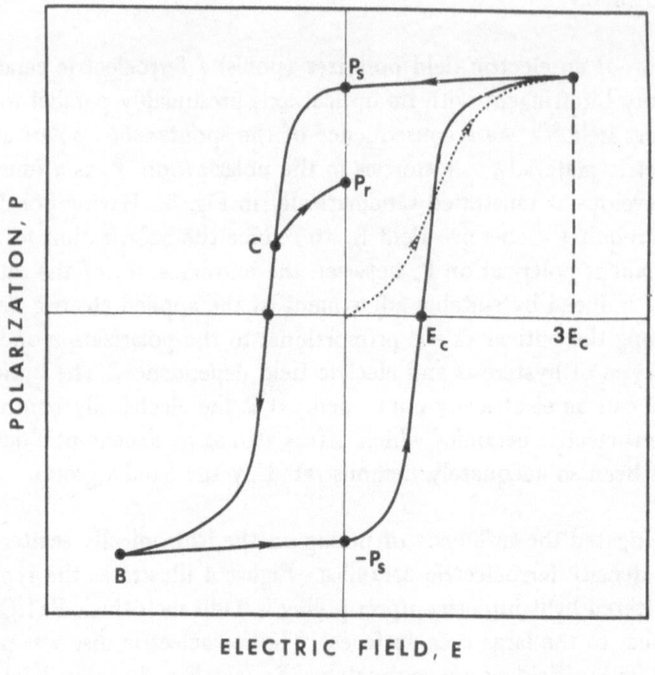

Fig. 3. Schematic hysteresis in the polarization vs electric field typical of ferroelectric ceramics.

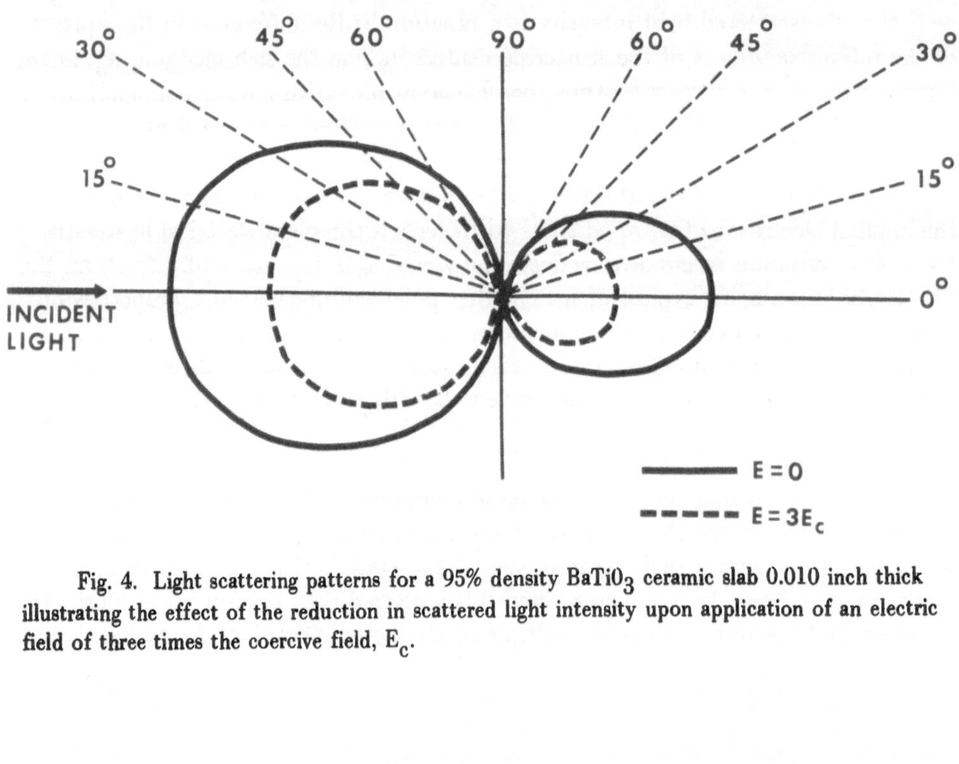

Fig. 4. Light scattering patterns for a 95% density $BaTiO_3$ ceramic slab 0.010 inch thick illustrating the effect of the reduction in scattered light intensity upon application of an electric field of three times the coercive field, E_c.

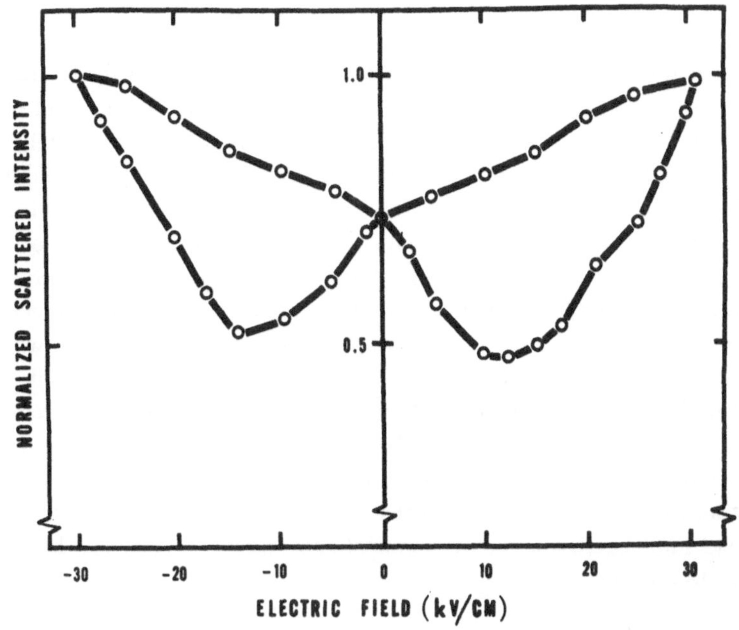

Fig. 5. Intensity of light scattered in the forward direction at ± 20° as a function of the applied electric field for a typical low-density $BaTiO_3$ ceramic slab.

of scattering centers one would expect that the influence of the change in refractive index due to the electric field would be large because of the large number of interactions per unit path length. However, as the scattering center density decreases, the number of scattering interactions decrease per unit path length thus decreasing the effect of the refractive index change and hence the electric field. The second point relates to very low angle scattering associated with grain boundaries and domain walls which we have not mentioned to this point. Such low-angle scattering has been observed to be an important consideration in the lead-zirconium-titanate (PZT) system since it has proven to be polarization dependent and thus offers possibilities of electronically controlling scattered light for small angular aperture optical systems.[1-4,8] We have not observed a similar effect in the $BaTiO_3$ system as yet and therefore will not discuss this type of scattering further.

ELECTROOPTICS OF $BaTiO_3$ CERAMICS

In this section we wish to discuss the work done in our laboratory to characterize the electrooptic properties of $BaTiO_3$ ceramics. We refer now to the unscattered transmitted component of light, it being understood that unless specifically stated otherwise the large-angle scattered component has been subtracted from the experimental results.

References pp. 166-167

EXPERIMENTAL PROCEDURE

Ceramic materials were prepared from a variety of commercially available $BaTiO_3$ powders of varying purity and particle sizes. Sources, purity, and particle size information is listed in Table I. A weighed quantity of powder (usually 10 to 20 grams) is poured into a cylindrical carbon die lined with pyrolytic graphite. The die is then inserted in a furnace, and the assembly positioned in a press which can provide up to 9 tons/in.^{2} pressure via a ram that mates with the interior of the die. A typical hot pressing sequence is to heat to 800°C, then apply desired pressure, resume heating to the predetermined sintering temperature, hold this temperature for one hour, cool to about 1050°C, remove pressure, and finally cool slowly to room temperature. The sintered specimen can then be ejected from the die and examined as desired. A subsequent oxidation at 1000°C is required to recover stoichiometry[24] and this can be done either before or after the shaping, lapping and polishing operations. It is more convenient to perform the oxidation step after these operations, however, since an anneal is necessary after such mechanical operations to relieve induced strains; i.e., the anneal and oxidation steps can be combined in a single heating cycle.

Two methods of preparing electrical contacts to the specimens were employed. In the cases where transparency of the electrodes were not a consideration, gold films were evaporated on the sample surfaces. When electrode transparency was required, RF sputtered tin oxide was used. As deposited, these tin oxide films typically exhibited high resistance, but subsequent annealing at about 475°C in air reduces the resistance to about 20 kilohms per square for films approximately 3000 Å thick. Two representative sample geometries are sketched in Fig. 6. The configuration shown in Fig.6a is used to pole specimens in the transverse mode, i.e., in the plane of the large-area faces of the sample. Figure 6b illustrates a geometry for

TABLE I

$BaTiO_3$ Powder Specifications

Supplier	Lot No.	Purity	Ave. Grain Size
TAM Division National Lead Co. Niagara Falls, N. Y.	CP 72677 220 P	99.5% (piezoelec. grade)	0.58 μ
Poly Research Corp. Great Neck, N. Y.	131-43	99.99%	1.8 μ

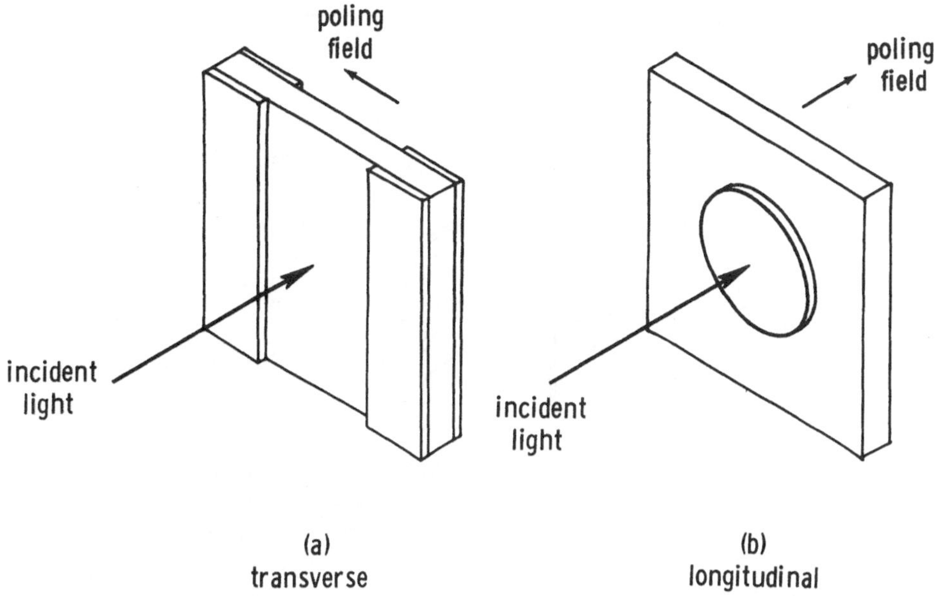

Fig. 6. Schematic illustration of sample electrode geometry defining the transverse and longitudinal poling configurations.

application of fields normal to the large-area faces, the longitudinal mode. It is this latter configuration that was employed exclusively for the determination of the electric field dependence of the isotropically scattered light previously discussed.

The experimental apparatus for observing the electrical and optical properties of the ceramics samples is illustrated schematically in Fig. 7. Light from a 1 mW helium-neon laser L is normally incident on a sample S which is between crossed polarizers P_1 and P_2. The light is modulated by the mechanical chopper C. The light transmitted (or scattered) through the sample is detected by the photomultiplier PM_1 after passing through the aperture A. The assembly of aperture, photomultiplier, and polarizer P_2 shown enclosed in the dashed lines of Fig. 7 could be rotated about a vertical axis through the center of the sample, thus facilitating detection of large angle scattering. The signal from the photomultiplier PM_1 was fed to the input of a phase-sensitive lock-in amplifier, which used a reference signal obtained from a second photomultiplier PM_2 that sensed the light from a small reference lamp L_r chopped at the same frequency as the incident laser light. Voltage was applied to the electrodes of the sample through a limiting resistor R and a series capacitor C. This facilitated the determination of the polarization of the sample by measuring the voltage across C with an electrometer.

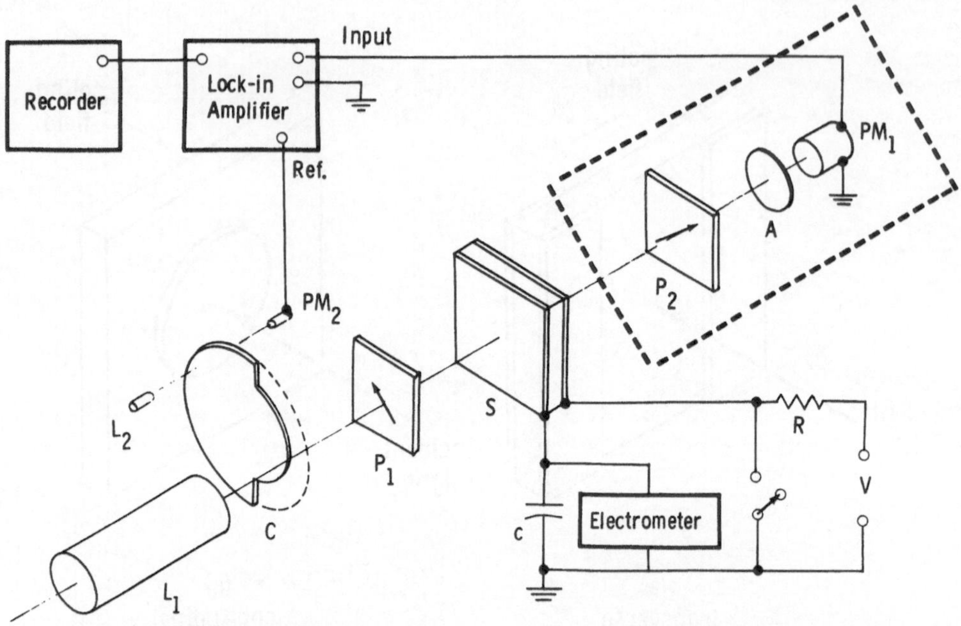

Fig. 7. Schematic representation of the experimental system employed in determining electrooptic coefficients and scattering patterns; L_1-helium-neon laser, L_2-neon lamp, C-chopper, P_1 and P_2-polarizers, S-electroded sample, A-aperture, PM_1 and PM_2-photomultipliers.

The electrooptic properties of the ceramics were obtained with the arrangement of Fig. 7 by determining the half-wave voltage. Briefly, the experiments are based on the following considerations. An apparent linear electrooptic coefficient can be defined in terms of the effective birefringence $\overline{\Delta n}$ induced by an applied electric field as

$$\overline{\Delta n} \equiv (n_1^3/2) r_{long} E_z L \qquad (1)$$

where r_{long} is the electrooptic coefficient for the longitudinal electrooptic effect (electric field applied in the same direction as the direction of light propagation), n_1 is the refractive index of the ordinary ray with zero applied field, E_z is the magnitude of the longitudinally applied electric field, and L is the thickness of the ceramic slab in the direction of light propagation.[25] Associated with $\overline{\Delta n}$ is a retardation $\Gamma = (\omega/c)\overline{\Delta n}$, where ω is the frequency of the light and c is the velocity of light in vacuum. A half-wave voltage is defined as that applied voltage which yields a retardation of π, that is

$$\Gamma = \pi(V/V_{1/2}) \qquad (2)$$

where $V = E_z L$ is the applied voltage and $V_{1/2}$ is the half-wave voltage. From Eq. (1) one finds that

$$V_{1/2} = \lambda/2n_1^3 r_{long} \quad (3)$$

where λ is the wavelength of the light. It can be readily shown that the fraction of incident light intensity transmitted at an arbitrary voltage is

$$I = \sin^2(\Gamma/2) = \sin^2(\pi V/2V_{1/2}) \quad (4)$$

and thus $V_{1/2}$ can be established by observing the variation of transmitted light intensity as a function of applied voltage.[11] The electrooptic coefficient r_{long} can then be obtained by Eq. (3). We have assumed in the above that $\overline{\Delta n}$ depends linearly on E_z which may not be the case. This is why we refer to r_{long} as an *apparent* electrooptic coefficient.

The procedure then for determining r_{long} is as follows: Place the sample in position as shown in Fig. 7 with the optical axis at 45° to both polarizer directions; determine the electric field dependence of the scattered light at some off-axis point for purposes of subtracting it out during subsequent analysis; next determine the on-axis variation of light intensity with applied voltage (polarization is simultaneously measured); subtract the scattered light and ascertain the half-wave voltage; and finally employ Eq. (3) to evaluate r_{long}.

EXPERIMENTAL RESULTS

Electrooptic coefficients of $BaTiO_3$ ceramics of varying densities have been determined, and there appears to be very little dependence of the coefficient on the density. A typical plot of smoothed intensity data as a function of electric field for a 0.0075 inch thick $BaTiO_3$ specimen of density 98.3% of theoretical is shown in Fig. 8. The half-wave voltage for this sample was determined to be 385 volts. Using a value of 2.4 for n_1 and with λ = 6328 Å, r_{long} was found to be 0.60 x 10^{-10} meter per volt. Similar observations of 65-35:2% La PZT resulted in the plot of Fig. 9, yielding a half-wave voltage of 275 volts and an r_{long} of 0.81 x 10^{-10} meter per volt. This latter value is approximately equal to the value determined by Land and Thacher (1.15 x 10^{-10} meter/volt) for similar material.[3]

These results for ferroelectric ceramic materials are listed and compared to similar values for several common electrooptic single crystals in Table II. It is evident that the magnitude of the coefficients for the ceramics is in a range consistent with practical use. In particular, it is of interest to note that the value of r_{long} determined for $BaTiO_3$ ceramics is substantially larger than the constant-strain value of

References pp. 166-167

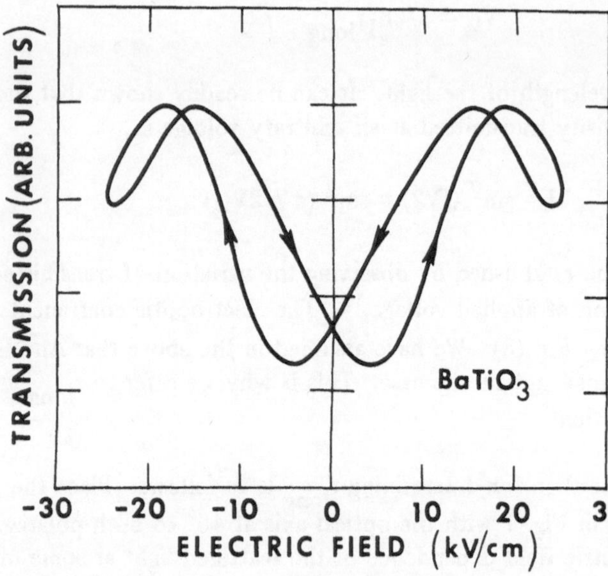

Fig. 8. Light transmission through a typical BaTiO$_3$ ceramic slab of relatively high density (transparent) between crossed polarizers as a function of electric field.

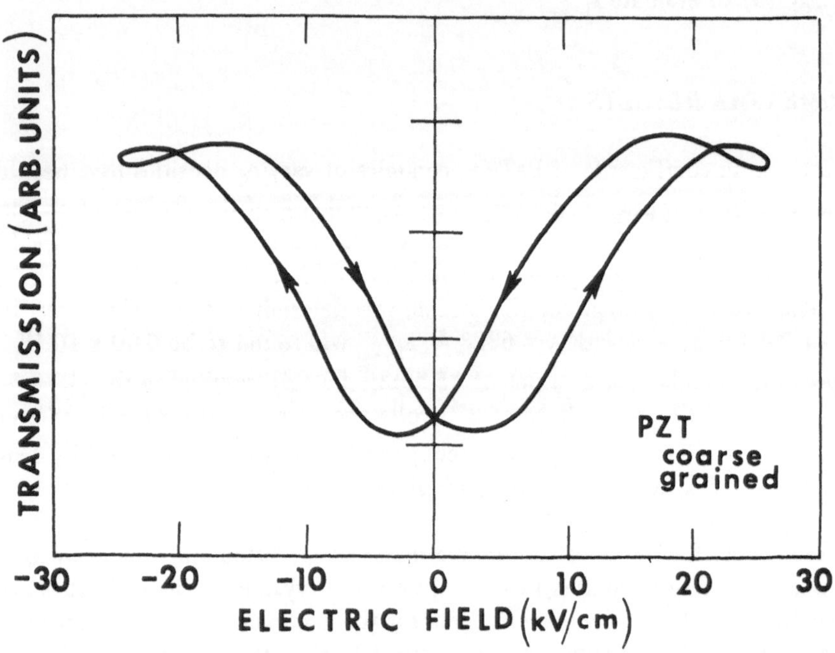

Fig. 9. Light transmission through coarse-grained lead-zirconate-titanate (PZT) with 2% lanthanum between crossed polarizers as a function of electric field. The PZT was obtained from the Clevite Corp.

TABLE II

Longitudinal Electrooptic Coefficients
For Several Ferroelectric Materials

	Material	r_{long} x 10^{10} meter/volt
Single Crystal	ADP (1)	0.17
	KDP (1)	0.214
	$LiNbO_3$	0.17
	$BaTiO_3$	0.23 (2)
	$(Sr_{0.75}Ba_{0.25})Nb_2O_6$	14.0
Ceramic	$BaTiO_3$	0.59
	PZT (3)	0.82

(1) $r_{long} = 2r_{63}$

(2) constant strain value

(3) 65-35 2% La coarse-grained

single crystal $BaTiO_3$. We shall return to this point later when considering the relation of the single crystal electrooptics to those of ceramics.

A most surprising, and perhaps significant, result was the observation of a longitudinal electrooptic effect in both $BaTiO_3$ and PZT samples that were poled in the longitudinal configuration (see Fig. 6b). Since the poling field in this case is in a direction normal to the large area faces of the sample one would expect that the induced optical axis associated with the birefringence to also be normal to the surfaces, while the optical properties in the plane of these surfaces would remain isotropic. Thus no electrooptic effect would be expected in the longitudinal configuration shown in Fig. 7. However, specimens poled in this manner exhibited an apparent electrooptic coefficient comparable in magnitude to that for the transverse-poled case listed in Table II. This observation was repeated several times with several different samples. We concluded that, in spite of the longitudinal poling mode employed, an anisotropy in the optical properties of the samples had to somehow develop that had a preferred direction in the plane of the large area surfaces, i.e., perpendicular to the poling direction.

Since longitudinal electrooptic effects are observed in both the transverse and longitudinally poled sample configurations, the optic axis induced by poling must be skew to the direction normal to the large area sample faces. Possible causes of such a skewness can be preferential orientation of grains and/or preferred directions of induced strain. Table III lists several possible contributions associated with the preparation of a ferroelectric ceramic specimen. X-ray diffraction (back-reflection Laue patterns) indicate that unpoled specimens that have been shaped and annealed do indeed exhibit some preferred crystallographic orientation. Also, observation of specimens during poling using a polarizing microscope suggest the possibility of induced directional strain, the direction of the strain being related to the sample and electrode geometry. These early results, however, are inconclusive and we are continuing to study this effect.

DISCUSSION AND CONCLUSIONS

The large value of the longitudinal electrooptic coefficient of $BaTiO_3$ ceramics relative to the comparable single crystal value can be accounted for directly. For a tetragonal crystal of 4mm symmetry, such as $BaTiO_3$ at room temperature, the electrooptic coefficient matrix is of the form[26]

$$r_{ij} = \begin{vmatrix} 0 & 0 & r_{13} \\ 0 & 0 & r_{13} \\ 0 & 0 & r_{33} \\ 0 & r_{42} & 0 \\ r_{42} & 0 & 0 \\ 0 & 0 & 0 \end{vmatrix} \qquad (5)$$

where for single crystal $BaTiO_3$ $r_{13} = 8.0$, $r_{33} = 23.0$, and $r_{42} = 820$, in units of 10^{-12} volts/meter.[27] The longitudinal electrooptic coefficient is[28]

$$r_{long} = (n_3^2/n_1^2)\, r_{33} - r_{13} \qquad (6)$$

where n_1 is the refractive index perpendicular to the optical axis and n_3 that parallel to the optical axis. The symmetry of the poled ferroelectric ceramic (an infinite-fold rotation axis along the poling direction) has the same electrooptic coefficient matrix as Eq. (5) above.[23] The corresponding values of the non-zero r_{ij}'s for the ceramic would be related to the single crystal r_{ij}'s through an appropriate statistical averaging over the random orientation of individual crystallites comprising the ceramic.[29] Thus the large single crystal value of r_{42} is mixed into the values of r_{13} and r_{33} of the ceramic, resulting in a larger r_{long} for the ceramic than for the single

TABLE III

Possible Sources of Preferred Orientation
in Ferroelectric Ceramics

During hot-pressing	Particle size and shape effects Influence of grain growth Effect of impurities Thermal gradients and heat transfer
During sample shaping	Directional strain induced by sawing, lapping and polishing Effect of sample geometry and mounting in annealing Cooling through the Curie temperature (phase transformation stress, etc.)
During poling	Electrode-sample geometry induced strain 90° vs 180° domain wall motion Space charge and surface effects

crystal. A similar explanation has been offered to account for the relatively large piezoelectric effect in $BaTiO_3$ ceramics compared to the single crystal.[30]

Several researchers have tried to relate the properties of ceramics to single crystals via statistical averaging over grains.[29-32] All of these calculations have been based on the assumption that the individual crystallites of the ceramic are randomly oriented about the uniaxial direction established by the poling field. We observe, however, that this assumption may not be valid, because there appears to be some preferred orientation of grains which give rise to the development of an optical axis skew to the poling field direction. Once the origin and nature of this orientation effect is established, an attempt to determine r_{long} for the ceramic from single crystal r_{ij}'s would be of value.

The possibility of oriented crystallites in ceramics suggests an interesting speculation. In principle it should be possible to induce a preferred orientation that would not only preserve the longitudinal poling effect (with its attendant device application advantages) but also optimize the contribution of the single crystal r_{42} to the longitudinal electrooptic effect of the ceramic. We thus could obtain a longitudinal electrooptic coefficient as large as 5×10^{-10} meter/volt which would yield a half-wave voltage (see Eq. (3)) of approximately 40 volts, a particularly attractive value for many device applications.

References pp. 166-167

The successful application of $BaTiO_3$ ceramics as electrooptic modulators of light is contingent on the fabrication of a transparent material. We have been able to obtain densities of $BaTiO_3$ ceramics as high as 99.3% of theoretical when employing the highest purity commercially available fine powder as starting material. This density still has sufficient residual porosity to scatter appreciable quantities of incident light. It appears that small quantities of additives that will improve densification are required but these additives must not degradate the ferroelectric properties of the ceramic.[18,33,34] There is no question that much more work is required before $BaTiO_3$ can be considered for possible device applications paralleling those already demonstrated for the PZT system (see the next paper in this volume by land and Thacher).[7]

We conclude by summarizing the salient points of our work to date. Ceramic $BaTiO_3$ of relatively high density has been prepared by hot-pressing techniques, the highest density so far realized approaching transparency but not to a point yet satisfactory for image processing. The scattered light associated with the residual porosity in the ceramics is consistent with an isotropic scattering model. This scattered light exhibits an electrooptic effect upon application of an electric field (a field-dependent intensity) which to the best knowledge of the authors has not previously been reported. The isotropically scattered light can be separated from the unscattered forward transmitted light enabling the study of electrooptic effects associated with the normal change in the birefringence induced by applied electric fields. A value of about 0.6×10^{-10} meter/volt was obtained for the longitudinal electrooptic coefficient of ceramic $BaTiO_3$, which could be interpreted in terms of the single crystal electrooptic coefficients. Finally, we observe for the first time a longitudinal electrooptic effect in ferroelectric ceramic samples which have been poled in the same direction as the incident light, and applied electric field, a quite unexpected result if the ceramics are truly composed of randomly oriented crystallites. It seems reasonable to conclude that $BaTiO_3$ ceramics warrant further attention and development as potential electrooptic light control devices.

ACKNOWLEDGMENTS

The authors express their appreciation of the many valuable contributions of C. E. Bleil, L. T. Klauder, J. G. Gay, D. M. Roessler, M. H. Brooks and D. S. Eddy. The fine technical assistance of L. Green in obtaining experimental data is gratefully acknowledged.

REFERENCES

1. *C. E. Land, 1967 International Electron Devices Meeting, Washington, D. C.; Sandia Laboratories Reprint SC-R-67-1219, October, 1967.*

2. *C. E. Land and P. D. Thacher, Proc. IEEE 57, 751(1969).*
3. *P. D. Thacher and C. E. Land, IEEE Trans. Elec. Devices ED 16, 515(1969).*
4. *C. E. Land and R. Holland, IEEE Spectrum 7, 71(1970).*
5. *C. E. Land and G. H. Heartling, J. Phys. Soc. Japan (suppl.) 28, 96(1970).*
6. *C. E. Land, International J. Nondestructive Testing 1, 315(1970).*
7. *C. E. Land and P. D. Thacher, this volume, p. 171.*
8. *R. E. Nettleton, J. Appl. Phys. 39, 3646(1968).*
9. *J. R. Maldonado and A. H. Meitzler, IEEE Trans. Elec. Devices ED 17, 148(1970).*
10. *A. H. Meitzler, J. R. Maldonado and D. B. Fraser, Bell System Tech. J. 49, 953(1970).*
11. *W. C. Stewart and L. S. Cosentino, Ferroelectrics 1, 149(1970).*
12. *J. R. Maldonado and A. H. Meitzler, to be published.*
13. *G. H. Heartling, Am. Ceram. Soc. Bull. 43, 875(1964).*
14. *G. H. Heartling and W. J. Zimmer, Am. Ceram. Soc. Bull. 45, 1084(1966).*
15. *G. H. Heartling, Am. Ceram. Soc. Bull. 49, 564(1970).*
16. *G. H. Heartling and C. E. Land, submitted to J. Am. Ceram. Soc.*
17. *W. D. Kingery, Introduction to Ceramics (John Wiley and Sons, Inc., New York, 1960).*
18. *R. W. Rice, "Hot Forming of Ceramics" in Ultrafine-Grain Ceramics, J. J. Burke, N. L. Reed and V. Weiss, Eds. (Syracuse University Press, Syracuse, New York, 1968), p. 211.*
19. *K. Okazaki and K. Takahashi, J. Phys. Soc. Japan (suppl.) 28, 329(1970).*
20. *H. Van de Hulst, Light Scattering by Small Particles (John Wiley and Sons, New York, 1957).*
21. *I. L. Fabelinski, Molecular Scattering of Light (Plenum Press, New York, 1968).*
22. *M. Born and E. Wolf, Principles of Optics (Pergamon Press, New York, 1970).*
23. *F. Jona and G. Shirane, Ferroelectric Crystals (MacMillan Co., New York, 1962).*
24. *V. J. Tennery and J. C. Venerus, Am. Ceram. Soc. Bull. 36, 59(1957).*
25. *See for example A. Yariv, Quantum Electronics (John Wiley and Sons, Inc., New York, 1967).*
26. *J. F. Nye, Physical Properties of Crystals (Oxford University Press, London, 1957).*
27. *I. P. Kaminov and E. H. Turner, Appl. Optics 5, 1612(1966).*
28. *I. P. Kaminov, "Electrooptic Materials" in Ferroelectricity, E. F. Weller, Ed. (Elsevier Publishing Co., Amsterdam, 1967).*
29. *See for example N. Uchida and T. Ikeda, Jap. J. Appl. Phys. 6, 1079(1967).*
30. *M. Marutake and T. Ikeda, J. Phys. Soc. Japan 12, 233(1957).*
31. *H. G. Baerwald, Phys. Rev. 105, 480(1957).*
32. *M. Marutake, J. Phys. Soc. Japan 11, 807(1956).*
33. *M. Deri, Ferroelectric Ceramics (MacLaren and Sons Ltd., London, 1966).*
34. *C. A. Miller, J. Materials Sci. 3, 463(1968).*

ELECTROOPTIC PROPERTIES OF Ba, Sn, AND La MODIFIED LEAD ZIRCONATE TITANATE CERAMICS†

C. E. LAND and P. D. THACHER

Sandia Laboratories, Albuquerque, New Mexico 87115

ABSTRACT

Electrooptic effects observed so far in ferroelectric lead zirconate titanate ceramic materials are transverse electrooptic effects (linear, quadratic, and memory), strain-biased longitudinal electrooptic memory effect, and electrically controlled scattering. The electrooptic memory effects are related to variations of ceramic birefringence with remanent polarization. All the above electrooptic effects are found in Ba, Sn, and La modified lead zirconate titanate ceramics except electrically controlled scattering. Both the character and magnitude of these effects are strongly dependent on the composition and grain size of the ceramic material as well as on the temperature and light wavelength of observation. The results for the La modified ceramics, which can be made highly transparent, are discussed in the greatest detail. The various electrooptic effects are shown to arise either from orientation of ferroelectric domains or from field-enforced ferroelectric distortion of certain paraelectric materials.

INTRODUCTION

Poled lead zirconate titanate - $Pb(Zr,Ti)O_3$ - ferroelectric ceramics are uniaxially birefringent on a macroscopic scale, and they have a number of useful electrooptic properties. Because they are polycrystalline, they tend to scatter light more than ferroelectric single crystals. In recent years, however, improved fabrication methods involving hot-pressing[1-3] increased the optical transparency of Bi doped* $Pb(Zr,Ti)O_3$ solid solutions to the extent that their electrooptic properties could be measured and evaluated.[4] Lanthanum doping further enhanced the transparency by reducing light scattering.[5,6] Haertling improved the switching characteristics by reducing the coercivity without adversely affecting the optical properties when he modified* the $Pb(Zr,Ti)O_3$ system first with Ba[7] to form $(Pb,Ba)(Zr,Ti)O_3$ or PBZT, and then with Sn[8] to form $Pb(Sn,Zr,Ti)O_3$ or PSZT. A further effort to improve both the switching and optical properties led Haertling[9] to modify the $Pb(Zr,Ti)O_3$ system with La to form $(Pb,La)(Zr,Ti)O_3$ or PLZT. The result was a dramatic increase in optical transparency and a fine compositional control of the electric and electro-optic properties.

As indicated above, the most important of the modified solid solution systems is PLZT because of its unusually high optical transparency between the wavelengths of 0.4 and 6 μm and its versatile composition-controlled electrooptic properties. Materials in the PLZT system can exhibit an electrooptic memory effect, a conventional quadratic (Kerr) electrooptic effect, or a linear (Pockels) electrooptic effect depending on La content and Zr/Ti ratio. Electrooptic memory effects are found in low-coercivity rhombohedral and tetragonal PLZT compositions. The Kerr quadratic electrooptic effect occurs in slim-loop** ferroelectric PLZT compositions for electric fields E in the range $0 \leq |E| < |E_S|$, where E_S is the value of electric field at which the ferroelectric polarization begins to saturate. A linear electrooptic effect - the first known to exist in ceramics - is found in high-coercivity tetragonal PLZT compositions. The PLZT ceramics are presently being evaluated for use in several electrooptic devices including light valves, shutters, modulators, spectral filters, memories, displays, and image storage devices.[8-12]

A method of obtaining a longitudinal electrooptic effect in ferroelectric ceramics was recently described by Maldonado and Meitzler.[12] A uniaxial biasing strain is applied to the ceramic in order to induce a birefringence with the principal axes of the optical indicatrix along the strain axes. When the strain-induced optic axis is normal to the light propagation direction, an electric field applied parallel to the light propagation direction causes a variation in the strain-induced birefringence. This technique has many important device implications, one of which is described in Reference 12.

* For purposes of this paper, the term "doped" refers to addition of less than 5 atom % of the doping element; the term "modified" refers to addition of 5 atom % or more of the modifying element.

** Slim-loop ferroelectric materials are characterized by a Curie point near room temperature, a field-enforced paraelectric to ferroelectric phase transition, and an almost anhysteretic polarization vs. electric field response (see Fig. 1(G)).

The remainder of this paper consists of a summary of the electrooptic effects in ferroelectric ceramics and a more detailed discussion of the electrooptic properties of PBZT, PSZT, and PLZT materials. The paper concludes with a phenomenological explanation of the basis of the various electrooptic effects.

REVIEW OF ELECTROOPTIC EFFECTS IN CERAMICS

BASIC ELECTROOPTIC EFFECTS

Two fundamentally different electrooptic effects have been observed in hot-pressed $Pb(Zr,Ti)O_3$ ceramics; (1) electrically controlled birefringence, and (2) electrically controlled light scattering. Fine-grained Bi or La doped $Pb(Zr,Ti)O_3$ ceramics with average grain size less than 2 μm have an effective birefringence $\overline{\Delta n}$ that can be varied either by partially switching the remanent polarization P_r (electrooptic memory effect) or by applying an external electric field E (conventional electro-optic effect).[4,6] Modified $Pb(Zr,Ti)O_3$ ceramics - PBZT, PSZT, and PLZT - with average grain sizes up to 10 μm also exhibit electrically controlled birefringence.[7-9] Coarse-grained Bi doped $Pb(Zr,Ti)O_3$ ceramics with average grain size greater than 2 μm have light scattering properties that can be varied by switching the orientation of the ferroelectric polarization (electrooptic scattering effect).

ELECTRICALLY CONTROLLED BIREFRINGENCE

A summary of the electrooptic effects involving variable birefringence in ceramics is shown in Fig. 1 for ceramics having two distinct P vs. E behaviors. Diagrams in Fig. 1 are idealized representations of the various effects which have been observed. Figure 1(A) shows a hysteresis loop in which polarization P is plotted vs. electric field E for a ferroelectric ceramic memory material.* The material is poled initially by applying an electric field of sufficient magnitude to cause saturation (curve a-b of Fig. 1(A)), and then removing the field (curve b-c). The remanent polarization at (c) is commonly referred to as saturation remanence P_R. In ceramics, remanent polarization can be switched incrementally by applying electric field pulses of controlled amplitude and duration[4] or by voltage-controlled partial switching.[10] Incremental switching is indicated in Fig. 1(A) between (c) at P_R and (d) at $P_r = 0$ and between (d) and (f) at $-P_R$.

* Designated "square-loop" ceramic elsewhere in this paper in order to contrast it with the hysteresis behavior of the slim-loop material.

References pp. 194-196

Electrooptic effects which have been observed in ferroelectric ceramic memory materials with P vs. E. characteristics similar to that shown in Fig. 1(A) are illustrated schematically in Figs. 1(B) through 1(D) for transverse effects and Figs. 1(E) and 1(F) for longitudinal effects. Figure 1(B) illustrates the electrooptic memory effect, i.e., the dependence of $\overline{\Delta n}$ on remanent polarization P_r, as P_r is switched incrementally from positive P_R at (c) through zero (d) to negative P_R at (f), then back (f-d-c). Birefringence measurements are made with E = 0. Figure 1(C) illustrates the quadratic electrooptic effect obtained as E is applied in the saturation direction (c-b). Considerable hysteresis is associated with this effect in some materials as indicated by the dashed lines, and in other materials the effect may become nearly linear.[9] Figure 1(D) illustrates the linear electrooptic effect observed in certain high coercivity tetragonal PLZT materials. This effect is essentially anhysteretic along the linear portions of the $\overline{\Delta n}$ vs. E characteristic.

Figure 1(E) shows the longitudinal electrooptic memory effect obtained by means of a static strain bias by Maldonado and Meitzler[12] in their image storage devices. The birefringence at (a) is due to the biasing strain. When the ceramic is poled with E parallel to the light propagation direction, $\overline{\Delta n}$ increases from (a) to (c). As the polarization is switched from positive P_R(c) through zero (d) to negative P_R(f), $\overline{\Delta n}$ increases to a maximum at (d) and then decreases to its value at (f),

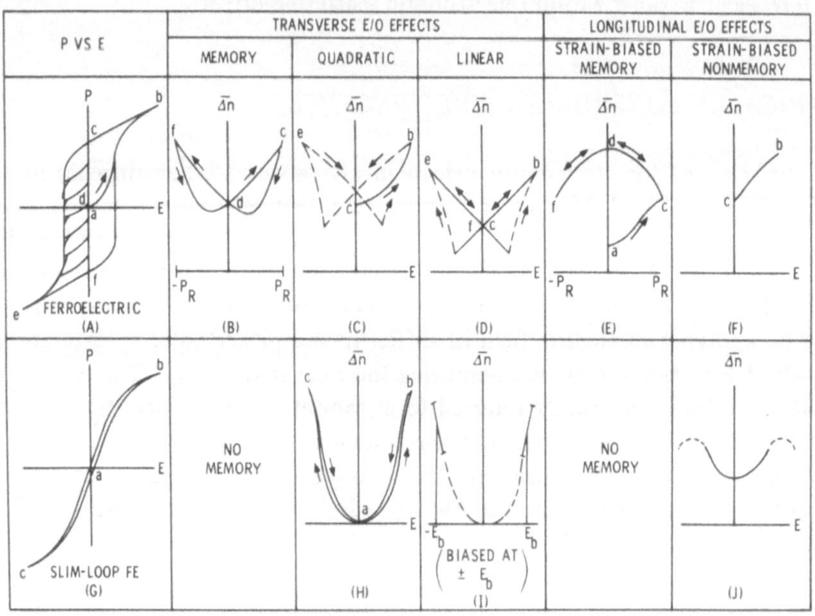

Fig. 1. Schematic representations of the various electrooptic effects involving variable birefringence in ferroelectric and slim-loop ferroelectric ceramics. Electrooptic effects (B) through (F) are associated with materials characterized by the hysteresis loop (A); electrooptic effects (H), (I), and (J) are associated with slim-loop ferroelectrics (G).

which is approximately equal to its value at (c). The effect is then reversible from (f) through (d) to (c). Figure 1(F) illustrates a strain-biased electrooptic effect without memory. After poling, the electric field is applied in the saturation direction (c-b) which produces a corresponding increase in $\overline{\Delta n}$. When the field is removed, $\overline{\Delta n}$ returns to its value at (c).

Figure 1(G) shows a P vs. E curve for a slim-loop ferroelectric ceramic. Electrooptic effects which have been observed in these materials are illustrated in Figs. 1(H) through 1(J). Note that electrooptic memory effects are not present in slim-loop materials. A quadratic (Kerr) electrooptic effect is illustrated in 1(H). The $\overline{\Delta n}$ vs. E curve is quadratic only until saturation begins to occur as shown by the slight inflections near (b) and (c) in Fig. 1(H). Nearly linear electrooptic effects may be obtained in slim-loop materials by applying biasing field E_b as shown in Fig. 1(I). The biasing field should be of sufficient magnitude to place the operating point near the inflection in the $\overline{\Delta n}$ vs. E curve if maximum sensitivity and linearity are desired. A quadratic longitudinal electrooptic effect without memory may be obtained by strain-biasing a slim-loop material as shown in Fig. 1(J).

ELECTRICALLY CONTROLLED LIGHT SCATTERING

Electrically controlled light scattering, as observed in coarse-grained Bi doped $Pb(Zr,Ti)O_3$, is illustrated schematically in Fig. 2. The upper half of Fig. 2 shows a ceramic plate which is forward-scattering light in a narrow pattern along the propagation direction of the incident light; the lower half of Fig. 2 shows the same plate scattering light in a broad pattern about the incident light propagation direction. The gap between the two electrode regions is the optically active area of the ceramic plate. If the ceramic polar axis in this region is parallel to the incident light propagation direction, light scattering similar to that shown in the upper half of Fig. 2 occurs. If the ceramic polar axis in the optically active region is normal to the light propagation direction, a scattering pattern similar to that in the lower half of Fig. 2 is obtained. Therefore, by switching the direction of the ceramic polar axis, one can change the amount of light incident on a detector of fixed aperture.[4,6] Light scattering phenomena in ferroelectric ceramics are discussed in considerable detail by Albers and Kaplit[13] in this volume.

ELECTROOPTIC PROPERTIES OF THE MODIFIED CERAMICS PBZT, PSZT, AND PLZT

GENERAL BEHAVIOR

A large number of intrinsic and extrinsic parameters play an important role in

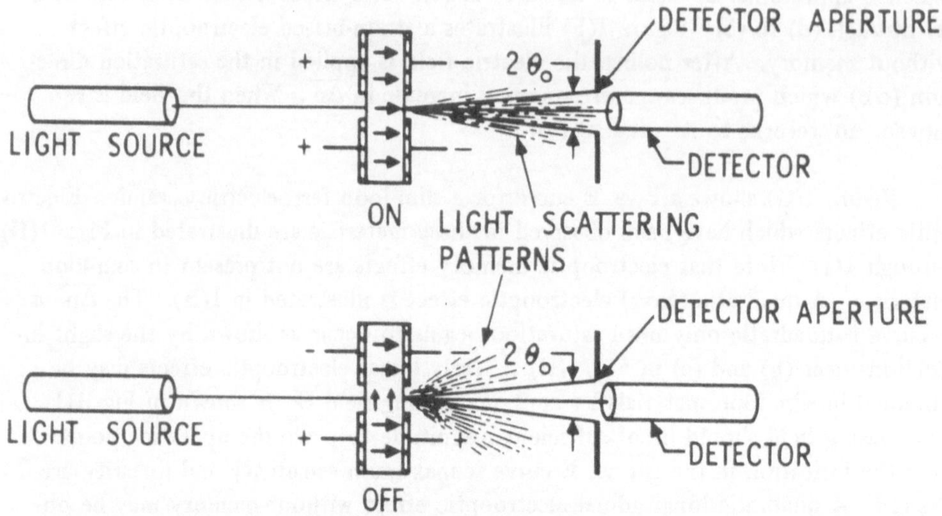

Fig. 2. Schematic illustration of electrically controlled scattering in a coarse-grained, Bi doped $Pb(Zr,Ti)O_3$ ceramic plate. Electrode configurations, switching polarities, and light scattering patterns are illustrated for two polarization states in the optically active area between the electrodes.

the magnitude and even the type of electrooptic effect observed in ceramic materials. Of the intrinsic parameters, composition (including doping and modification) and grain size are among the most easily measured, and effects of these parameters on various ceramic properties are discussed. Other important intrinsic parameters are internal stress, grain boundary structure, domain pinning points, etc. The exact role of these parameters is difficult to assess, but they affect the light scattering and switching behavior of ceramics in general, and internal stress modifies the intrinsic birefringence of domains in square-loop (Fig. 1(A)) ceramic material. The latter parameters will not be discussed further except to say that they and the grain size all depend on the exact process by which the ceramic is fabricated.[1-3,9] Extrinsic parameters of importance to the electrooptic effects in ceramics are light wavelength, temperature, and applied electric or stress fields. Effects of the latter two parameters are discussed in the second part of this section. The remainder of this part is devoted to a general discussion of the effects of composition, grain size, light wavelength, and temperature on electrooptic properties.

COMPOSITION DEPENDENT PROPERTIES

Types of Electrooptic Properties. Modification of the basic $Pb(Zr,Ti)O_3$ system with Ba, Sn, or La determines first of all whether the ceramics will exhibit the square-loop or the slim-loop hysteresis behavior. The most thorough work has been carried out on La modified ceramics because of their high transparency and variety of electrooptic effects. This composition dependence of the electrooptic properties of the PLZT system is indicated in the room temperature phase diagram, Fig. 3. Points on the figure represent the compositions studied. This study permitted mapping of compositional areas on the phase diagram in which the various electrooptic effects shown in Fig. 1 are known to occur. For example, the area ABC includes compositions which have electrooptic memory characteristics (Fig. 1(B)); area BDEC includes compositions with high coercivity which have linear electrooptic characteristics at saturation remanence (Fig. 1(D)); and area EFGH includes compositions which have slim-loop quadratic electrooptic characteristics (Fig. 1(H)). Comparison of electrooptic properties measured for compositions in Fig. 3 has proved helpful in predicting properties of other compositions in the figure.

Magnitude of Electrooptic Effects. Increasing the Ba, Sn, or La content lowers the temperature of the ferroelectric-paraelectric phase boundary in a composition-temperature phase diagram. This lowers the Curie point of ferroelectric compositions with constant Zr/Ti ratio. This effect is illustrated for the PLZT system in Fig. 4. A lowered Curie point affects both the intrinsic birefringence of domains in square-loop ceramics and the amount of electric field induced birefringence in slim-loop ceramics. The very existence of slim-loop behavior, indicated by the area EFGH of Fig. 3, is also dependent on the La content of the modified ceramic. Reduction of the Curie point has important effects on the electrical properties of PLZT as well.[9] Increasing Ba, Sn, or La content is accompanied by a corresponding reduction in coercivity, and, near the ferroelectric-paraelectric phase boundary, a reduction in remanent polarization. In applications of the memory mode of operation, reduction in coercivity is usually highly desirable because memory states can be changed with lower switching voltages.

GRAIN SIZE DEPENDENT PROPERTIES

Light Scattering. Light depolarizing by multiple scattering is significantly reduced in the Ba, Sn, and La modified ceramics when compared to either Bi or La doped $Pb(Zr,Ti)0_3$. In the unpoled state, multiple light scattering is even found to decrease with increasing grain size up to 10 μm or more. This behavior is in contrast to Bi doped $Pb(Zr,Ti)0_3$ whose primary electrooptic effect for grain sizes greater than 2 μm is an electrically controlled scattering.[4,6] It has been observed, however, as average grain size increases in the modified materials, the amount of multiple scattering in both square-loop and slim-loop ceramics becomes increasingly dependent on the magnitude of electric field and polarization.

Birefringence. The birefringence $\overline{\Delta n}\,(P_R,0)$ of all the modified ceramics which

References pp. 194-196

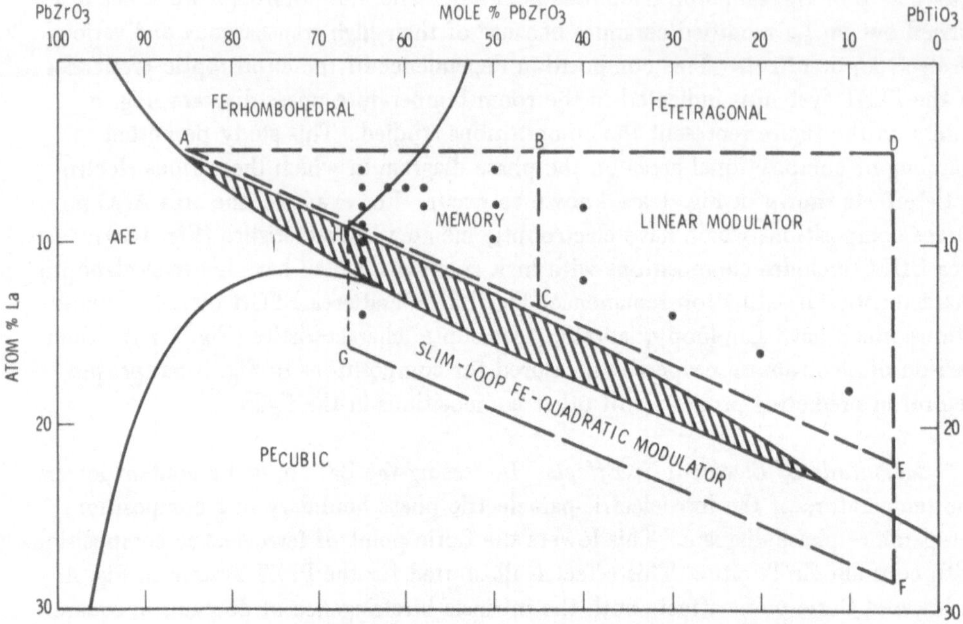

Fig. 3. Room temperature phase diagram of the PLZT system showing compositional areas particularly applicable to electrooptic memory (ABC), linear modulator (BDEC), and slim-loop ferroelectric quadratic modulator (EFGH) behavior. The shaded region represents mixed phases.

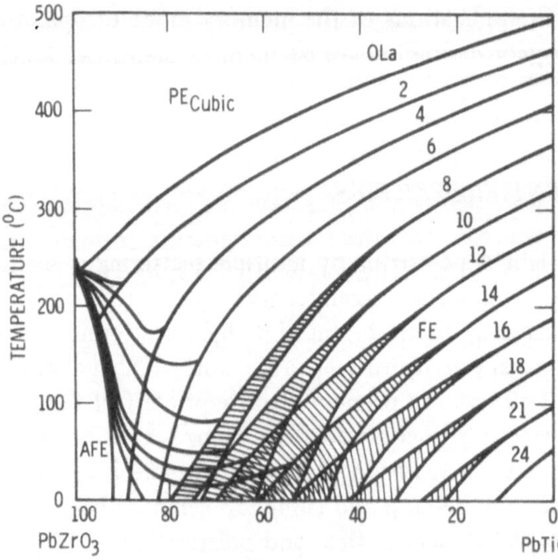

Fig. 4. Phase diagram of the PLZT system showing various constant La concentration levels. The shaded regions represent mixed phases.

have been studied increases with increasing grain size. A possible explanation may be the increased domain alignment allowed at P_R in the larger grain size materials. This explanation is supported by the fact that P_R for a given modified composition also increases with increasing grain size. The effect of grain size on $\overline{\Delta n}(P_R,0)$ for PLZT 8/65/35 (8 atom % La; 65/35 Zr/Ti ratio) is shown in Fig. 5. As grain size increases from 2 to 3 μm, $\overline{\Delta n}(P_R,0)$ increases from -0.0044 to about -0.008 and then remains nearly constant as grain size is further increased to 10 μm. The smaller value of $\overline{\Delta n}(P_R,0)$ of PLZT 8/65/35 as compared to PLZT 2/65/35[5] may be explained, in part, by the decreased Curie points of materials with increased La content.

Electrooptic Memory Effect. The range of $\overline{\Delta n}(P_r,0)$ as a function of P_r for electrooptic memory materials increases with increasing grain size. This effect is discussed in more detail later.

WAVELENGTH DEPENDENT PROPERTIES

Light Transmission. The optical transmittances vs. wavelength λ of Bi and La doped and Ba, Sn, and La modified $Pb(Zr,Ti)O_3$ have certain similarities.[4,5,9] For

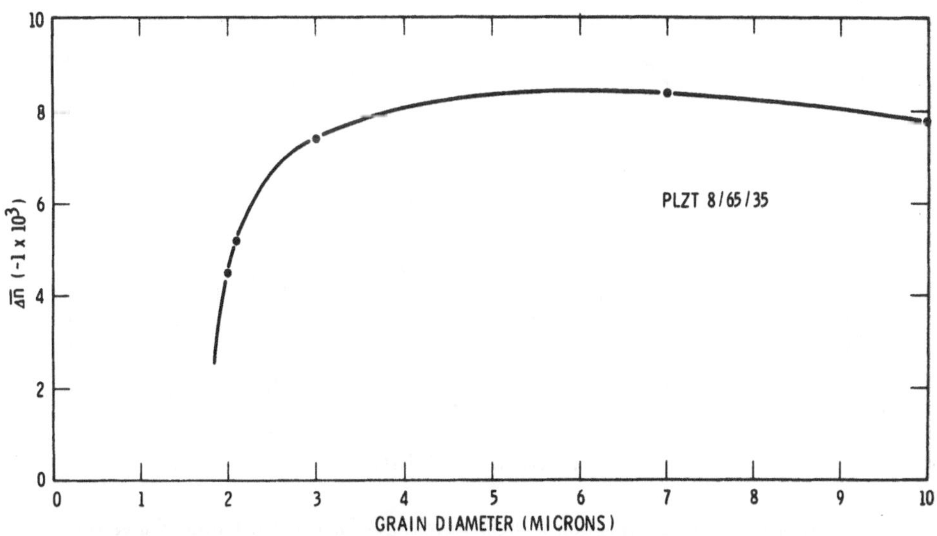

Fig. 5. Effective birefringence $\overline{\Delta n}(P_R,0)$ vs. grain size for PLZT 8/65/35.

all these materials, the transmittances rise above the fundamental absorption edge at between .3 and .4 μm and increase with increasing λ to some maximum value in the infrared between 4 and 6 μm. The transmittances then decrease to zero at 12 to 14 μm. The transmittance behavior of PBZT and PSZT is quite similar to that of La doped $Pb(Zr,Ti)O_3$.[5] The transmittance of PLZT, however, is much higher. Transmission losses of polished plates of PLZT 8/65/35 up to 0.25 mm thick consist primarily of the reflection losses between 0.5 and 6 μm. This is illustrated by the curves of Fig. 6. Curve (a) shows transmittance vs. λ for the range 0 to 16 μm, curve (b) shows transmittance vs. λ from .3 to .7 μm. The transmittance increases abruptly between .37 and .4 μm and more gradually between .4 and .55 μm, where it reaches its maximum value. In thicker plates (up to several mm thick) the transmittance increases gradually with increasing λ throughout the visible and reaches a maximum value in the infrared.

The reduction in transmittance with decreasing λ in the visible is due almost entirely to scattering in Bi and La doped $Pb(Zr,Ti)O_3$.[4,5] Although scattering measurements have not been made on PBZT, PSZT, and PLZT, it is believed that losses in these materials at the lower end of the visible spectrum are also due primarily to scattering.

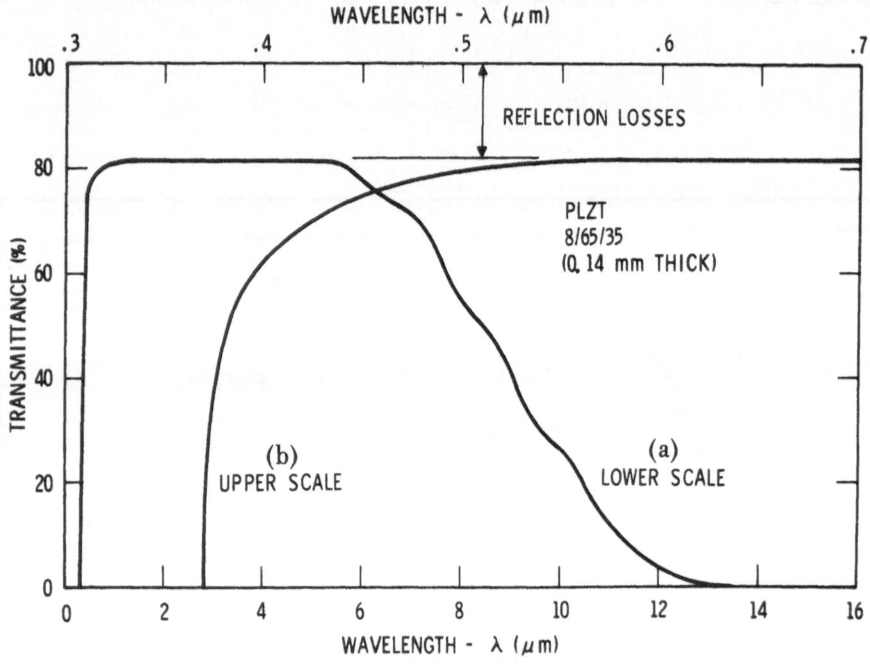

Fig. 6. Transmittance as a function of light wavelength for PLZT 8/65/35. Curve (a) - lower scale - shows transmittance from 0 to 16 μm; curve (b) - upper scale - shows response from 0.3 to 0.7 μm on an expanded scale. Front surface reflection losses are approximately 18%.

Refractive Index. The refractive indexes of modified ceramics decrease with increasing λ. Figure 7 shows the refractive index vs. λ for slim-loop PBZT 29/71/29 (29 atom % Ba; 71/29 Zr/Ti ratio), and PLZT 11/65/35 as well as for unpoled (isotropic) square-loop PLZT 2/65/35. Variation of La content appears to have little effect on the refractive index of compositions with constant Zr/Ti ratio, but variation of that ratio has a significant effect. This is due to the difference between the refractive indexes of $PbZrO_3$[14] and $PbTiO_3$[15,16] indicated in Fig. 7.

Birefringence. Birefringence as well as the refractive indexes themselves are dispersive in the modified ceramics as illustrated in Fig. 8. Curves are shown of $\overline{\Delta n}(P_R,0)$ for PLZT 2/65/35 and PLZT 12/40/60 and of $\overline{\Delta n}(E = 10\ kV/cm)$ for PLZT 11/65/35. The shape of the $\overline{\Delta n}$ vs. λ curves seems to be essentially independent of La content. This is emphasized by comparing $\overline{\Delta n} \times 10^{-1}$ for PLZT 2/65/35 with $\overline{\Delta n} \times 4$ for PLZT 11/65/35. However, the magnitude of $\overline{\Delta n}$ at a given wavelength depends strongly on both La content and Zr/Ti ratio.

TEMPERATURE DEPENDENCE OF THE BIREFRINGENCE

The temperature dependence of the birefringence of $Pb(Zr,Ti)O_3$ single crystals was first studied by Fushimi and Ikeda.[17] They found that birefringence generally

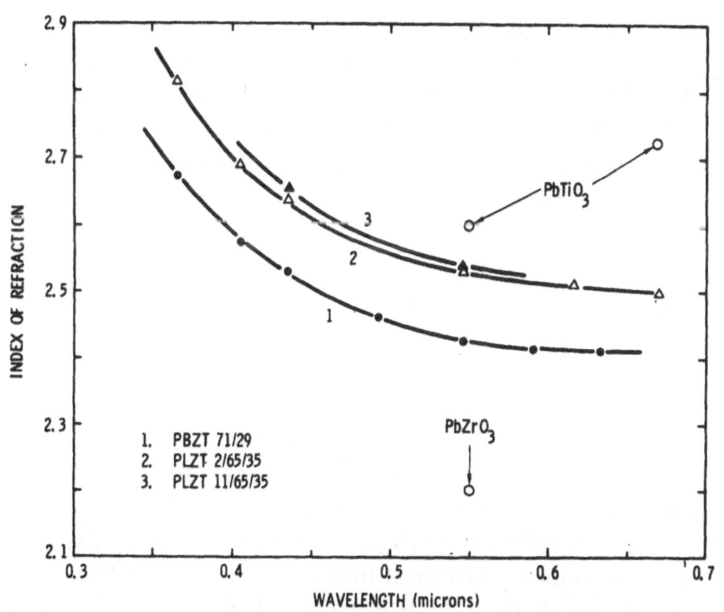

Fig. 7. Refractive index vs. wavelength for PBZT 29/71/29, PLZT 2/65/35 and PLZT 11/65/35. Points also indicate reported values for single crystal $PbTiO_3$ and $PbZrO_3$. (See Refs. 14-16.)

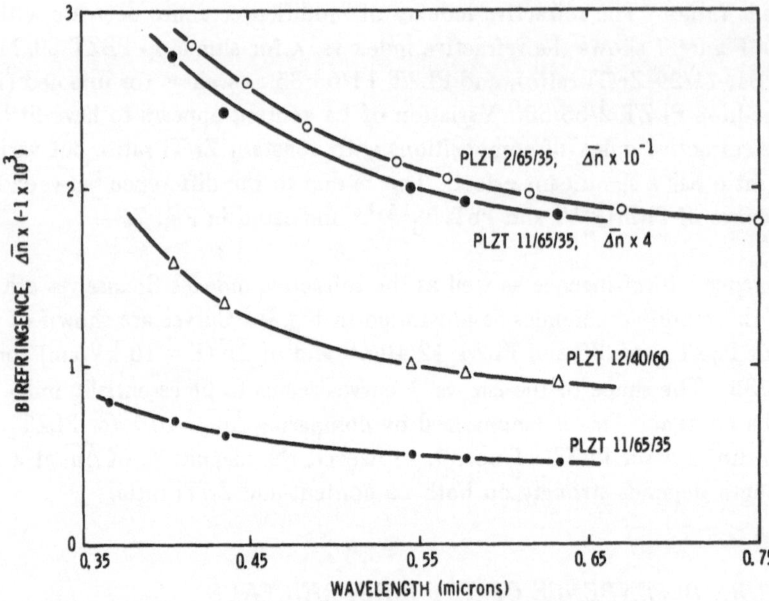

Fig. 8. Effective birefringence $\overline{\Delta n}$ $(P_R,0)$ vs. wavelength for PLZT 2/65/35 and PLZT 12/40/60. Also, $\overline{\Delta n}$ (E = 10 kV/cm) vs. wavelength for PLZT 11/65/35.

decreases slowly with increasing temperature until the more rapid decrease to zero at the Curie point. Similar behavior has been observed in coarse-grained Bi and La doped $Pb(Zr,Ti)O_3$ ceramics[8] and Ba modified ceramics.[8] In fine-grained ceramics, however, the rate of change of $\overline{\Delta n}$ with increasing temperature is usually somewhat larger. The $\overline{\Delta n}$ $(P_R,0)$ vs. temperature behavior is illustrated for PBZT 20/80/20 ceramic of various grain sizes in Fig. 9. All samples were initially poled to saturation remanence at room temperature.

ELECTROOPTIC EFFECTS

ELECTROOPTIC MEMORY EFFECT

PLZT Ceramics. The electrooptic memory effect occurs in compositions indicated in area ABC of Fig. 3. This memory effect is shown for PLZT 8/65/35 of 2 μm average grain size in Fig. 10, which is a plot of $\overline{\Delta n}$ $(P_r,0)$ as a function of normalized remanent polarization $P = P_r/P_R$. The hysteresis of the $\overline{\Delta n}$ curve is considerably larger and more asymmetrical than that observed for PLZT 2/65/35.[5] The asymmetry of the $\overline{\Delta n}$ $(P_r,0)$ vs. P characteristic always depends on the polarity of the original poling field, though the characteristic is repeatable within measurement accuracy over at least several switching cycles.

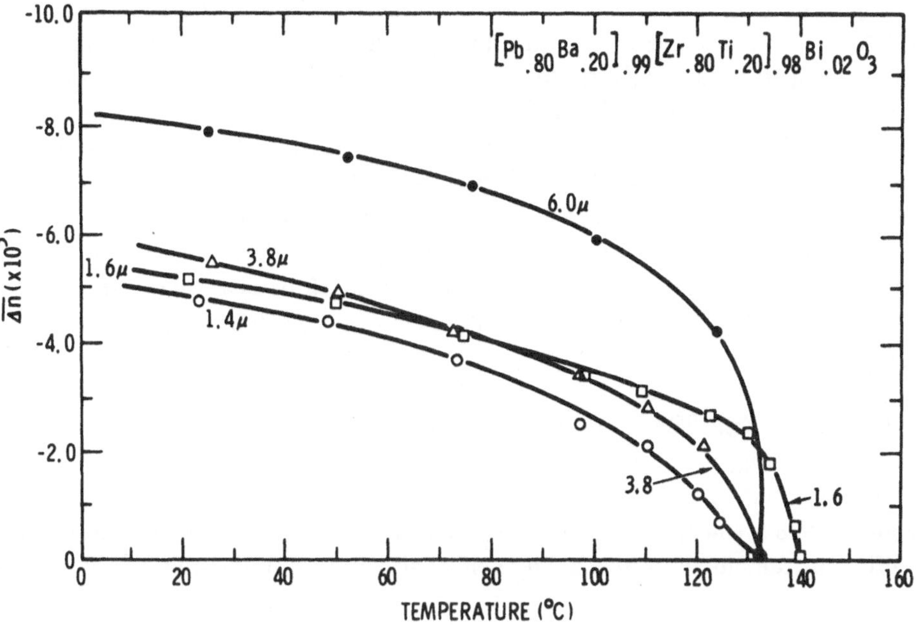

Fig. 9. Effective birefringence $\overline{\Delta n}\,(P_R,0)$ vs. temperature for 1.4, 1.6, 3.8, and 6 μm average grain size PBZT 20/80/20.

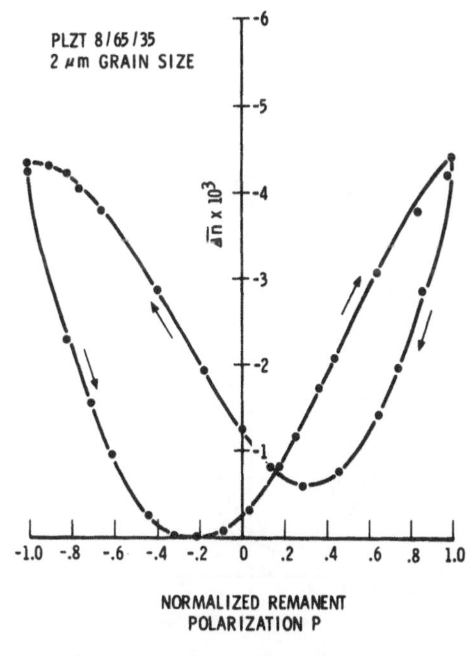

Fig. 10. Effective birefringence $\overline{\Delta n}\,(P_r,0)$ vs. P for 2 μm average grain size PLZT 8/65/35. Arrows indicate switching directions. The ceramic was originally poled in the positive direction.

Figure 11 shows the effect of grain size on curves of $\overline{\Delta n}\,(P_r,0)$ vs. P for PLZT 8/65/35. As grain size increases from 2 to 3 μm, $\overline{\Delta n}\,(P_R,0)$ increases from -0.0044 to -0.008, and the range of $\overline{\Delta n}\,(P_r,0)$ increases from 0.0044 to about 0.007. Only the 2 μm grain size ceramic achieves $\overline{\Delta n}\,(P_r,0) = 0$ for some state of remanent polarization. This fine-grained material is the first ever observed to be optically isotropic on a macroscopic scale at some P near zero. The range of $\overline{\Delta n}\,(P_r,0)$ for this ceramic is therefore 100% of the value of $\overline{\Delta n}\,(P_R,0)$, whereas for PLZT 2/65/35[5] the range is only about 45% of the value of $\overline{\Delta n}\,(P_R,0)$. However, the magnitude of the change in $\overline{\Delta n}\,(P_r,0)$ is still larger in PLZT 2/65/35 because $\overline{\Delta n}\,(P_R,0)$ at $\lambda = 0.633$ μm is -0.0044 for PLZT 8/65/35 as compared to -0.012 for PLZT 2/65/35.

PBZT Ceramics. The electrooptic memory effect in PBZT 20/80/20 ceramic is illustrated in Fig. 12. The behavior is similar in many respects to that of PLZT 8/65/35. As grain size increases from 1.4 to 6 μm, $\overline{\Delta n}\,(P_R,0)$ increases from -0.003 to about -0.0083, and the range of $\overline{\Delta n}\,(P_r,0)$ increases from about 0.0015 to 0.0038. The effect of the initial positive poling field is also evident in the asymmetry of the curves.

PSZT Ceramics. Electrooptic memory effects in two PSZT compositions, each with 50 atom % Sn in the B(Sn,Zr,Ti) site of the ABO_3 lattice, are shown in Fig. 13.

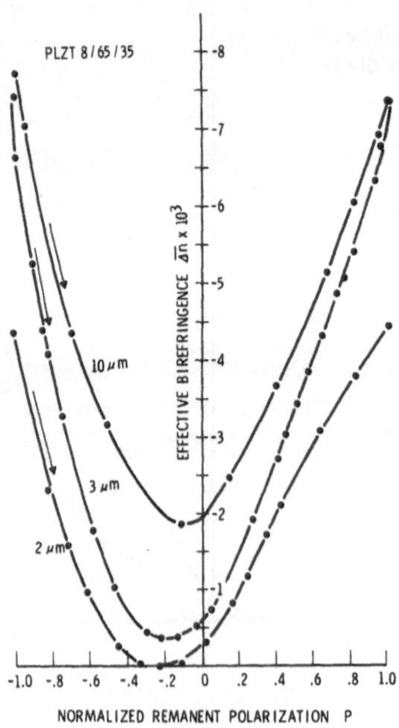

Fig. 11. Effective birefringence $\overline{\Delta n}\,(P_r,0)$ vs. P for 2, 3, and 10 μm average grain size PLZT 8/65/35. The ceramics were originally poled in the positive direction.

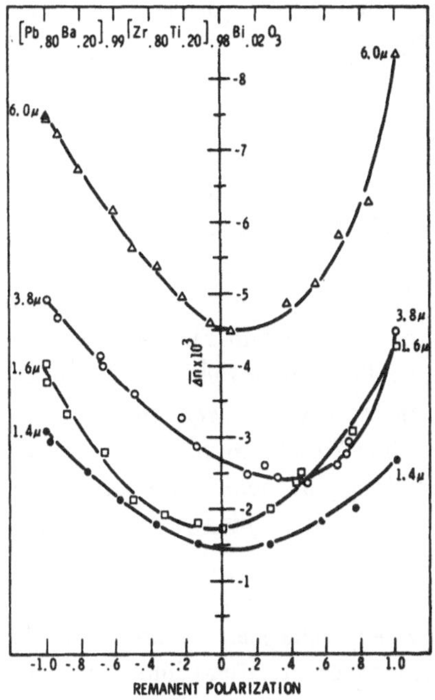

Fig. 12. Effective birefringence $\overline{\Delta n}\,(P_r,0)$ vs. P for 1.4, 1.6, 3.8, and 6 μm average grain size PBZT 20/80/20. The ceramics were originally poled in the positive direction.

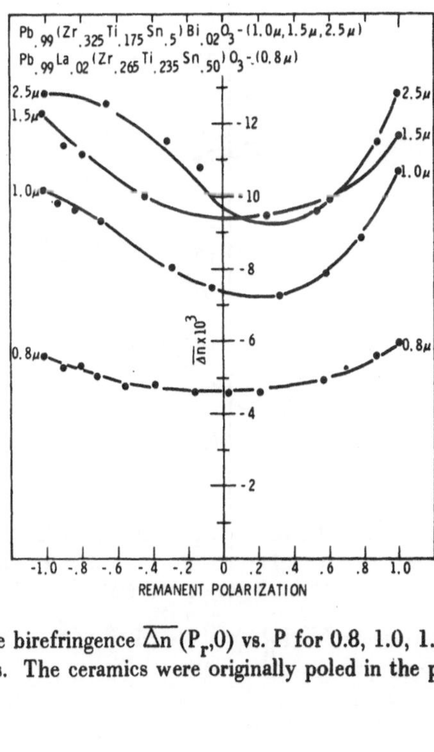

Fig. 13. Effective birefringence $\overline{\Delta n}\,(P_r,0)$ vs. P for 0.8, 1.0, 1.5, and 2.5 μm average grain size PSZT compositions. The ceramics were originally poled in the positive direction.

References pp. 194-196

Again the grain size dependence of $\overline{\Delta n}(P_R,0)$ and the range of $\overline{\Delta n}(P_r,0)$ are characteristic of the modified solid solutions.

LINEAR ELECTROOPTIC EFFECT

High coercivity tetragonal PLZT compositions in the area BDEC of the phase diagram, Fig. 3, have linear electrooptic characteristics after they have been poled. These are the first ceramics known to exhibit a linear electrooptic effect. Examples are shown for three PLZT compositions in Fig. 14. These materials were poled to positive saturation remanence P_R using a field $E = 2E_c (\approx 30$ kV/cm) prior to measuring $\overline{\Delta n}$. The measured $\overline{\Delta n}$ vs. E curves were then found to remain linear for positive applied fields exceeding 20 kV/cm and for negative applied fields which do not exceed the domain switching threshold. The curves also showed very little hysteresis below the switching threshold. This threshold is indicated by the abrupt change of slope of the curves.

A convenient and physically significant comparison of the linear electrooptic effect in PLZT ceramics with those of several linear electrooptic crystals may be made by examining the respective transverse electrooptic coefficients r_c defined in the equation,[18,19]

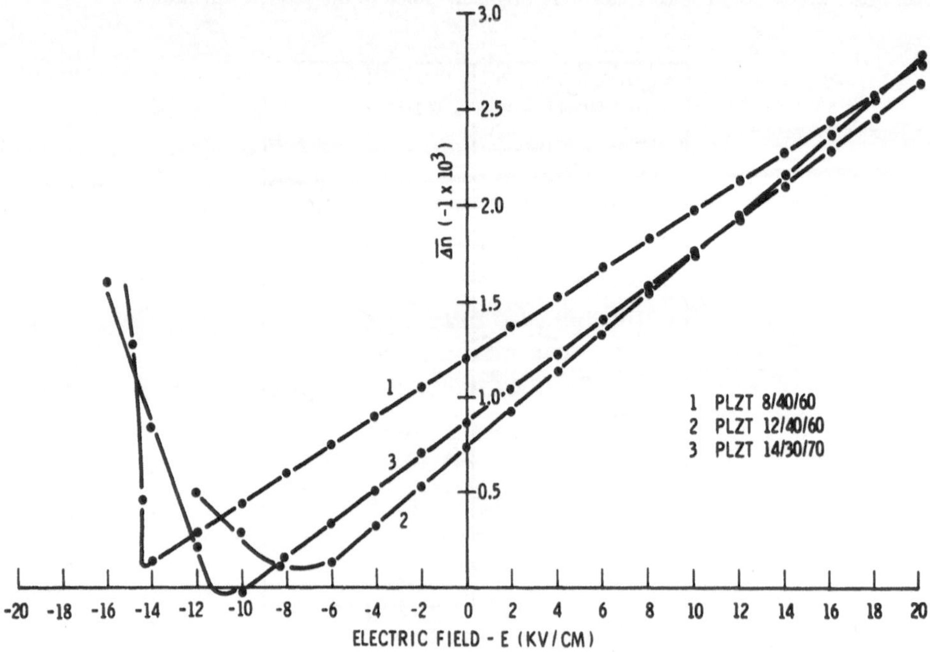

Fig. 14. Effective birefringence $\overline{\Delta n}(P_R,E)$ vs. electric field E for three PLZT high coercivity ceramics exhibiting linear electrooptic effects.

$$\overline{\Delta n} = -(1/2) n_1^3 r_c E_3 , \quad (1)$$

where n_1 is the refractive index normal to the applied field E_3. From Eq. (1) it can be seen that r_c is simply a measure of the slopes of the curves in Fig. 14. Values of r_c for several electrooptic crystals and two of the PLZT ceramics of Fig. 14 are listed in Table I.

SLIM-LOOP QUADRATIC ELECTROOPTIC EFFECT

PBZT Ceramics. The quadratic electrooptic effect characteristic of slim-loop ferroelectric ceramics was first observed in PBZT 29/71/29.[20] The addition of 29 atom % Ba in the A(Pb,Ba) site of the ABO_3 lattice reduces the Curie point to slightly below 25°C. This composition is, therefore, paraelectric at room temperature. When an electric field is applied, the material becomes ferroelectric with rhombohedral symmetry. In this field-enforced ferroelectric phase, the ceramic is birefringent, and its birefringence is proportional to E^2 for E less than E_S. At E_S saturation effects begin to occur. The $\overline{\Delta n}$ vs. E curve for PBZT 29/71/29 is shown in Fig. 15. For this material E_S is about 14 kV/cm.

TABLE I

Electrooptic Coefficient r_c

Material	$r_c \times 10^{10}$ m/V
$LiNbO_3$	0.16[a]
$LiTaO_3$	0.22[a]
$Ba_2NaNb_5O_{15}$	0.38[a]
KD_2PO_4	0.53[a,b]
PLZT 8/40/60	1.02
PLZT 14/30/70	1.12
$(Sr_{.5}Ba_{.5})Nb_2O_6$	2.10[c]
$(Sr_{.75}Ba_{.25})Nb_2O_6$	14.00[c]

[a] *J. T. Milek and S. J. Welles, Linear Electrooptic Modulator Materials, Electronic Properties Information Center, Hughes Aircraft Co., Culver City, California, 1970.*

[b] $r_c = 2r_{63}$ *measured at* $\lambda = 0.546$ *μm.*

[c] *E. G. Spencer, P. V. Lenzo, and A. A. Ballman, "Dielectric Materials for Electrooptic, Elastooptic, and Ultrasonic Device Applications," Proc. IEEE, Vol. 55, 2074-2108, Dec. 1967.*

References pp. 194-196

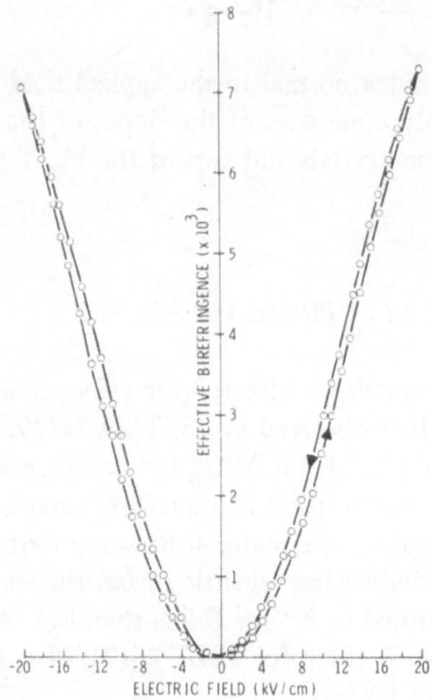

Fig. 15. Effective birefringence $\overline{\Delta n}$ (E) vs. electric field E for the slim-loop ferroelectric PBZT 29/71/29.

PLZT Ceramics. The $\overline{\Delta n}$ vs. E characteristics for three of the PLZT slim-loop compositions are shown in Fig. 16. The dramatic change in electrooptic sensitivity of these compositions as the atom % La is varied from 9 to 11 is illustrated in Fig. 17 which shows $\overline{\Delta n}$ plotted as a function of E^2.

A convenient method of comparing the quadratic electrooptic effects in slim-loop PBZT and PLZT materials with those of other quadratic electrooptic materials is by means of the transverse quadratic electrooptic coefficients R and g. These coefficients are determined from the relations

$$\overline{\Delta n} = -(½)n_1^3 R E_3^2 \tag{2}$$

and

$$g \approx R/(\epsilon_o \epsilon_{33}^T)^2 . \tag{3}$$

It is apparent that R may be calculated directly from Eq. (2) and the curves of Fig. 17. It is then possible to calculate g using Eq. (3) provided the free static dielectric constants $\epsilon_o \epsilon_{33}^T$ of the PBZT and PLZT compositions are known. The results of such calculations for the materials of Figs. 15, 16, and 17 are given in Table II along with the result for single crystal KTN.

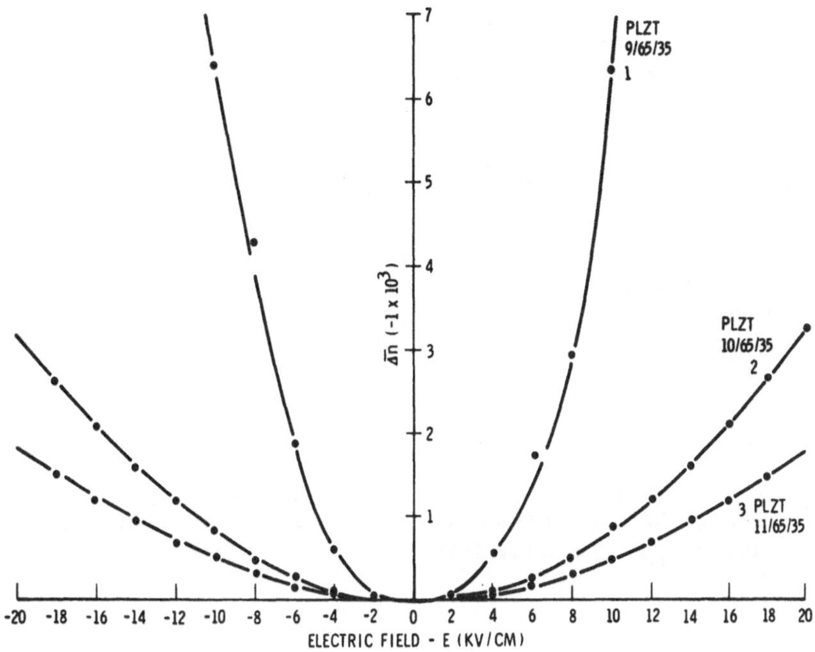

Fig. 16. Effective birefringence $\overline{\Delta n}$ (E) vs. electric field E for three slim-loop ferroelectric PLZT ceramics.

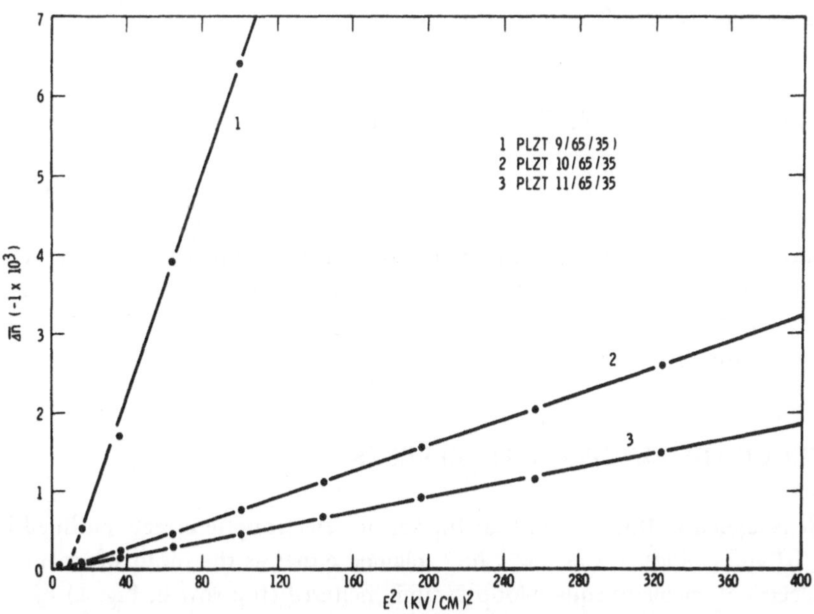

Fig. 17. Effective birefringence $\overline{\Delta n}$ (E) vs. electric field squared E^2 for the compositions of Fig. 16.

TABLE II

Electrooptic R and g Coefficients

Material	$R \times 10^{16} (m^2/V^2)$	$g(m^4/c^2)$
PBZT 29/71/29[a]	3.20	0.016
PLZT 9/65/35	9.12	0.018
PLZT 10/65/35	1.07	0.010
PLZT 11/65/35	0.60	0.014
$KTa_{0.65}Nb_{0.35}O_3$	5.3[b]	0.16[b]

[a] *Reference 20.*

[b] *Calculated from data given by P. H. Smakula and P. C. Claspy, "The Electro-Optic Effect in* $LiNbO_3$ *and KTN," Trans. Met. Soc. AIME, Vol. 239, pp. 421-424, Mar. 1967.*

Values of g for the PLZT compositions in Table II are in good agreement with those of several PBZT compositions and that of a single PLZT composition already reported.[20] The value of R for PLZT 9/65/35 is 9.1 compared to a maximum value of 3.2 for the best PBZT composition in Ref. 20.

SUMMARY OF THE ELECTROOPTIC EFFECTS

A summary of the various electrooptic effects found in ferroelectric ceramics is presented in Table III. The intent of this Table is to correlate all of the ceramic electrooptic effects reported in the literature with their schematic representation in Fig. 1 and with the ceramic materials in which the effects have been observed. Also indicated in the Table are effects which are possible but which have not yet been reported in the literature.

ORIGIN OF THE ELECTROOPTIC EFFECTS

It is argued in this section that the various electrooptic effects outlined in Table III and in Figs. 1 and 2 can be explained either as the result of orientation of ferroelectric domains in square-loop ceramic material (top row in Fig. 1) or as the result of variations in the amount of ferroelectric distortion in slim-loop material (bottom row in Fig. 1). For purposes of the discussion, it will be assumed that

TABLE III

Summary of Electrooptic Effects in Ferroelectric Ceramics

Material	Transverse Electrooptic Effects: Memory[a]	Transverse Electrooptic Effects: Quadratic[b]	Transverse Electrooptic Effects: Linear[c]	Longitudinal E/O Effects: Strain-Biased Memory	Longitudinal E/O Effects: Strain-Biased Non-Memory	Electrooptic Scattering Effect
(1) Coarse-grained Bi Doped PZT						Fig. 2[d] (3,4,6,13)[e]
(2) Fine-grained Bi Doped PZT	Fig. 1(B)[d] (3,4,6)[e]	Fig. 1(C)[d] (4,6)[e]				
(3) Fine-grained La Doped PZT	Fig. 1(B) (5,6,7,8,10)	Fig. 1(C) (5,6,7,8)		Fig. 1(E) (11,12)	Fig. 1(F) X[f]	
(4) Ferroelectric PBZT	Fig. 1(B) (7,8)	Fig. 1(C) (7,8)		Fig. 1(E) X	Fig. 1(F) X	
(5) Slim-loop PBZT		Fig. 1(H) (8,20)	Fig. 1(I) X		Fig. 1(J) X	
(6) Ferroelectric PSZT	Fig. 1(B) (8)	Fig. 1(C) (8)		Fig. 1(E) X	Fig. 1(F) X	
(7) Rhombohedral PLZT	Fig. 1(B) (9)	Fig. 1(C) (9)		Fig. 1(E) (11,12)	Fig. 1(F) X	
(8) Slim-loop PLZT		Fig. 1(H) (9,20)	Fig. 1(I) X		Fig. 1(J) X	
(9) Tetragonal PLZT			Fig. 1(D) (9)			

[a] $\overline{\Delta n}(P,E) = \overline{\Delta n}(P_r,0)$

[b] $\overline{\Delta n}(P,E) = \overline{\Delta n}(P_R,0) + \lambda KE^2$ *(Refer to Fig. 1(C))*

$\overline{\Delta n}(P,E) = \lambda KE^2$ *(Refer to Fig. 1(H))*

[c] $\overline{\Delta n}(P,E) = \overline{\Delta n}(P_R,0) \pm \lambda KE$ *(Refer to Fig. 1(D))*

[d] *Figure which gives a schematic representation of the effect.*

[e] *Numbers in parentheses refer to references.*

[f] *Effects possible but either not yet observed or not yet reported.*

temperature, wavelength, and strain are held constant. This is not a presumption that temperature pulses, wavelength shifts or strain waves may not provide interesting optical effects in ceramics but rather an attempt to limit discussion to the electrooptic effect alone. For the same reason of simplification, optical effects caused by variations in composition, grain size, or other intrinsic parameters will not be considered.

TRANSVERSE ELECTROOPTIC MEMORY EFFECT

When the birefringence of a ceramic plate is measured, one actually is measuring an effective birefringence $\overline{\Delta n}$ which is both proportional to the intrinsic birefringence Δn of a domain and dependent on an average over all domain orientations in the ceramic. If E defines the z direction, the orientation of the polarization P of a single domain can be defined in spherical coordinates by the angle θ measured from the z axis and the azimuthal angle φ measured from the x axis. The x direction will in turn be defined by the incident light direction so that, in particular, the transverse electrooptic effect is under discussion. The effective birefringence of the domain measured with respect to the electric field direction is then given by

$$\Delta n(\theta,\varphi) = \Delta n(\cos^2 \theta - \sin^2 \theta \sin^2 \varphi) \ . \tag{4}$$

If azimuthal symmetry is assumed, $\langle \sin^2 \varphi \rangle = ½$, and the θ dependence of the effective birefringence is

$$\Delta n(\theta) = \Delta n(3/2 \cos^2 \theta - ½) \ . \tag{5}$$

The total $\overline{\Delta n}$ is now given by multiplying $\Delta n(\theta)$ by the volume of domains with P along θ, integrating over θ, and dividing by the total volume. The resulting integral can be worked out simply in two limiting cases: Isotropic domain orientation (thermally depoled ceramics), and maximum possible alignment of the domains along E. In the former case $\overline{\Delta n} = 0$, and in the latter $\overline{\Delta n} = (2/\pi)\Delta n$ for domains of rhombohedral symmetry and $\overline{\Delta n} = (\sqrt{3}/\pi)\Delta n$ for tetragonal symmetry.[21]

The actual $\overline{\Delta n}$ achieved in ceramics at P_R is found to fall considerably below the maximum values given above because of incomplete domain alignment. The only measurement on which a comparison can be based was made by Fushimi and Ikeda[17] on a single crystal of rhombohedral $Pb(Zr_{0.85}Ti_{0.15})O_3$ or PZT 85/15. They obtained $\Delta n = -0.042$ which can be compared with $\overline{\Delta n} = -0.011$ extrapolated for PZT 85/15 from data in Ref. 8. In this case $\overline{\Delta n} = 0.26\ \Delta n$ rather than the maximum possible value of $\overline{\Delta n} \approx 0.64\ \Delta n$.

According to Uchida and Ikeda,[21] the component of the intrinsic ferroelectric strain perpendicular to an applied field is given by

$$S_1 = -\tfrac{1}{2}\,\delta(3/2\cos^2\theta - \tfrac{1}{2}) \tag{6}$$

where θ is the angle between P and E, and where $(3/2)\delta = (c/a) - 1$ for tetragonal ferroelectrics and $\delta = (\pi/2) - \alpha$ for rhombohedral ferroelectrics, with α the rhombohedral angle. The effective strain $\overline{S}_1$ measured in a ceramic is obtained from Eq. (6) in the same way $\overline{\Delta n}$ is obtained from Eq. (5). If measurements are made with field applied, the strain $\overline{S}_1$ will include a piezoelectric contribution and the birefringence $\overline{\Delta n}$ an electrooptic contribution, but it will be assumed that such contributions are small compared to those of orienting domains. In that case, since Eq. (6) has the same form as Eq. (5), $\overline{S}_1$ should have the same form as $\overline{\Delta n}$. This fact has been noted experimentally[22] and is indicated schematically in rows B and C of Fig. 18 for square-loop ceramics. The resulting linear relation between $\overline{\Delta n}$ and $\overline{S}_1$ is shown adjacent to these curves in order to emphasize the fact that, though the intrinsic properties Δn and δ need bear no relationship to one another, the measured effective quantities $\overline{\Delta n}$ and $\overline{S}_1$ are related through a common dependence on domain orientation processes. These same arguments show that the reason $\overline{\Delta n}$ is not uniquely related to P is because the ceramic polarization is given by an average over $\cos\theta$ rather than over the $(3/2)(\cos^2\theta - 1/3)$ appearing in Eq. (5).[21]

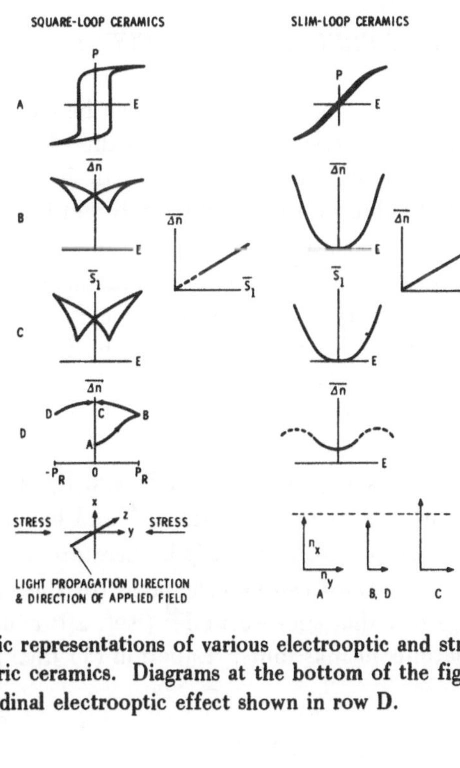

Fig. 18. Schematic representations of various electrooptic and strain effects in square-loop and slim-loop ferroelectric ceramics. Diagrams at the bottom of the figure are used to explain the strain-biased longitudinal electrooptic effect shown in row D.

The slope of the curve of $\overline{\Delta n}$ vs. $\overline{S}_1$ should be

$$\overline{\Delta n}/\overline{S}_1 = -2\Delta n/\delta \qquad (7)$$

from Eqs. (5) and (6), and the two sides of this equation can be separately calculated. The birefringence of PZT 85/15 has already been noted[17] to be $\Delta n = -0.042$, and data by Sawaguchi[23] indicate $\delta = 15' = 4.4 \times 10^{-3}$ radians for PZT 85/15. The right hand side of Eq. (7) is thus equal to 19. Measurements on ceramic PZT 85/15 have not been carried out, but Berlincourt[22] has deduced $\overline{\Delta n}/\overline{S}_1$ for PZT 80/20 from curves for $\overline{\Delta n}$ vs. E and $\overline{S}_1$ vs. E. He obtained $\overline{\Delta n}/\overline{S}_1 = 18$. This excellent agreement is a convincing argument in favor of explaining the electrooptic effects in square-loop ceramics as a domain orientation process.

As an aside, it should be noted that Eq. (7) also presents a means of determining the intrinsic birefringence of a ferroelectric material available in ceramic rather than single crystal form. The left side of Eq. (7) consists only of measurements of ceramic parameters, and δ can be obtained from x-ray work on ceramic material. Even though the determination of $\overline{\Delta n}/\overline{S}_1$ is not a trivial measurement, the total effort can still be far less than that of obtaining a good single crystal of the material.

TRANSVERSE QUADRATIC ELECTROOPTIC EFFECT

Electrooptic effects in slim-loop ceramics can also be explained by means of Eq. (7). The increase in ferroelectric distortion (and therefore polarization) with increasing field in slim-loop material leads to a quadratic increase in both $\overline{\Delta n}$ and $\overline{S}_1$. This is indicated for slim-loop ceramics in rows B and C of Fig. 18. Shown adjacent to these curves is the resulting linear relation between $\overline{\Delta n}$ and $\overline{S}_1$. The reason $\overline{\Delta n} \propto E^2$ is because $\Delta n \propto P^2$ and $P \propto E$ for reasonably low fields.[20] At higher fields curves of $\overline{\Delta n}$ vs. E tend to saturate, but curves of $\overline{\Delta n}$ vs. P are truely quadratic. It must also be true that the domain distribution is reasonably constant as a function of E or else $\langle\cos^2\theta\rangle$ would vary with E and modify the observed quadratic dependence of $\overline{\Delta n}$ on E.

The strain $\overline{S}_1$ is also a quadratic function of P (and E) because $\delta \propto P^2$.[24] This relation has been known in $BaTiO_3$ for some time,[25] and has recently been found in $PbTiO_3$ as well.[26] The term $\Delta n/\delta$ in Eq. (7) is therefore a constant independent of E or P. A value of this constant can be calculated for the slim-loop ceramic PBZT 29/71/29 from the fact that $\Delta n = -0.11\ P^2$ (Ref. 20) and $\overline{S}_1 = -7.5 \times 10^{-3}\ P^2$ (Ref. 22), with both relations in mks units. Equation (7) then gives $-2\Delta n/\delta = 15$, about the same value as found for PZT 85/15. This similarity of values should not be surprising since the compositions themselves are rather similar.

LONGITUDINAL ELECTROOPTIC EFFECTS

If a stress is applied to the ceramic perpendicular to the direction of light propagation, the longitudinal electrooptic effect shown in Fig. 1 and in row D of Fig. 18 can be observed.[12] In view of the close relation between birefringence, strain, and domain orientation just developed for the case of zero stress, it is not surprising that a qualitative explanation of the strain-biased electrooptic effect can also be made in terms of domain orientation. The explanation will be made for the variation of $\overline{\Delta n}$ with P_r shown for square-loop ceramics in row D of Fig. 18, but the explanation can easily be extended to the other longitudinal electrooptic effects shown in Fig. 1. In the interest of clarity, the explanation will also be made for positive uniaxial materials ($\Delta n > 0$) rather than for the negative uniaxial materials found in the $Pb(Zr,Ti)O_3$ system.

At the bottom of Fig. 18 is shown a coordinate system useful to the discussion of the strain-biased electrooptic effect. A compressive stress is indicated along the y axis, but tensile stress along the same axis leads to a similar explanation and behavior. With a compressive stress along the y axis, the polarization of the individual domains in a thermally depoled ceramic tends to be oriented in the xz plane because of unit cell elongation along the direction of polarization. A birefringence can thus be observed in the ceramic with light propagating along z because more birefringent domains are oriented along x than y. This birefringence is noted at point A on the $\overline{\Delta n}$ vs. P_r curve, and the effect on the indexes n_x and n_y is given in exaggerated form in diagram A at the bottom of Fig. 18. (Note that both n_x and n_y arise from an average over domain orientations so that in relation to the ordinary (n_o) and extraordinary (n_e) indexes of an individual domain, it must be true that $n_e > n_x > n_y > n_o$ in this experiment.)

If an electric field is now impressed along z, domains will tend to switch from the xy plane to the z direction; but because of the stress along y, the polarization components of more domains will switch from y to z than from x to z. The effect on the indexes is shown in diagram B. The index n_x is reduced somewhat, but n_y is reduced much more. The resulting increase in birefringence at P_R is noted at point B on the $\overline{\Delta n}$ vs. P_r curve. If the electric field is now reversed to achieve $-P_R$, symmetry dictates that, disregarding hysteresis effects, the indexes in diagrams B and D and the birefringences at points B and D be the same. For intermediate values of P_r, however, the birefringence is increased. This behavior arises because during the switching process the polarization components of more domains orient along x than y due to the compressive stress. As noted in diagram C, both n_x and n_y increase during switching, but n_x increases more.

LIGHT SCATTERING EFFECTS

The effect of electrically controlled light scattering[4,13] shown in Fig. 2 can also be related to domain orientation, though the precise mechanism cannot be isolated with any certainty from data in Ref. 4. The first explanation of the scattering was given by Nettleton.[27] His theory traced the polarization dependence of the scattering to Rayleigh scattering from periodic groups of 180° domain walls whose orientation would depend on the polarization direction. Two further mechanisms can be inferred from data in Ref. 4. The data first of all show the increase in light scattering obtained by rotating P_R away from the light propagation direction. Qualitatively similar behavior arises either from increasing the sample thickness or decreasing the light wavelength. Scattering increases in the former case because of an increase in the number of scattering events and in the latter case because of an increase in the magnitude of the scattering at each event. In both cases, scattering is presumed to arise from index of refraction discontinuities at domain walls and grain boundaries - the only optically related feature other than the internal stress distribution which varies with P_R. The reason for the scattering increase with decreasing wavelength comes from the pronounced increase in Δn at shorter wavelengths. This evidence leads to two other mechanisms which can give polarization dependent scattering: Domains are elongated in the direction of poling such that fewer domain walls are encountered by light propagating in the direction of polarization; and index discontinuities at domain walls are smaller for light propagating in the polarization direction. The validity and relative importance of these mechanisms awaits experimental verification.

ACKNOWLEDGMENTS

We wish to thank G. H. Haertling for fabrication of all ceramic materials, I. D. McKinney and B. C. Kayate for sample preparation, and J. C. Crawford and G. H. Haertling for valuable discussions.

REFERENCES

† *This work was supported by the U. S. Atomic Energy Commission.*

1. *G. H. Haertling, "Hot-pressed Lead Zirconate-Lead Titanate Ceramics Containing Bismuth," Am. Ceram. Soc. Bull., Vol. 43, 875-879, Dec. 1964.*

2. *G. H. Haertling and W. J. Zimmer, "An Analysis of Hot-pressing Parameters for Lead Zirconate Lead Titanate Ceramics Containing Two Atom Percent Bismuth," Am. Ceram. Soc. Bull., Vol. 45, 1084-1089, Dec. 1966.*
3. *G. H. Haertling, "Hot-pressed Ferroelectric Lead Zirconate Titanate Ceramics for Electro-optical Applications," Am. Ceram. Soc. Bull., Vol. 49, 564-567, June 1970.*
4. *C. E. Land and P. D. Thacher, "Ferroelectric Ceramic Electrooptic Materials and Devices," Proc. IEEE, Vol. 57, 751-768, May 1969.*
5. *P. D. Thacher and C. E. Land, "Ferroelectric Electrooptic Ceramics With Reduced Scattering," IEEE Trans. on Electron Devices, Vol. ED-16, 515-521, June 1969.*
6. *C. E. Land and R. Holland, "Electrooptic Effects in Ferroelectric Ceramics," IEEE Spectrum, Vol. 7, 71-78, Feb. 1970.*
7. *C. E. Land and G. H. Haertling, "Optical Properties of Ferroelectric Ceramics," Proc. 2nd Intl. Mtg. on Ferroelectricity, Kyoto, Japan, Supplement to J. Phys. Soc. Japan, Vol. 28, 96-99, 1970.*
8. *C. E. Land, "Recent Developments in Electrooptic Ceramics," Intl. J. of Nondestructive Testing, Vol. 1, 315-336, 1970.*
9. *G. H. Haertling and C. E. Land, "Hot-pressed (Pb,La)(Zr,Ti)0_3 Ferroelectric Ceramics for Electrooptic Applications," J. Am. Ceram. Soc., Vol. 54, 1-11, Jan. 1971.*
10. *J. R. Maldonado and A. H. Meitzler, "Ferroelectric Ceramic Light Gates Operated in a Voltage-controlled Mode," IEEE Trans. on Electron Devices, Vol. ED-17, 148-157, Feb. 1970.*
11. *A. H. Meitzler, J. R. Maldonado, and D. B. Fraser, "Image Storage and Display Devices Using Fine-grain, Ferroelectric Ceramics," Bell System Technical Journal, Vol. 49, 953-967, July-Aug. 1970.*
12. *J. R. Maldonado and A. H. Meitzler, "Strain-biased Ferroelectric Picture Devices," Presented at the IEEE Electron Device Research Conference, Seattle, Wash., June 1970; Proc. IEEE, Vol. 59, 368-382, March 1971.*
13. *W. A. Albers and M. Kaplit, "Visible Light Scattering in Ferroelectric Ceramics," this volume, p. 151.*
14. *F. Jona, G. Shirane, and R. Pepinsky, "Optical Study of $PbZr0_3$ and $NaNb0_3$ Single Crystals," Phys. Rev., Vol. 19, 1584-1590, March 1955.*
15. *Referred to in Ref. 17. The wavelength is assumed to approximate $\lambda = 0.55$ μm.*
16. *M. L. Sholokhovich and E. G. Fesenko, "Preparation and Structure of Crystals of Some Lead-containing Ferroelectric Substances and Their Solid Solutions." Bull. Acad. Sci. USSR, Phys. Ser. 24, 1244-1247, 1960. Their data were taken in an S-Se mixture which approximates data at $\lambda = 0.67$ μm according to M. H. Francombe and B. Lewis [Acta Cryst. 11, 696(1958)].*
17. *S. Fushimi and T. Ikeda, "Optical Study of Lead Zirconate Titanate," J. Phys. Soc. Japan, Vol. 20, pp. 2007-2012, Nov. 1965.*
18. *I. P. Kaminow and E. H. Turner, "Electrooptic Light Modulators." Proc. IEEE, Vol. 54, 1374-1390, Oct. 1966.*
19. *I. P. Kaminow, "Electro-optic Materials," in Ferroelectricity, E. F. Weller, Ed., (Elsevier, New York, 1967).*
20. *P. D. Thacher, "Electrooptic g Coefficients of Pb-containing Oxygen-octahedra Ferroelectrics: Ceramic (Pb,Ba)(Zr,Ti)O_3," J. Appl. Phys., Vol. 41, 4790-4797, Nov. 1970.*
21. *N. Uchida and T. Ikeda, "Electrostriction in Perovskite-type Ferroelectric Ceramics," Jap. J. Appl. Phys., Vol. 6, 1079-1088, Sept. 1967.*
22. *D. A. Berlincourt, Clevite Corp. (private communication).*
23. *E. Sawaguchi, "Ferroelectricity Versus Antiferroelectricity in the Solid Solutions of $PbZr0_3$ and $PbTi0_3$," J. Phys. Soc. Japan, Vol. 8, 615-629, Sept. - Oct. 1953.*

24. *F. Jona and G. Shirane, Ferroelectric Cyrstals (MacMillan, New York, 1962).*

25. *W. J. Merz, "The Electric and Optical Behavior of $BaTiO_3$ Single-domain Crystals," Phys. Rev., Vol. 76, 1221-1225, Oct. 1949.*

26. *G. A. Samara, "Pressure and Temperature Dependence of the Dielectric Properties and Phase Transitions of the Ferroelectric Perovskites: $PbTiO_3$ and $BaTiO_3$," submitted to Ferroelectrics.*

27. *R. E. Nettleton, "Polarization-dependent Rayleigh Scattering in a Coarse-grained Ferroelectric Ceramic," J. Appl. Phys., Vol. 39, 3646-3654, July 1968; "Domain Structure and Scattering of Visible Light in a Perovskite Ceramic," IEEE Trans. Electron Devices, Vol. ED-16, 602, June 1969.*

LIGHT SCATTERING PROPERTIES OF NEMATIC LIQUID CRYSTALS†

P. S. PERSHAN

Division of Engineering and Applied Physics
Harvard University, Cambridge, Massachusetts

ABSTRACT

The fundamental properties of simple liquid crystal systems will be reviewed. In particular the elastic and hydrodynamic properties will be discussed in terms of a new, simple, rigorous theory that is more general than previous theories. The light scattering properties observed by Chatelain, and more recently by the Orsay group, follows simply from this theory. Basically this differs from previous theories in that the hydrodynamics and elastic properties are treated purely in terms of the usual conserved quantities (momentum, energy and mass) without the need to postulate a separate equation for the director. The introduction of this separate equation, that has been common, requires assumptions that are not usually necessary in rigorous hydrodynamics and that are not in fact needed to explain the existing observations on liquid crystals.*

INTRODUCTION

The central theme of this paper will be the unusual light scattering properties of the simplest nematic liquid crystals. Fundamental experiemental studies of this phenomenon have been carried out by Chatelain[1] and more recently by the "Orsay Liquid Crystal Group".[2,3] Interpretation of these experiments have been in terms of an existing hydrodynamic theory for liquid crystals.[4] Recently an alternative hydrodynamic theory has been proposed,[5] and we will interpret the light scattering properties of nematic liquid crystals in terms of this theory.

* *See "Note added in proof" on p. 206.*
References p. 206

NEMATIC LIQUID CRYSTALS

BACKGROUND

Considering the availability of general review papers on liquid crystals we will not attempt a comprehensive discussion of their various properties.[6] Rather we will introduce only those properties that are essential for the phenomena under consideration. The simplest type of liquid crystals, the ones called nematics, are invariably formed from organic molecules that are much longer in one dimension than any other. Typically cigar or sausage shaped molecules can form nematics. Optically active long molecules often form a different type of liquid crystal, called cholesteric, rather than nematic. In this article we will speak only of the nematic type, ignoring both the cholesteric and a third type of liquid crystal called smectic. At high temperatures (typically greater than 50 to 100°C) the cigar shaped molecules are randomly oriented and one has a simple organic liquid. However, as the temperature is lowered one often observes a latent heat, characteristic of a first order phase transition.[7] In the low temperature phase the material has many properties characteristic of simple fluids but also some distinctly different ones. Firstly, the low temperature nematic phase, is optically anisotropic. The simplest liquid crystalline phases are uniaxial and the symmetry is ∞/mm. In this phase the molecules are more or less all oriented parallel to each other, thus defining a symmetry axis. [The alignment is not complete, but we will not go into that detail.] In contrast to the long range order characterizing molecular orientation the molecular centers of mass have only short range order. This last fact is part of the reason for the liquid like properties of liquid crystals.

A second important difference between simple isotropic liquids and nematic liquid crystals is that the latter appear to be able to sustain a static shear stress if that static shear is such as to cause an inhomogeneous variation in the direction of the local symmetry axis. For example, if a unit vector $\tilde{n}(r)$ describes the direction of the local symmetry axis, stresses that would induce a non-vanishing $\partial\tilde{n}(r)/\partial r_i$ can exist for long times. Frank[8] has discussed an elastic theory of liquid crystals using these derivatives and the existing hydrodynamic theory[9] derives, in part, from Frank's elastic theory.

The rigorous justification for any hydrodynamic theory follows from the existence of microscopic conservation laws. The usual hydrodynamic equations,[10] for example, correspond to the conservation laws for mass, momentum and energy,

$$\partial\rho/\partial t + \nabla\cdot\mathbf{g} = 0 \tag{1a}$$

$$\partial\mathbf{g}/\partial t + \nabla\cdot\mathbf{T} = 0 \tag{1b}$$

$$\partial E/\partial t + \nabla\cdot\mathbf{J}^{\epsilon} = R \tag{1c}$$

where ρ is the mass density, **g** is the total momentum density, **T** is the stress tensor, E is the energy density, $\mathbf{J}^{\epsilon}$ is the energy current density and R is a function that describes the energy dissipation due to viscous effects. The existing hydrodynamic theory for liquid crystals introduces a fourth equation, for which there does not appear to be any separable conservation law, in an attempt to describe $\partial \bar{n}(r)/\partial t$.* The argument, sometimes made, that this fourth equation derives from conservation of "internal angular momentum" appears to be fallacious, since it is hard to understand why it should be necessary for a liquid crystal but not for the isotropic fluid. Also conservation of angular momentum is actually contained within the equations for linear momentum density. In particular, if $m^{\alpha,i}$ and $\mathbf{R}^{\alpha,i}$ are the mass and position of the αth particle on the ith molecule the momentum density $\mathbf{g} = \sum_{\alpha,i} m^{\alpha,i}\, \dot{\mathbf{R}}^{\alpha,i}\, \delta(\mathbf{r} - \mathbf{R}^{\alpha,i})$ may be expanded in various moments about the molecular centers of mass. Take $\mathbf{P}^i = \sum_{\alpha} m^{\alpha,i}\, \dot{\mathbf{R}}^{\alpha,i}$ and $\ell^i = \sum_{\alpha} m^{\alpha,i}\, (\mathbf{R}^{\alpha,i} - \mathbf{R}^i) \times \dot{\mathbf{R}}^{\alpha,i}$ to represent respectively the total linear momentum of the ith molecule and the angular momentum of the ith molecule about its own center of mass $\mathbf{R}^i$. Then the first two terms in a Taylor expansion of $\delta(\mathbf{r} - \mathbf{R}^{\alpha,i})$ about the center of mass $\mathbf{R}^i$ obtain $\mathbf{g} = \mathbf{P} + \frac{1}{2} \nabla \times \ell + \ldots$ where $\mathbf{P} = \sum_i \mathbf{P}^i\, \delta\, (\mathbf{r}\text{-}\mathbf{R}^i)$ and $\ell = \sum_i \ell^i\, \delta\, (\mathbf{r}\text{-}\mathbf{R}^i)$ are the linear momentum density associated with motion of the molecular centers of mass and the density corresponding to the internal angular momentum of the molecules about their centers of mass. Conservation of total angular momentum $[\mathbf{L} = \int d^3r(\mathbf{r} \times \mathbf{g}) = \int d^3r(\mathbf{r} \times \mathbf{p}) + \int d^3r(\ell)$ + a surface term$]$ follows from Eq. 1b. This is most easily demonstrated by dropping surface terms and taking $T_{ij} = T_{ji}$. The internal angular momentum ℓ is not a separately conserved quantity but rather the total angular momentum, the sum of ℓ and $\mathbf{r} \times \mathbf{p}$ is the conserved quantity and this follows from the equation for conservation of linear momentum. We argue that a fourth equation is not necessary for a hydrodynamic theory of liquid crystals.

The same point may be argued from a different approach. The elastic or hydrodynamic properties of anisotropic single crystals, including molecular crystals, are discussed without regard for the number or types of optical phonon branches that may be present. The essential fact is that for elastic deformations varying slowly in both space and time, in the absence of external fields, the internal structure of the unit cell follows the low frequency acoustic deformations instantaneously. The existence of optical phonon branches certainly influences the values and structure of the elastic constants of the crystal but it is not necessary to include dynamical equations for the optical phonons in a long wavelength, low frequency elastic theory of crystals. Similarly, for simple molecular fluids one does not include dynamical equations for the internal molecular coordinates in a hydrodynamic theory. The assertion we make here is that the three equations, Eqs. (1a), (1b), (1c), should be sufficient for a hydrodynamic theory of liquid crystals. The difference between crystals, liquids and liquid crystals should lie solely in the constitutive relations defining the stress tensors **T**, (the i, j components of **T** are T_{ij}).

* *See "Note added in proof" on p. 206.*

References p. 206

ELASTIC-HYDRODYNAMIC THEORY: NORMAL MODES

The full elastic hydrodynamic theory that is proposed will be discussed elsewhere.[11] Here we will present only an approximate version obtained by making certain simplifying assumptions. This will be sufficient to understand the physical origin of the observed phenomena.

Consider small deformations of a uniform elastic solid of ∞/mm symmetry. That is of a non-polar uniaxial system. These deformations are described in terms of strains $\mu_{ij} = \frac{1}{2}(\partial_i\mu_j + \partial_j\mu_i)$. Expanding the elastic energy density as a power series in these strains and keeping only the leading terms obtains[12]

$$F = \tfrac{1}{2}(\mu_{11} + \mu_{22})K_1(\mu_{11} + \mu_{22}) + \tfrac{1}{2}\mu_{33}K_2\mu_{33} + \mu_{33}K_3(\mu_{11} + \mu_{22})$$
$$+ \tfrac{1}{2}[(\mu_{11} - \mu_{22})K_4(\mu_{11} - \mu_{22})] + 2\mu_{12}K_4\mu_{12} \qquad (2)$$
$$+ 2\mu_{13}K_5\mu_{13} + 2\mu_{23}K_5\mu_{23}$$

where the 3-axis is the symmetry axis and the other two are arbitrary orthogonal axes. For an isotropic solid $K_1 = K_2 - K_4 = K_3 + K_4$ and $K_4 = K_5$. The common statement about the elastic energy of a fluid is that in addition to the above $K_4 = K_5 = 0$, so that $F = \frac{1}{2}K_1(\mu_{11} + \mu_{22} + \mu_{33})^2$. This assures that there is no elastic energy associated with either *uniform* shear or *uniform* tension (i.e. uniform means $\mu_{i,j}$ = constant). The argument we make here is that this last statement is not sufficient to obtain liquid like behavior. Equation (2) represents only the leading terms in F and the vanishing of K_4 and K_5 still leaves open the possibility that there can be elastic energy associated with *non-uniform* shears. For example, if there are additional terms in F proportional to $[\partial_i\mu_{jk}]^2$ these would supply such an energy. For a solid these terms are completely dominated by the lower order terms and one need never worry about them. Normal liquid behavior, however, requires that these terms also must vanish. The assertion is that liquid crystals are materials for which some of the K's vanish but for which F contains some terms proportional to $[\partial_i\mu_{jk}]^2$.

Specializing to an incompressible nematic liquid crystal and assuming that the only elastic deformations are those contained in the Frank elastic theory,[8] [see Eq. (4) below] the following form for the elastic energy density is obtained

$$F = \tfrac{1}{2}M_6[\partial_3(\mu_{11} + \mu_{22})]^2 + \tfrac{1}{2}L_6\left\{[\partial_1(\mu_{11} + \mu_{22})]^2 + [\partial_2(\mu_{11} + \mu_{22})]^2\right\}$$
$$+ \tfrac{1}{2}M_4\left\{[\partial_3(\mu_{11} - \mu_{22})]^2 + 4(\partial_3\mu_{12})^2\right\} \qquad (3)$$
$$+ 2M_5[(\partial_3\mu_{13})^2 + (\partial_3\mu_{23})^2]$$

subject to the additional restriction, due to the assumption of incompressibility, that $\mu_{11} + \mu_{22} + \mu_{33} = 0$. The more complete theory[5,11] allows for an additional elastic constant, to be associated with compressibility and other non-uniform shears. Present experimental data suggests that these other terms may be present but are less importnat than the ones included here.

Since elastic deformations will induce local variations in the local symmetry axis, Eq. (3) can be related to the Frank elastic theory.[8] A simple assumption is

$$\delta n_1 = \partial_3 \mu_1$$

$$\delta n_2 = \partial_3 \mu_2. \tag{4}$$

Strictly speaking one requires a slightly more general relation than Eq. (4);[11] however it is approximately correct and for our purposes it will suffice. Using Eq. (4) it is possible to establish the following relations between the elastic constants in Eq. (3) and the Frank elastic constants $K_{ii}^{(F)}$
$M_4 = K_{22}^{(F)}$, $M_5 = K_{33}^{(F)}$, $M_6 = K_{11}^{(F)} - K_{22}^{(F)} + 2K_{33}^{(F)}$, and $L_6 = - K_{33}^{(F)}$. The negative value of L_6 is consistent with elastic stability.[11]

The constitutive relation for T_{ij} [see Eq. (1c)] contains two parts, an elastic, or non-dissipative part, and a viscous or dissipative part. The elastic part of T_{ij} is obtained from elastic energy density by a variational argument;[13]

$$T_{ii}^{(1)} = - \partial F/\partial\mu_{ii} + \partial_k(\partial F/\partial(\partial_k \mu_{ii})) \tag{5a}$$

and if $i \neq j$

$$T_{ij}^{(1)} = - \frac{1}{2}\, \partial F/\partial\mu_{ij} + \frac{1}{2}\, \partial_k(\partial F/\partial(\partial_k \mu_{ij})) \tag{5b}$$

In view of the incompressibility and other assumptions leading to Eq. (3), $\partial F/\partial\mu_{ii} = \partial F/\partial\mu_{ij} = 0$ in the present discussion. The viscous part of $T_{ij}^{(2)}$ is obtained by exactly the same methods one would employ for a solid of the same symmetry. The function R, introduced in Eq. (1c) describes the rate of energy dissipation due to viscous effects. For an incompressible system of ∞/mm symmetry there are only 3 viscosity coefficients, and

$$R = \frac{1}{2}\eta_6(\dot\mu_{11} + \dot\mu_{22})^2 + \frac{1}{2}\eta_4[(\dot\mu_{11} - \dot\mu_{22})^2 + 4\dot\mu_{12}{}^2] + 2\eta_5[\dot\mu_{13}{}^2 + \dot\mu_{23}{}^2]\,. \tag{6}$$

The viscous parts of T_{ij} are obtained from $T_{ii}^{(2)} = - \partial R/\partial\mu_{ii}$ and $T_{ij}^{(2)} = - \frac{1}{2}\,\partial R/\partial\mu_{ij}$ for $i \neq j$. For example, the viscous parts of T_{11} and T_{12} are $T_{11}^{(2)} = - (\eta_6 + \eta_4)\mu_{11} - (\eta_6 - \eta_4)\mu_{22}$ and $T_{12}^{(2)} = - 2\eta_4\mu_{12}$. The equation of motion for a normal mode propagating in the 1-3 plane, i.e. $\mathbf{q} = (q_1,0,q_3)$ with displacements in the 2 direction follows from Eq. (lb)

$$\rho\left\{\partial^2/\partial t^2 + [\eta_4 q_1^2 + \eta_5 q_3^2]\,\partial/\partial t + M_5 q_3^4 + M_4 q_1^2\right\}\mu_2 = 0 \qquad (7)$$

where we have used $\mathbf{g} = \rho\mu$ and ρ is the mass density. Experimentally it appears that the values of $\{\eta_i\}$, $\{M_i\}$, and ρ are such that $\eta^2 >> 4M\rho$. Using this inequality the eigenfrequencies are approximately

$$\omega_F \simeq -i(\eta_4 q_1^2 + \eta_5 q_3^2)/\rho$$

$$\omega_s \simeq -iq_3^2\,[M_5 q_3^2 + M_4 q_1^2]\,[\eta_5 q_3^2 + \eta_4 q_1^2]^{-1}. \qquad (8)$$

The solution ω_F corresponds to the usual diffusive-like behavior of low frequency transverse waves in a simple liquid. The above inequality is such that $\omega_F >> \omega_s$. For a simple liquid M_4, M_5 vanish and $\omega_s = 0$, corresponding also to the behavior of simple fluids. The finite value of ω_s and its relation to Frank-like constants M_4 and M_5 is a property unique to liquid crystals. Typically ω_s is in the range from 0 to 10^6 sec $^{-1}$ for q's of $\sim 10^4$ cm^{-1}. The light scattering measurements made by the Orsay group[3] obtain ω_s for the nematic liquid crystal (PAA). The resulting values correspond to $M_5/\eta_{55} \simeq 6.8 \times 10^{-5}$ in cgs units, $M_4/M_5 \simeq 0.45 \pm 0.15$ and $\eta_4/\eta_5 = 0.9 \pm 0.3$. Independently measured values of M_5, M_4,[14] and η_5[15] obtain $M_4/M_5 = 0.28$, $M_5/\eta_5 = 6.1 \times 10^{-5}$ cgs units. The agreement is satisfactory.

Equation (8) predicts that if q_1 remains finite as $q_3 \to 0$, $\omega_s \to 0$. Although the Orsay measurements do not have this feature, they do indicate a very small value of ω_s for this condition. If this is not an experimental artifact, it can be explained by removing some of the assumptions leading to Eq. (3).[5,11] Away from $q_3 = 0$ agreement between Eq. (8) and the experiment is good.

Light Scattering from Normal Modes

The ∞/mm symmetry of the nematic phase admits the possibility that the optical frequency dielectric tensor has the form

$$\epsilon = \begin{pmatrix} \epsilon_1 & 0 & 0 \\ & \epsilon_1 & 0 \\ 0 & 0 & \epsilon_3 \end{pmatrix}. \qquad (9)$$

The largest interaction between the hydrodynamic normal modes and light waves results from the fact that the deformations $\partial_i\mu_j$ rotate this tensor. For simplicity we assumed, in the previous section, that the rotation was described by Eq. (4). To remain consistent with the assumption we now take

$$\delta\epsilon_{13} = \delta\epsilon_{31} = (\epsilon_3 - \epsilon_1)\partial_3\mu_1$$
$$\delta\epsilon_{23} = \delta\epsilon_{32} = (\epsilon_3 - \epsilon_1)\partial_3\mu_2. \tag{10}$$

The differential cross section per cm^3 for light scattering is given by

$$d\sigma/d\omega d\Omega = \pi^2 \lambda^{-4} \int d^3r \exp[-i\,\mathbf{q}\cdot\mathbf{r}]<\delta\epsilon_{so}\delta\epsilon_{so}(r,\omega)> \tag{11}$$

where $(\tilde{e}_s)_i\, \delta\epsilon_{ij}(\tilde{e}_o)_j \equiv \delta\epsilon_{so}$ if $\tilde{e}_s$ and $\tilde{e}_o$ are the unit vectors describing the polarizations of the scattered and incident light respectively, λ is approximately the wavelength of the incident light in the liquid crystal, and $\mathbf{q}$ is the difference between the wave vectors of the incident and scattered light in the medium. Employing Eqs. (10) and (11) the differential cross-section may be expressed in terms of the autocorrelation functions of the $\{\mu_i\}$. These autocorrelation functions are obtained, in the classical limit, by applying the fluctuation dissipation therorem to the equation[5] of motion; that is, to Eqs. (lb). As an explicit example from Eq. (7) one obtains

$$(2\pi)^{-1} \int_{-\infty}^{\infty} < \mu_2(0,\tau)^* \mu_2(\tau,q)> \exp(i\omega\tau)d\tau =$$

$$kT/\pi\rho^2\, [\eta_4 q_1^{\,2} + \eta_5 q_3^{\,2}] \left\{(\omega^2 + |\omega_F|^2)^{-1}(\omega^2 + |\omega_s|^2)^{-1}\right\}. \tag{12}$$

For the special case of incident light propagating along the 1-direction and polarized in the 3-direction, the differential cross-section per cm^3, for light scattered into the near forward direction and polarized in the 2-direction is given by

$$d\sigma/d\omega d\Omega = [\pi kT(\epsilon_3 - \epsilon_1)^2/\rho^2\lambda^4]\,[\eta_4 q_1^{\,2} + \eta_5 q_3^{\,2}]\, q_3^{\,2}/[(\omega^2 + |\omega_F|^2)(\omega^2 + |\omega_s|^2)].$$

For frequencies $\omega << |\omega_F| = \rho^{-1}[\eta_4 q_1^{\,2} + \eta_5 q_3^{\,2}]$, by substituting [from Eq. (8)]

$$q_3^{\,2}[\eta_4 q_1^{\,2} + \eta_5 q_3^{\,2}] = |\omega_s|\, [M_4 q_1^{\,2} + M_5 q_3^{\,2}]^{-1}$$

one obtains

$$d\sigma/d\omega d\Omega \approx [\eta kT(\epsilon_3 - \epsilon_1)^2/\lambda^4(M_4 q_1^{\,2} + M_5 q_3^{\,2})]\; [|\omega_s|/(\omega^2 + |\omega_s|^2)]. \tag{13}$$

The experimental geometry appropriate to Eq. (13) is shown in Figure 1. The values of q_1 and q_3 are approximately given by

$$q_1^{\,2} \simeq (2\pi/\lambda_o)^2\, [\epsilon_3^{1/2} - \epsilon_1^{1/2}]^2$$
$$q_3^{\,2} \simeq (2\pi/\lambda_o)^2\, \epsilon_1 \sin^2\theta \cos^2\psi \tag{14}$$

References p. 206

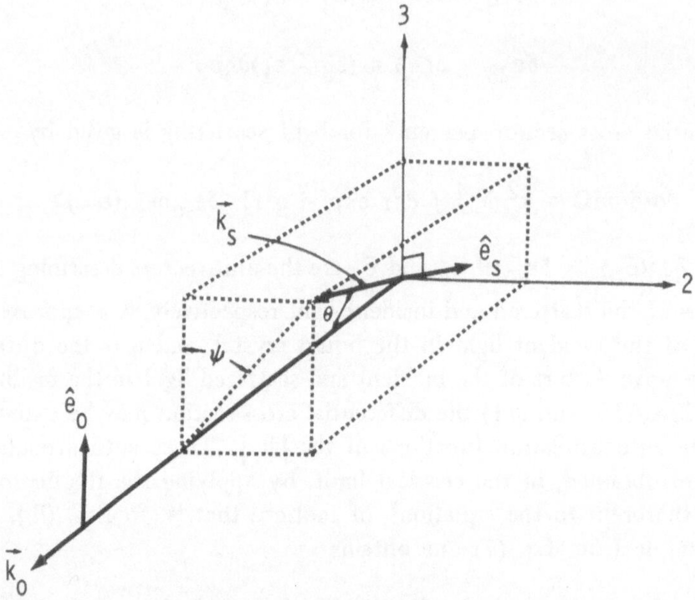

Fig. 1. The geometry appropriate to Eq. (13). Incident light of wave vector k_o in the 1-direction has polarization $\tilde{e}_o$ in the 3-direction. Scattered light is examined for wave vector $\vec{k}_s$ as shown with polarization $\tilde{e}_s$ in the 1-2 plane. The 3 axis is the direction of uniaxial symmetry. Nematics typically have $\epsilon_3 > \epsilon_1$ so that $|k_o| > |k_s|$.

where λ_o is the free space wavelength of the incident light wave, $|k_o| = 2\pi\lambda_o^{-1}\,\epsilon_3^{1/2}$. The determination of ω_s from the Orsay measurements mentioned at the end of the previous section follows from the light scattering spectrum predicted by Eq. (13). The Orsay group observed Lorentzian spectra with half widths that can fit Eq. (8) except near $q_3 = 0$. The discrepancy near $q_3 = 0$ is small and can be corrected by a slight generalization of Eq. (3). The discussion at the end of the last section pertains to this point. For quantitative intensity, rather than line shape, comparison one must make further small corrections to the cross-section in order to account for polarization factors. These are straightforward.

Chatelaine's earlier measurements[1] of total scattered light intensity into a given solid angle can be interpreted by

$$\frac{d\sigma}{d\Omega} = \int \frac{d\sigma}{d\omega d\Omega}\, d\omega = \pi^2(\lambda)^{-4}kT(\epsilon_3 - \epsilon_1)^2[M_5q_3{}^2\ 3\ M_4q_1{}^2]^{-1}. \qquad (15)$$

The cross-section would blow up in the forward direction were it not for the fact that for the chosen polarizations $\epsilon_1 \neq \epsilon_3$ keeps $q_1{}^2 > 0$ as $\theta \to 0$. This formula is only approximately applicable to the Chatelain observations in that we have not included scattering due to a second fluctuation, with the same q vector, but polarized mostly in the 1-3 plane. This corresponds to a second transverse mode, also displaying diffusive type behavior, and which has not been discussed in the present article. Although a complete description of the Chatelaine experiement requires inclusion of the effects of the second mode, the dominant contribution is probably described by Eq. (15).

SUMMARY OF LIGHT SCATTERING DISCUSSION

The most important optical property of nematic liquid crystals is the uniaxial symmetry of its dielectric tensor. A homogeneous, static, nematic would behave optically like a uniaxial single crystal. The physical fact is that one never has a homogeneous, static nematic. In the absence of externally induced inhomogeneities, thermal fluctuations induce sizeable variations in the uniaxial direction and these in turn lead to quite strong light scattering. In particular the light scattering in the near forward direction is particularly strong and also time dependent. Typical frequencies are in the range of $10^3 \rightarrow 10^6$ sec^{-1}. External disturbances can certainly produce other inhomogeneities above the thermal limit and increase the light scattering cross-section by orders of magnitude. On the other hand, at least in principle, different external influences may also reduce the thermal scattering. For example, if a strong external magnetic field could provide an anisotropy energy larger than kT, ω_s would be significantly raised and $(\partial_3\mu_2)^2$ decreased. It remains to be seen whether this type of effect is practical.

Similar types of light scattering effects occur in cholesteric and smectic crystals; however, their analysis requires a more complicated elastic energy than described by Eq. (3). Also, not all smectic systems are uniaxial and it is not clear that a director (i.e. $\bar{n}$) is sufficient to describe the local symmetry.

A further complication that arises in practice with nematic liquid crystals is the appearance of static "disinclinations" that are analogous to dislocations in single crystals. The effect is that unless special precautions are taken a given sample of nematic liquid crystal will have static variations in its local symmetry axis far in excess of the thermal variations discussed here. External forces can often eliminate these.

ACKNOWLEDGEMENT

We would like to emphasize that this paper is a condensed description of work done in collaboration with P. C. Martin and Jack Swift. In the course of that work we benefited from conversations with R. Meyer, D. Turnbull, B. Halperin, D. Litster, M. Stephen, and a brief exchange with P. G. de Gennes.

References p. 206

REFERENCES

* Supported in part by the Advanced Research Projects Agency Under Contract DAHC-15-67-C-0219, by the Joint Services Electronics Program (U. S. Army, U. S. Navy, and U. S. Air Force) Under Contract NOOO14-67-A-0298-0006, and by the Division of Engineering and Applied Physics, Harvard University.

1. P. Chatelaine, Acta. Cryst. 1, 315(1948).
2. M. Pierre-Gilles de Gennes, C. R., Acad. Sc., Paris 266, 15(1968).
3. Orsay Liquid Crystal Group. Phys. Rev. Lett. 22, 1361(1969).
4. Groupe d'Etudes des Cristaux Liquides (Orsay), J. Chem. Phys. 51, 816(1969).
5. P. C. Martin, P. S. Pershan, and Jack Swift, Phys. Rev. Lett. 25, 844(1970).
6. One excellent review article on liquid crystals is by I. G. Chistyakov, Usp. Fiz, Nauk (USSR) 89, 563(1966). [Sov. Phys. - Usp. 9, 551(1967).]
7. Some interesting pretransition effects have been observed, J. D. Litster and T. W. Stinson, J. Appl. Phys. 41, 996(1970), and T. W. Stinson and J. D. Litster, Phys. Rev. Lett. 25, 503(1970).
8. F. C. Frank, Dics. Faraday Soc. 25, 19(1958).
9. J. L. Eriksen, Arch. Ratl. Mech. Anal. 4, 231(1960), F. M. Leslie, Quart. Jour. Mech. & Appld. Math. 19, 358(1969), and M. J. Stephen (to be published).
10. For our purposes the "usual hydrodynamic equations" are as treated by L. D. Landau and E. M. Lifshitz, Fluid Mechanics, Pergamon, London, 1959.
11. P. C. Martin, P. S. Pershan, and Jack Swift (in preparation).
12. See, for example, L. D. Landau and E. M. Lifshitz, Theory of Elasticity, Addison-Wesley, Reading, Ma. (1959), p. 39.
13. See, for example, the argument of L. D. Landau and E. M. Lifshitz, The Classical Theory of Fields (revised second edition), Addison-Wesley, Reading, Ma. (1962), p. 87.
14. A. Saupe, Z. Naturforsch. 15a, 815(1960).
15. M. Miesowicz, Nature 158, 27(1946). His η_1 and η_3 and our η_5 and η_4 respectively.

Note added in proof: The assertion made in this paper that Eqs. (1) are sufficient to describe the observed properties of nematic liquid crystals is in error. This is discussed fully by Dieter Forster, Tom C. Lubensky, Paul C. Martin, Jack Swift and P. S. Pershan, Phys. Rev. Letters 26, 1016 (1971). The single omission made in the present manuscript, and also in Ref. 5, was to omit the fact that the director (see Eq. (4)) does not strictly follow the displacements but follows it with a phase lag. Although this phase lag vanishes in the long wavelength limit, it must be included in the hydrodynamics. The small discrepancy between the experiments of Ref. 11 and Eq. (8) in the limit that $q_3 \to 0$ is removed when this relaxation, or phase lag, is taken into account.

LIQUID-CRYSTAL LIGHT-CONTROL EXPERIMENTS

R. A. SOREF

Sperry Rand Research Center, Sudbury, Massachusetts 01776

INTRODUCTION

Liquid crystals are sensitive to a number of external influences: Pressure, temperature, mechanical shear, low-frequency electric fields, low-frequency magnetic fields, ultraviolet irradiation, and chemical vapors. Any of these stimuli can produce a measurable change in the liquid crystal's light-scattering properties. Light "control" is possible because the external stimuli modify the optical scattering pattern. They control the amplitude, phase, and angular direction of light emerging from the liquid crystal.

Of the above external influences, electric and magnetic fields are easiest to manipulate in the laboratory. Thus, the most convenient techniques for light control are voltage control and current control. Accordingly, we shall confine our discussion to electro-optic and magneto-optic effects.

This paper will survey effects recently reported. Two electro-optic effects - turbulence and helix pitch dilation - are singled out because of their prominence in current research. New experimental results on both topics are given here.

Before reviewing light-control phenomena, let us clarify some basic practical matters: (1) The typical experimental geometry; (2) the influence of boundary surfaces; and (3) the definitions of regular and anomalous molecular alignment.

BACKGROUND DISCUSSION

DEVICE GEOMETRY

Figure 1 shows the basic electro-optic device configuration. A thin film of liquid crystal is held between parallel plates, and the faces in contact with the liquid are coated with a transparent electrode material (e.g., indium oxide). Voltage is applied as shown. The device operates in a reflective mode when one electrode is made metallic.

The film's thickness in the **z** direction is generally 10 to 100 optical wavelengths (5 to 50 μm), which ensures large-scale changes in the film's light scattering when the field is turned on. The 5-to-50 μm thickness also makes it possible to achieve fields of 10^4 to 10^5 V/cm in the liquid with 10 to 100 V driving voltages. These low voltages are an important advantage in applications.

Most experiments have been done with $\mathbf{E} \parallel \mathbf{z}$, although the field $\mathbf{E} \perp \mathbf{z}$ can also be applied with electrodes on the x-y surfaces. From a practical standpoint, $\mathbf{E} \perp \mathbf{z}$ would be undesirable because wall effects (see below) would counteract the field-induced alignment. Also, the optical filling factor would be low and the drive voltages comparatively high.

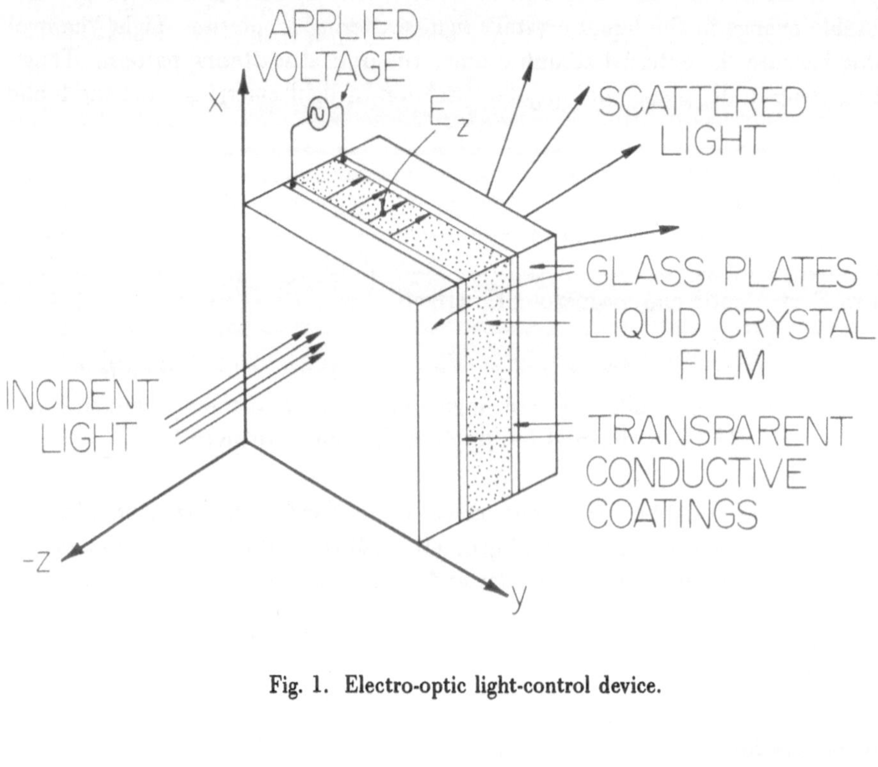

Fig. 1. Electro-optic light-control device.

In magneto-optic experiments, a thin film is employed, as in Fig. 1, and the sample is immersed in a magnetic field H_x or H_y. If provision is made for optical transmission, H_z can be used.

WALL EFFECTS

The electrode surfaces or "walls" have a significant effect upon the molecular alignment at zero field. For example, in nematic liquids, three boundary effects are found that depend upon surface treatment. First, if the walls are rubbed in a particular direction, the nematic molecules align nearly homogeneously with their optic axes in the x-y plane (Fig. 1), parallel to the rubbing direction. Second, when the walls are untreated, the molecules group into internally parallel clusters, the cluster axes being oriented randomly in the x-y plane. Third, the molecules align perpendicular to the x-y plane when the walls are specially cleaned.

The first kind of alignment is analogous to the ordering of ferroelectric domains that takes place when a ferroelectric ceramic is poled (see the paper by Land and Thacher elsewhere in this volume). The walls serve the same function as a poling field applied perpendicular to **z** . Not only do the walls give initial ordering, but they often determine the final order that emerges after the field E_z is shut off. However, in some liquid crystals, the field E_z perturbs the wall alignment irreversibly.

Although wall effects have been known for more than twenty years,[1] information about them is still fragmentary and unsystematized. What factors determine the penetration depth of alignment? What influence does the electrode composition have? Does it make a difference whether the sample cell is filled by capillary action or by compression of a liquid drop? These are just a few of the unanswered questions that exist.

Without information of this kind, it is difficult to predict and to reproduce wall ordering; yet, in device applications, it is important to control both the uniformity and the extent of initial alignment across the device. Thus, future progress in light control will be linked to progress in the control of wall effects.

REGULAR AND ANOMALOUS ALIGNMENT

The electric and magnetic susceptibilities of liquid-crystal molecules are, in general, anisotropic. The dielectric anisotropy is said to be positive when $\epsilon_{\parallel}/\epsilon_{\perp} > 1$ and negative when $\epsilon_{\parallel}/\epsilon_{\perp} < 1$, where $\epsilon_{\parallel}$ and $\epsilon_{\perp}$ are the dielectric constants parallel to and perpendicular to the long axis of the molecule. Similar definitions of the positive and negative diamagnetic anisotropy apply to $\chi_{||}$ and $\chi_{\perp}$, the diamagnetic susceptibility components.

In the presence of an external field, the molecules tend to align regularly or anomalously. Alignment is regular when positive anisotropy molecules prefer to be parallel to the field and negative anisotropy molecules prefer to be perpendicular to the field. Anomalous alignment refers to the tendency of positive anisotropy molecules to align perpendicular to the field and negative anisotropy molecules to align parallel to the field. Examples of this are given in the next section.

SURVEY OF ELECTRO-OPTIC AND MAGNETO-OPTIC EFFECTS

A large number of electro-optic and magneto-optic effects have been reported in the past few years. Several effects have already become the basis for light-control devices and the remainder have real but unexploited potential for this application. The profusion of effects is not surprising since one out of every 200 organic compounds has a liquid-crystal phase.

The effects range across all three structural categories of liquid crystals: Nematic, smectic, and cholesteric. In some cases, mixed liquid-crystal systems are involved, and, in others, combined E and H fields are used.

Table I summarizes recent experimental observations taken from the literature. Effects are categorized according to molecular organization. A thin film geometry, like that of Fig. 1, is used in each effect except for surface deformation. In this instance, the thin film has one free surface and the deformation fields are applied Xerographically.

The first four effects involve continuous motion of the liquid crystal in the steady-state field-on condition. Storage, which is the fourth effect, denotes emulsification of cholesteric material dissolved in the nematic "host". Two types of phase transformation are listed. They are reversible transitions since the initial phase reappears when the field is shut off.

Erasure and quenching are both regular alignment effects that occur at relatively high audio frequencies. The anomalous alignment phenomena take place at low frequencies. The guest-host interaction, a color-switching effect, uses the cooperative alignment of dichroic, nonionic impurities added to a regularly alignable nematic.

Historically, the terminology "disturbed" and "undisturbed" for cholesteric materials arose from how the films were prepared on a substrate. A film made by

TABLE I

Electro-optic and Magneto-optic Effects in Liquid Crystals

Effect of field upon liquid crystal	Mesophase structure at zero field	Field	Frequency	References
domain formation	nematic smectic	E_z E_z	a a	17, 18, 19 27
turbulence	nematic smectic	E_z E_z	a a	15, 25 this work, 27
oscillatory domain motion	nematic smectic	E_z E_z	a a	37 this work
creation of stored texture	nematic+cholesteric smectic	E_z E_z	a a	38 this work
erasure of stored texture	nematic+cholesteric smectic	E_z E_z	b b	38 this work
smectic-to-nematic phase transformation	smectic	E_z	a	28
cholesteric-to-nematic phase transformation	cholesteric cholesteric+nematic cholesteric	E_z E_z H_x	a a a	39 40 11, 41
quenching of turbulence	nematic	E_z	b	25
regular alignment	nematic nematic	E_z or H_x $E_x + H_y$	a, b a, b	10, 42 43
anomalous alignment	nematic smectic	$E_z + H_x$ $E_z + H_x$	a a	44, 45, 46 14
induced Grandjean-plane texture	nematic+cholesteric	E_z	b	2
induced undisturbed texture	nematic+cholesteric	E_z	a	3
guest-host interaction	nematic+dye(s)	E_z	a, b	47
dilation of helix pitch	undisturbed cholesteric	E_z or H_z	a, b	7, 31, 48
contraction of helix pitch	disturbed cholesteric	E_z or H_x	a, b	7, 49, 50
disturbed-to-undisturbed conversion	disturbed cholesteric	E_z	a	31
surface deformation	nematic smectic cholesteric	H_x or H_z E_z E_z	a a a	13 51 51, 52

a. dc, or low audio frequency
b. audio mid-range frequency

References pp. 230-231

solvent evaporation was "undisturbed", and the same film, subjected to mechanical shear, became "disturbed". In a Fig. 1 - cell, the initial alignment tends to be disturbed, which means the helical ordering axes are perpendicular to the walls. When a strong electric field E_z is applied, the liquid breaks up into helix domains that rotate 90°. The screw axes of the domains line up parallel to the walls. Then the alignment is undisturbed.

Once formed, the undisturbed state persists for many hours at zero field. An applied field dilates the pitch up to the cholesteric-to-nematic conversion point. Conversely, if one applies E_z to a disturbed-state liquid, the field contracts the pitch prior to disturbed-to-undisturbed conversion. It is assumed here that the cholesteric material has $\epsilon_{\|} > \epsilon_{\perp}$. If not, the electric field reduces any imperfections that exist in the helicoidal structure, but it does not change the pitch.

In a nematic-cholesteric system with negative anisotropy, a Grandjean plane texture can be induced.[2] When the nematic solvent has a positive dielectric anisotropy, it should be possible to induce undisturbed alignment of the cholesteric component at low frequencies without dynamic scattering. Carr[3] obtained evidence of induced undisturbed alignment in a nematic-cholesteric mixture using microwave dielectric-loss measurements. In future experiments, optical detection should offer more direct proof of the induced undisturbed state.

Let us now examine the optical interactions that occur in the Table I effects. Table II summarizes the molecular and optical characteristics of representative electro-optic effects selected from Table I. To show how the field reorders the molecules and changes their light scattering, the field-off and field-on states are compared in Table II. (Magneto-optic effects are discussed later in this section.)

The table shows that a large, diverse group of optical interactions is available. These include diffuse scattering, scattering with memory, optical modulation (using polarizers), induced wavelength-selective absorption, and tunable diffraction. Applications of these unique effects were reviewed recently.[4,5] The applications story is just now beginning to unfold and is far from complete.

Shelton and Shen[6] used the circular birefringence of a cholesteric mixture at $E = 0$ to attain phase-matched optical harmonic generation. Phase matching should also be possible through application of a low-frequency electric field. The field would modify the circular birefringence sufficiently for matching, provided that the initial helix pitch (determined by the constant operating temperature) were properly selected.[7] The cholesteric compound is one for which $\epsilon_{\|} > \epsilon_{\perp}$.

To conclude this section, we briefly review liquid-crystal magneto-optic effects, beginning with nematics. Magnetic alignment of a nematic may be thought of as the

TABLE II

Electro-optic Light-control Effects in Liquid Crystals (Fig. 1 Geometry)

Field Effect (and initial structure)	Molecular Ordering		Optical Properties	
	at zero field	at a high field	at zero field	at a high field
domain formation (nematic)	primarily ∥ to walls	liquid rotation; cylindrical vortex motion	transparent; birefringent (uniaxial positive)	parallel array of cylindrical lenses
turbulence (nematic)	primarily ∥ to walls	nonlaminar flow	transparent; birefringent (uniaxial positive)	intense, diffuse wide-angle scattering
turbulence (smectic)	smectic-A texture; fans or polygons	nonlaminar flow	translucent; birefringent (biaxial)	intense, diffuse, wide-angle scattering
oscillatory domain motion (nematic)	primarily ⊥ to walls	domain vibration in direction ⊥ to domain-lines	transparent; birefringent (uniaxial positive)	medium-intensity diffuse scattering
creation of stored texture (nematic-cholesteric mixture)	primarily ∥ to walls	disruption of cholesteric solute; nematic turbulence	transparent; birefringent (uniaxial positive)	intense, diffuse, wide-angle scattering
erasure of stored texture (nematic-cholesteric mixture)	disordered texture, resembling focal-conic	aligned ⊥ to field	diffuse scattering	transparent; birefringent (uniaxial positive)
cholesteric-to-nematic transformation (cholesteric)	primarily the Grandjean plane texture	aligned ∥ to field	iridescent scattering	transparent; isotropic in the z-direction

References pp. 230-231

TABLE II. (Continued)

Field Effect (and initial structure)	Molecular Ordering		Optical Properties	
	at zero field	at a high field	at zero field	at a high field
quenching of turbulence (nematic)	nonlaminar flow	aligned ⊥ to field	subsiding, diffuse scattering	transparent; birefringent (uniaxial positive)
regular alignment, using E_z (nematic)	randomly oriented clusters aligned ∥ to walls	aligned ∥ to field	transparent; depolarized scattering	transparent; isotropic in z-direction
regular alignment, using $E_x + H_y$ (nematic)	aligned ∥ to H_y	partially re-aligned ∥ to E_x	transparent; birefringent; optical activity*	Pockel's effect*
guest-host interaction (nematic doped with dichroic dye)	randomly oriented clusters aligned ∥ to walls	aligned ∥ to field	linear dichroism (wavelength-selective absorption)	transparent; (colorless), with isotropy along z-axis
dilation of helix pitch (undisturbed cholesteric)	domains ∥ to walls within focal-conic texture	molecules in layers aligned more ∥ to field	birefringent (uniaxial negative) Bragg or Raman-Nath diffraction	decreased birefringence; decreased diffraction angles
contraction of helix pitch (disturbed cholesteric)	Grandjean plane texture	conically perturbed (tightened) helices	optical activity; selective circular reflection; Bragg diffraction	decreased rotary power; decreased diffraction angles
surface deformation (cholesteric)	Grandjean plane texture	helix reorientation, and mass transport along field gradients	Bragg diffraction (dispersive reflection)	altered reflection from deformed areas

* *for light propagating at an angle to the z-axis*

interaction of the field with self-ordered macroscopic regions of the liquid. Experiments support this picture. For example, Chistyakov and Chiakovskii[8] found that a 500 G field gave nearly complete alignment of randomly oriented volume elements; yet, increasing the field to 16 kG had little effect on the mutual molecular ordering within regions. Another view of the magnetic interaction can also be taken, namely, that the field interacts with individual molecules by modifying the phase-organizing forces.[9]

The Freedericksz transition[10] is a good example of a magneto-optic interaction in a nematic. Here, the birefringence for light propagating in the z direction (Fig. 1 cell) decreases in the following way. A field H_z is applied to nematic molecules that have their axes parallel to the glass plates. Above a critical field strength, the molecules reorient. Their angular position rotates progressively towards the z axis as one travels away from each glass plate into the liquid. The critical field strength is typically 10 kG.

Optical properties are also a key indicator of the interaction between a cholesteric liquid crystal and an external magnetic field. For example, the optical "disinclination lines" in a wedge-shaped cholesteric layer give a visual readout of pitch dilation when the liquid is subjected to a magnetic field.[11] The field, applied normal to the helix axes, unwinds the helicies and transforms the disturbed cholesteric to a nematic when $H \approx 10$ kG. The individual molecules, which have $\chi_{\parallel} > \chi_{\perp}$, align parallel to H as expected. In another experiment, however, a mixture of cholesteryl chloride and cholesteryl myristate aligned with its helical-ordering axes parallel to a 25 kG field.[12] Whether this mixture had $\chi_{\perp} > \chi_{\parallel}$ is unknown; therefore, one cannot say definitely that the alignment was anomalous.

The foregoing statements imply that high fields are needed to observe magneto-optic effects. Generally, this is true, but there are special situations where moderate fields suffice. de Gennes,[13] for example, predicts that the free surface of a nematic film will deform under a 3 kG field into an optically observable domain structure.

As a rule, magnetic fields are less effective than electric fields in producing molecular alignment. For example, in some nematics,[14] 10 kG is the alignment equivalent of 2.7 kV/cm. Similarly, it is found that electro-optic devices can function with slightly less than 10 kV/cm, while magneto-optic devices require about 10 kG. The generation and control of 10 kV/cm, as in Fig. 1, is easy, but the production and manipulation of 10 kG is more cumbersome. Therefore, magnetic light-control devices appear less practical than their electrical counterparts.

References pp. 230-231

ELECTRIC-FIELD-INDUCED TURBULENCE

Work has been under way in a number of laboratories to clarify the mechanism of turbulence. Recent developments in turbulence, including our experimental results, will be presented in this section.

Heilmeier[15] coined the name "dynamic scattering" to denote the diffuse wide-angle light scattering associated with field-induced turbulence. In effect, the scattering comes from centers 1-to-5 μm in diameter and is not highly wavelength dependent in the visible. The scattering, which has a high front/back ratio, can be excited either with dc fields or ac fields, whose frequency lies below a critical value.

Experimental evidence indicates that two conditions must be fulfilled before a nematic can exhibit dynamic scattering: its conductivity must be greater than a critical value around 10^{-10} ohm^{-1} cm^{-1}, and its dielectric anisotropy must be negative.

Helfrich[16] recently proposed that a conduction-induced torque arises in nematics whose conductivity is anisotropic. This volume torque supplements the dielectric-anisotropy torque and the elastic torque due to distortion of the molecular orientation pattern. Helfrich assumed a continuum model for the nematic, unlike Heilmeier,[15] who analyzed the field-induced disruption of aligned swarms. Both authors, however, use the concepts of material flow and shear. Helfrich predicted that the conduction-induced distortion would result in turbulence or in stationary distorted orientation patterns (domains).

Domain formation in nematics has been known for many years, but the hydrodynamic structure underlying the domains was recognized only recently. Our experiments in conductive nematics with $\epsilon_\perp > \epsilon_\parallel$ illustrate how the domain features are linked to surface preparation. If the sample cell plates are stroked with a clean cloth before the cell is filled, the domain lines form perpendicular to the rubbing direction. An example of this is shown in Fig. 2(a) for the room-temperature nematic p-methoxybenzilidene-p-n-butylaniline (MBBA). For the Fig. 2(b) cell, the walls were untreated. Here, a more randomized domain pattern results, although the domains retain their characteristic width ~ L and wavelength ~ 2L, where L is the sample thickness. The excitation voltage was 6.6 V rms at a frequency of 30 Hz for both cells in Fig. 2. This voltage is approximately equal to the domain threshold V_c, first reported by Williams.[17] To first order, V_c is independent of L.

Seemingly, domains are at rest. A closer inspection by Penz[18] and Durand et al.[19] revealed that a cylindrical vortex motion exists within domains. Between the static domain boundaries, the liquid rotates in circular paths. The paths lie in

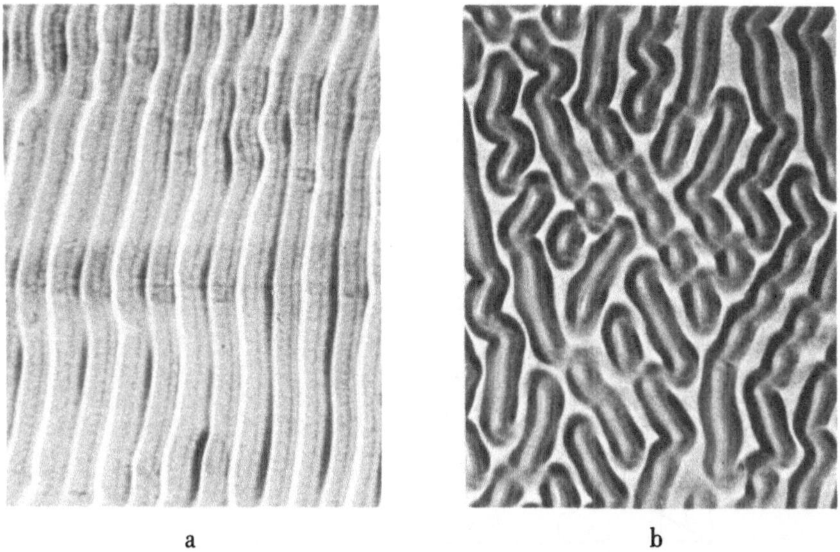

Fig. 2. Domain formation in two room-temperature nematics: (a) p-methoxybenzylidene-p-n-butylaniline; (b) 48.8% p-methoxyformyloxybenzylidene-p-n-butylaniline, 48.8% p-ethoxybenzylidene-p-n-butylaniline, 2.4% p-n-butoxybenzoic acid. Microscope photographs taken with transmitted unpolarized light. Area shown is 350 μm x 450 μm. Cell thickness = 25 μm. Applied field ⊥ page.

planes perpendicular to the boundaries. In adjacent domains, the vorticity is antiparallel and the domain lines are real or virtual images of the incident illumination. These observations verify many aspects of Helfrich's theory.[16]

A complex sequence of events takes place as the applied voltage V is raised above V_c. Domains begin to move and subdivide with a regular oscillatory motion. Here the velocity v of small particles in the liquid increases as $v = a(V^2 - V_c^2)$ according to Durand et al.,[19] and the light scattering builds up from its small initial value. With increased voltage, domain motion gives way to nonlaminar flow as a second hydrodynamic instability threshold is reached. The term "dynamic scattering" aptly describes optical behavior in the turbulent regime, but it seems inappropriate for the light focusing that occurs in the laminar-flow regime near V_c.

Figure 3 illustrates the voltage dependence of optical scattering[20] and correlates the measured contrast ratio with the liquid's internal structure. Electrical current is also plotted for a later discussion. The scattering ordinate in Fig. 3 is the off/on ratio of diffuse transmittance within a large solid angle. As the detector's acceptance angle diminishes, the contrast ratio goes up and approaches a limiting value (assuming that collimated coherent light is incident on the cell). In principle, very high ratios are attainable using spatial filtering.[21] Off/on ratios of 10^4 have been measured.[22]

References pp. 230-231

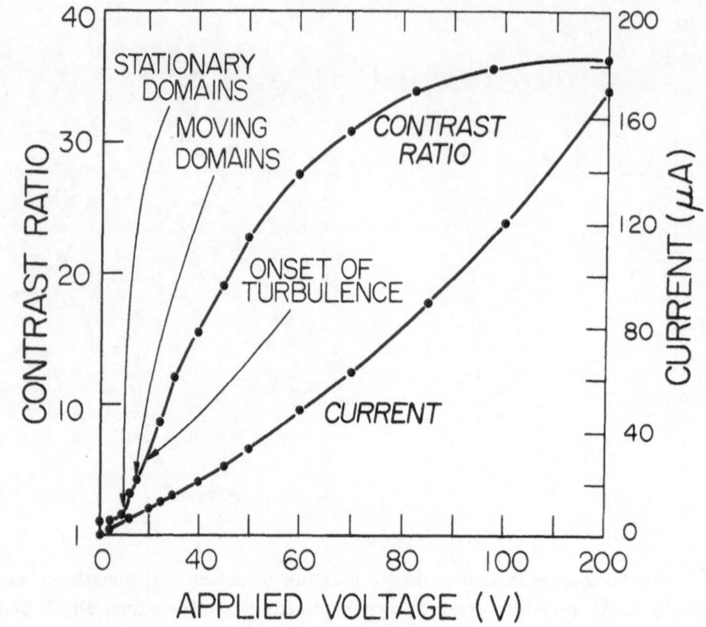

Fig. 3. Optical scattering and dc electrical current vs voltage for MBBA at 25°C [after Jones, Creagh and Lu, Appl. Phys. Letters *16*, 61(1970)]. Cell thickness = 25 μm.

While most authors agree that the buildup of space charge is essential for turbulence, they disagree on whether the charge comes from conductivity anisotropy in an E field. For example, Koelmans and van Boxtel[23] assert that unipolar space-charge injection in an isotropic liquid causes turbulence. This question certainly warrants further study.

A model which has been suggested for charge injection is Schottky emission at the cathode of the liquid crystal cell.[15] This model gives a current dependence of

$$I = A \exp[(B/kT)E^{1/2}] .$$

The current data of Fig. 3 has this dependence at the higher voltage levels where $I \propto \exp(c\ V^{1/2})$. Also, Lu and Jones[24] obtained a good fit to the Schottky emission model with the E - I values found in a thick nematic cell. The nonuniform field distribution peaked at the electrodes, indicating the presence of heterocharges.

Heilmeier's observations[25] suggest that water molecules promote electron injection at the cathode. Presumably, these electrons are captured by neutral molecules, transported as negative ions, and oxidized at the anode, thus freeing molecules to repeat the cycle. The contribution of positive ions and other ionic impurities is unknown at present.

Above a critical excitation frequency f_c, the conduction-induced torque falls below its dc value and tends toward zero because of incomplete charge buildup.[16] The frequency f_c is governed by the space-charge relaxation time t_s as follows:

$$f_c = t_s^{-1} = \sigma/2\pi\epsilon\epsilon_o$$

where ϵ_o is the free-space permittivity and σ is the liquid's conductivity. For $f > f_c$, turbulence is not produced because the space charge density can no longer follow the applied field. Quenching and erasure (Table I) occur at frequencies where conduction-induced torque is suppressed. Here the dipolar nature of neutral molecules becomes significant as the dielectric alignment forces take command.

During the past two years some understanding has been gained of the transient response of turbulence. Koelmanns and van Boxtel,[23] for example, investigated t_r, the risetime of turbulence. From several experiments in MBBA, they constructed the following empirical law

$$t_r \propto \eta/\epsilon\epsilon_o\, E^2$$

where η is the liquid-crystal viscosity. They further interpret t_r as the sum of t_h and t_v, where t_h is the time to develop hydrodynamic instability and t_v is the time to accelerate the liquid to maximum velocity. Their data indicate that $t_h \approx t_v$. They do not discuss the minimum possible risetime. The minimum time may be limited by the arc-over (breakdown) field of the sample cell.

Heilmeier[15] studied the falltime t_f, the natural relaxation time for turbulence after E is turned off. Experimentally, he found that

$$t_f \propto \sigma^{-1/2}\, L^2$$

and he interprets this on the premise that the walls "regain control", that is, the reorientation of initially turbid regions starts at the walls and propagates into the bulk at a diffusion-controlled rate.

Heilmeier[25] also reported that quenching can be done with a high-frequency field or with a unipolar pulse whose duration is less than t_s. One can understand the quenching process as follows. Disruption due to ions-in-transit is negligible, and the dielectric torque, now dominant, aligns molecules perpendicular to the field, as do the walls. The minimum quench times are about 2 msec,[25] and are approximately equal to the RC time of the sample cell.

References pp. 230-231

Little has appeared in the literature about turbulence in smectic liquid crystals. We have produced both turbulence and optical storage in certain smectic materials and find the electro-optic behavior rather different from that described above for nematic-phase materials.

Experiments were performed in a mixture of 11% by weight of p-n-butoxy-benzaldehyde (an isotropic liquid at room temperature), and 89% of p-n-butoxy-benylidene-p-n-butylaniline. The mixture has a nematic Schlieren texture from 29°C to the isotropic point at 49°C. A smectic phase exists from 21°C to 29°C. We tentatively identify this as the "smectic A" phase of Sackmann and Demus[26] because it has either a simple fan texture or a polygon texture that arises following a mild mechanical shear. Broken fans are occasionally seen in the texture, so the phase may be smectic C. Below 21°C, one observes a mosaic texture, smectic B. We did not investigate whether a third smectic phase exists at low temperatures.

The molecules have $\epsilon_{\parallel} > \epsilon_{\perp}$, judging from the regular alignment that was produced in the nematic phase with low frequency electric fields. Between crossed polarizers, the aligned nematic extinguished light propagating along z (Fig. 1). Some weak turbulence was seen at high fields around 10^5 V/cm.

Over a narrow temperature range, from 21°C to 23°C, turbulence was induced in the smectic liquid with dc fields or low frequency ac fields (f < 600 Hz). Starting at V = 0, domains did not form with increasing V; instead, an irregular material flow began at the interstices between fans. This is shown in the photomicrographs of Fig. 4, which compare the V = 0 and V = 38 V rms conditions. Also, the number of radial lines within each fan increased, and these "blades" rocked back and forth in a manner reminiscent of nematic domains, parametrically excited by ac drive (Tables I and II). While the blade oscillation was going on, the turbulence spread from its nucleation sites at the fan edges.

After field removal, a polygon texture forms. This texture, like the initial fan texture, remains for hours at E = 0 and scatters light. However, either texture can be erased rapidly with an ac field whose frequency exceeds 600 Hz. When the field is on, most of the texture becomes homeotropic (optically isotropic) along E, although small areas rotate slowly about the z axis in vortex fashion.

The smectic-phase electro-optic effects are summarized in Fig. 5, which presents the diffuse optical reflectance vs applied voltage for two excitation frequencies. The ordinate in Fig. 5 is the intensity of light back-scattered from the sample, which was measured with an integrating sphere. The turbulence threshold (24 V rms at f = 30 Hz) is three times higher than for the nematic of Fig. 3. At V = 130 V rms, the contrast between the clear and translucent states is comparable to that found in nematics (see Fig. 3). The reflectance of the stored state (Fig. 5) exhibits some

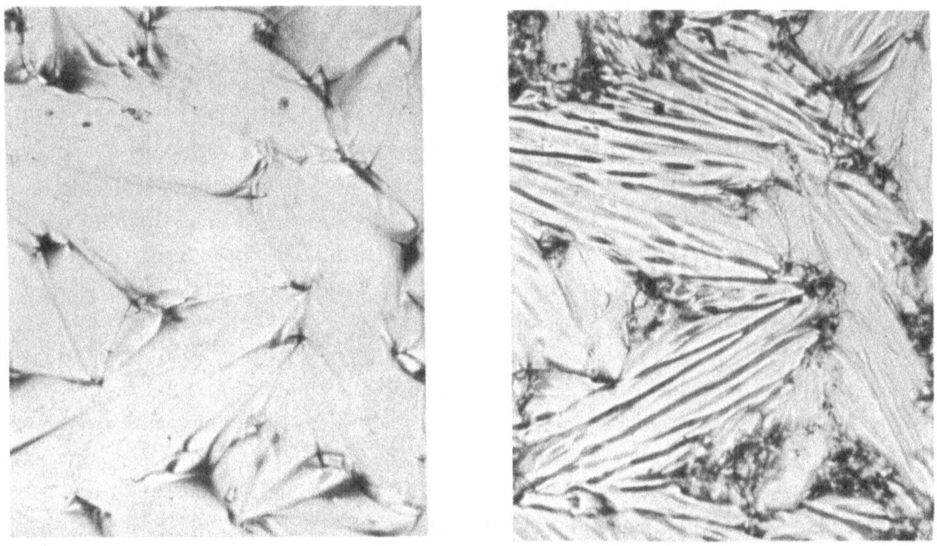

Fig. 4. Electric field-induced turbulence in the smectic phase of 11% p-n-butoxybenzaldehyde, 89% p-n-butoxybenzylidene-p-n-butylaniline at 23°C; (a) V = 0, (b) V = 38 V rms at f = 50 Hz. Photomicrographs taken with transmitted unpolarized light. Area shown = 350 μm x 450 μm. Cell thickness = 25 μm. Applied field ⊥ page.

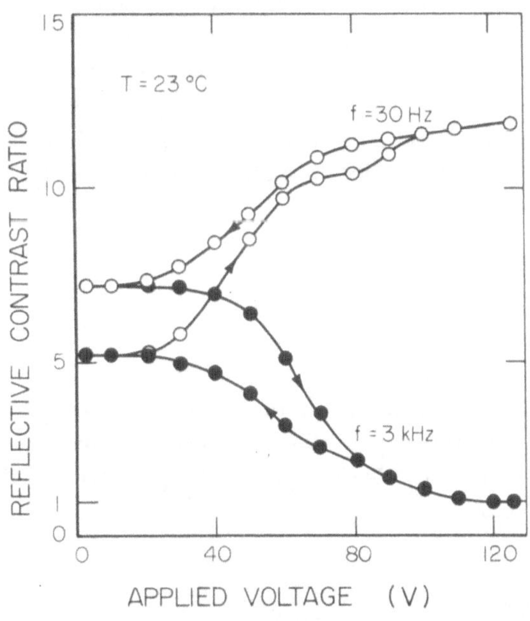

Fig. 5. Optical scattering vs applied rms voltage for two excitation frequencies 30 Hz and 3000 Hz. The smectic mixture, the temperature, and the cell thickness are the same as in Fig. 4.

References pp. 230-231

hysteresis because it depends slightly upon the initial $V \neq 0$ state.

Insight into the preceeding smectic-phase results can be gained from Carr's experiments[14] on an analogous liquid crystal EMC. Like the 89/11 mixture, EMC has $\epsilon_{\|} > \epsilon_{\perp}$. The smectic-A phase of EMC aligns anomalously with long axes perpendicular to E when it is subjected to a low-frequency field whose amplitude is less than the turbulence threshold in Fig. 5. To explain this, Carr[14] invokes conductive-torque theory. Because the smectic-A phase of EMC has anisotropic conductivity, Carr postulates that polarization charges arise at the interface of aligned molecular clusters when the field is applied.

A similar interpretation seems reasonable for our smectic mixture, namely, that a destabilizing conduction-induced torque is dominant for excitation frequencies less than 600 Hz. Since charges congregate on the smectic-fan boundaries, one expects a weaker conduction-induced torque within fans, which may explain the oscillatory blade instability described above (Fig. 4(b)).

The absence of domains (Fig. 4(b)) may be correlated with the positive dielectric anisotropy. Vistin and Kapustin[27] report field-induced domains in the smectic-A phase of p-(n-heptyl)oxybenzoic acid, but the nematic phase of their compound formed domains, unlike the nematic phase of our compound.

With high dc fields, Castellano and McCaffrey[28] converted the smectic-B phase of p-butoxybenzlidene-p′-aminopropiophenone to a homeotropic liquid crystal. They believe that a smectic-to-nematic phase transformation took place because, according to their estimates, Joule heating in the liquid exceeded $C_V \Delta T$, where C_V is the liquid's specific heat and ΔT is the temperature width of the smectic-B phase. This phase-transition hypothesis does not seem valid for the electro-optic response presented in Figs. 4 and 5.

FIELD-INDUCED HELIX PITCH DILATION

Light-control applications of helix pitch dilation are in an earlier stage of development than those of turbulence. In this section, new results on pitch dilation are given that pertain to three applications: Information display, light modulation and light-beam deflection. The basic experimental properties of this electro-optic effect are described first.

BASIC PROPERTIES

The cholesteric liquid is contained in a sandwich cell as in Fig. 1. The liquid consists of several cholesteric compounds mixed together such that $\chi_{\parallel}^{e} > \chi_{\perp}^{e}$, where $\chi_{\parallel}^{e}$ and $\chi_{\perp}^{e}$ are the electric susceptibility components parallel to and perpendicular to the molecular alignment at any point in the liquid. Initially, the film has a disturbed texture, and the applied field is parallel to **h**, the helix screw axis. Since $\chi_{\parallel}^{e} > \chi_{\perp}^{e}$, the field destroys the disturbed ordering, reorienting helix domains 90° and producing the undisturbed state. Reorientation requires a field of about 0.8 E_c (defined below).

Now, because E is perpendicular to **h**, the helix pitch dilates with increasing E. As E approaches a critical field E_c, the pitch becomes infinite, diverging logarithmically in the manner predicted by de Gennes[29] and Meyer.[30] At $E = E_c$, the molecules align parallel and the phase changes from cholesteric to nematic. Kahn[31] has experimentally verified the relation given by de Gennes[29] and Meyer[30] for the critical field:

$$E_c = (\pi^2/2p_o)[(k_{22})/(\chi_{\parallel}^{e} - \chi_{\perp}^{e})]^{1/2} .$$

Here k_{22} is an elastic modulus of the liquid crystal and p_o is the zero field pitch. A similar formula applies to the magnetic-field case.[29,30]

The optical scattering geometry for the undisturbed film is given in Fig. 6. Each helix domain, indicated by parallel lines, is a local site for Bragg scattering.

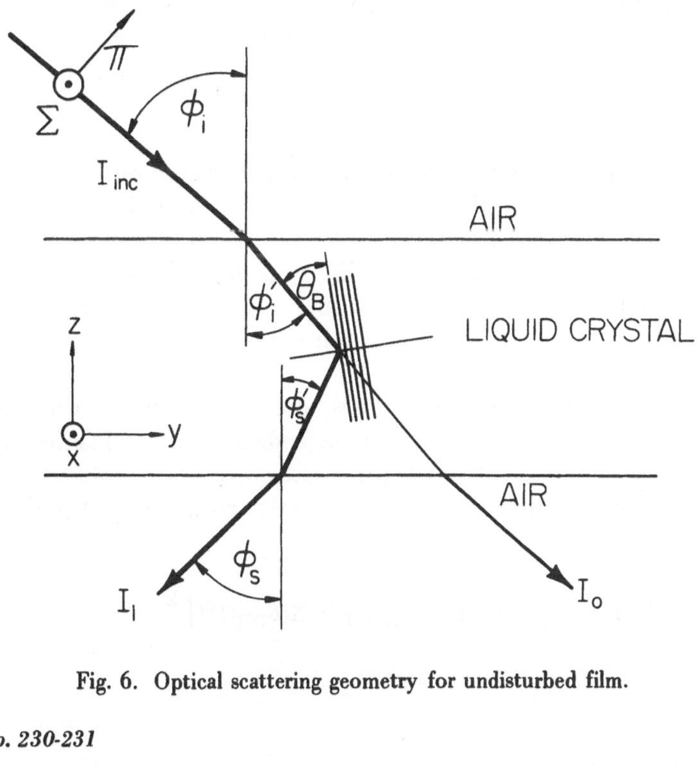

Fig. 6. Optical scattering geometry for undisturbed film.

A light beam of intensity I_{inc} is incident upon the film at an angle ϕ_i with respect to the normal. Part of this beam is diffracted and exits the film at an angle ϕ_{s1} with an intensity I_1. The undeviated portion leaves the film with intensity I_o. As shown in Fig. 6, the input beam can be polarized in the plane of incidence (π polarization) or normal to the plane of incidence (Σ polarization). Figure 6 shows transmitted I_0 and I_1. Alternatively, reflective diffraction is obtained by placing a mirror at the second interface.

The most intense light scattering comes from helix domains that satisfy the Bragg condition, defined as specular reflection off the helix layers (see Fig. 6). Such domains often have **h** tilted slightly out of the x-y plane. Fulfillment of the Bragg condition leads to the well-known relation between the scattered optical wavelength λ and the angles of incidence and scattering:

$$m\lambda = 2np \sin ½ [\sin^{-1} (\sin\phi_i/n) + \sin^{-1} (\sin\phi_{sm}/n)] \quad (1)$$

where n is the average refractive index of the liquid and p is the helix pitch. This relation takes account of multiple diffraction, m being the spectral order number and ϕ_{sm} the scattering angle of the mth-order beam.

Since p increases with E, one sees from Eq. (1) how E influences λ and ϕ_{sm}. For incident white light, the color components are dispersed in angle in accordance with Eq. (1) and the diffraction angles $\phi_{sm}(\lambda)$ are field controlled. When p_o is within the 0.3 to 0.6 μm range, visible colors (I_1) are observed at convenient viewing angles. At some vantage points, the observed color shifts from blue to red with increasing E.

In practice, the helix-domain axes **h** are not randomly oriented in the x-y plane. We found this to be so when the electrode surfaces were rubbed in the **x** direction before the cholesteric liquid was added. The resulting wall orientation persisted in the undisturbed state and many domains lined up along **x** (with **h** $\parallel$ **y**) as shown in Fig. 7. We deduced these orientations from measurements of I_1 when **k** $\perp$ **x** and when **k** $\perp$ **y**, where **k** is the propagation vector of incident light.

The theoretical dependence of I_1 on ϕ_i and ϕ_{s1} has not yet been worked out; however, a useful model for the intensity dependence can be obtained from the acousto-optic interaction (see Dixon's paper elsewhere in this volume). For ultrasonic light diffraction, Cohen and Gordon[32] derive an intensity dependence of the form

$$I_1/I_{inc} \propto \left\{\sin[½ K(\theta-\theta_B)W]/½ K(\theta-\theta_B)W\right\}^2 \quad (2)$$

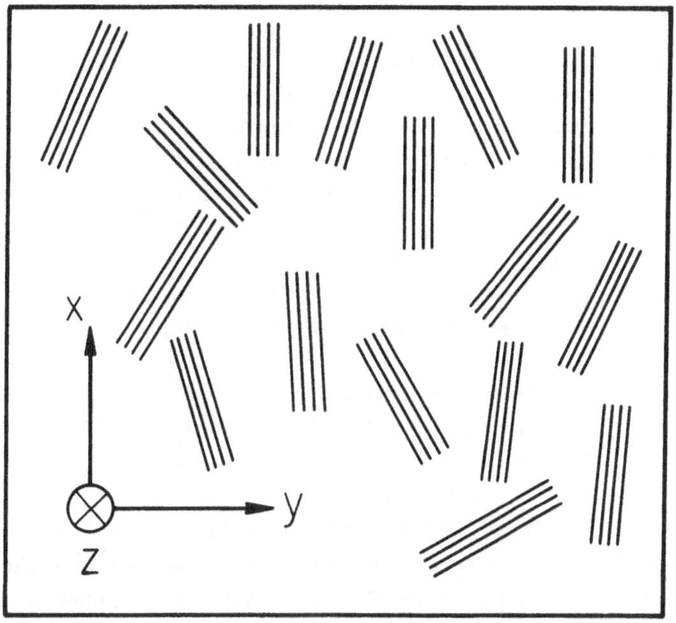

Fig. 7. Distribution of helix domains in an undisturbed film when the walls are rubbed in the x direction. The view perpendicular to the x-y plane is shown.

where K is the acoustic propagation constant, θ the incidence angle of light on the acoustic equiphase surface, θ_B the Bragg angle, and W the acoustic beamwidth. To check the relevance of this model, measurements were made of I_1 vs (ϕ_i, ϕ_{s1}) in several undisturbed cholesteric compounds. In direct analogy to the Cohen-Gordon result (Eq. (2)), we found that the following relation gave a good fit to the experimental data:

$$I_1/I_{inc} \propto \left\{ \sin[\beta(\phi_i' - \phi_{s1}')]/\beta(\phi_i' - \phi_{s1}') \right\}^2 \tag{3}$$

where β is a constant, and, from Snell's law, $\sin \phi_i = n \sin \phi_i'$ and $\sin \phi_{s1} = n \sin \phi_{s1}'$ Equation (3) includes the well-known result that I_1 peaks when $\phi_i = \phi_{s1}$, which implies that most sites in Fig. 6 have **h** $\perp$ **z**, as expected.

Second-order diffraction was observed at 24°C in the following undisturbed mixtures: 25% cholesteryl chloride, 64% cholesteryl geranyl carbonate, 11% cholesteryl myristate with p_0 = 5400 Å, and 45% cholesteryl chloride (CC), 31% cholesteryl oleyl carbonate (COC), 24% cholesteryl nonanoate (CN) with p_0 = 4600 Å. The second-order diffracted colors had maximum intensity when the applied field E_z was adjusted such that $\phi_{s2} = \phi_i$, a result analogous to Eq. (3). To our knowledge, this is the first report of I_2 in undisturbed films, although Berremen and

References pp. 230-231

Scheffer[33] reported second-order Bragg reflection in disturbed materials.

Under the microscope, the cholesteric film exhibits a focal-conic texture illustrated in Fig. 8. The interference colors that are seen when the film is viewed between crossed polarizers demonstrate that the film consists of localized birefringent regions, several microns in size. We have been calling these birefringent areas "helix domains" (they are, of course, different entities than nematic domains). For the moment, let us call them "grains". At high fields ($E > 0.6\ E_c$), the grains act as follows: Grain boundaries distort, some grains coalesce, and there is mass transport in the irregular regions that separate grains. The irregular areas give rise to diffuse light scattering that competes with light diffraction from grains.

It has not been recognized that the diffraction efficiency I_1/I_{inc} can be high. For example, we found efficient diffraction in an undisturbed film of 45% CC, 35% COC, 20% CN at 24°C. In our measurement, the illumination was collimated 6328 Å laser light incident at 45°. The dc field was set at 9.3 V/μm (E_c = 10 V/μm). This field shifted ϕ_{S1} to 45°, which maximized I_1 and gave a crescent-shaped diffraction pattern. The photodetector recorded the total light intensity within the crescent. The pitch was 1.25 p_o and p_o = 4500 Å. Measurements were made as a function of L for both the Σ and π input polarizations, with the results shown in Fig. 9. The stronger π interaction means that the π polarization sees a larger spatial "ripple" in the liquid's refractive index.

For device applications, the attractiveness of the 2%-to-9% efficiencies in Fig. 9 is mitigated by the wide angular spread in the beam I_1. The linewidth $\Delta\phi_{S1}$ of I_1 was measured under conditions identical to those in Fig. 9 ($\Delta\phi_{S1}$ is the separation between angles where I_1 (ϕ_{S1}) is down 3 dB from its maximum). A rapid increase of $\Delta\phi_{S1}$ with L was found. For L = 13, 25, 38, and 50 μm, the $\Delta\phi_{S1}$ values were 7°, 9°, 10°, and 16°, respectively. These values are consistent with the value $\Delta\phi_{S1}$ = 9° reported by Adams et al.[34] for a disturbed CC-CN film.

At present, little is known about transient response. The only empirical data have been given by Kahn,[31] who found a 1 msec electro-optic risetime for pitch dilation and a 2 msec falltime (p_o = 2330 Å).

INFORMATION-DISPLAY APPLICATIONS

A recent paper outlined display applications of pitch dilation.[35] The basic display concept is to apply the field to small, selected areas of the film. The localized field application gives a spatially modulated diffraction pattern - a pattern of contrasting colors. Matrix addressing is feasible because the nonlinear λ-vs-E curve has a color-shift threshold.[35] But the matrix has this drawback: Its color contrast

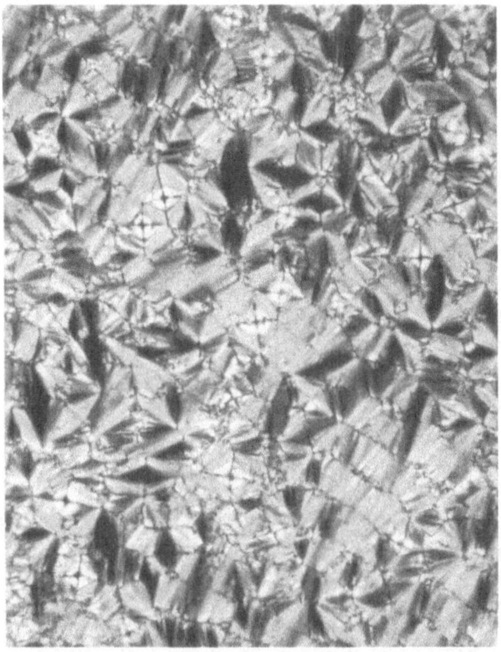

Fig. 8. Undisturbed 25 μm layer of 45% CC, 35% COC, 20% CN at 24°C with E = 0. Area shown is 350 μm x 450 μm. Photomicrograph taken using crossed polarizers.

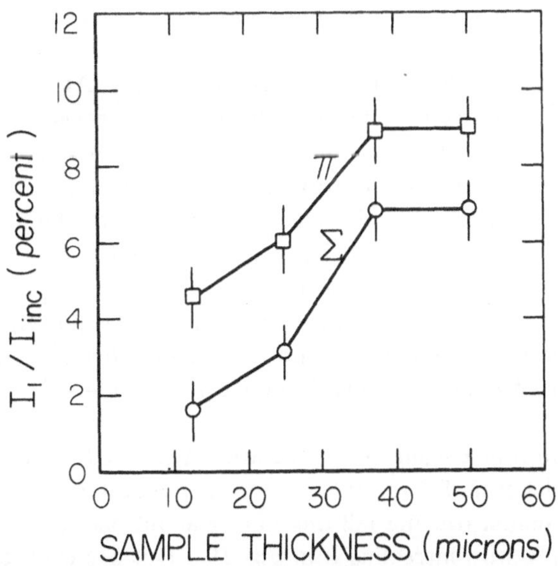

Fig. 9. I_1/I_{inc} vs L for Σ and π input polarizations. The liquid crystal is the same as in Fig. 8. Other experimental conditions are listed in the text.

References pp. 230-231

cannot be maintained in situations where the color-shift persistence is shorter than 20% of the frametime.

We also constructed a pitch-dilation display that had individually addressed elements. These elements did not have to be refreshed frequently because of their RC storage time. Multisecond storage exists since the cholesteric's resistivity ranges typically from 10^{13} to 10^{15} ohm-cm. The Fig. 9 efficiencies were useful, and the diffuse scattering did not present a serious problem. Both displays, however, had a narrow range of viewing angles, and they required collimated back illumination for optimum performance.

LIGHT MODULATION

The Fig. 6 experiment becomes equivalent to the Raman-Nath effect[36] in the limit $np/\lambda >> 1$, and $\phi_i = 0$. Then, from Eq. (1), the scattering angles are $\sin \phi_{sm} = m\,\lambda/p$ as in ultrasonic diffraction.[36]

Light modulation takes place in the acousto-optic interaction as follows. The fixed-frequency acoustic power controls the peak refractive index perturbation Δn, thereby governing the light intensity I_m/I_{inc} through the Bessel function relation $J_m^2(2\pi\Delta nL/\lambda)$. A similar light modulation would seem to hold for the liquid-crystal system of Fig. 6, but this is not the case because the applied field E_z does not, to first order, change the index corrugation Δn, although it does alter p and ϕ_{sm}. However, by decreasing ϕ_{s1}, the field does modulate I_1 in accordance with Eq. (3).

Modulation of I_o is also found in undisturbed films. At zero field, off-axis scattering robs the undeviated beam of intensity as it propagates through the liquid, while at high fields ($E \approx 0.9\ E_c$), intensity depletion is less severe. Consequently, the field can swing I_o between two distinct levels. We documented this modulation by measuring I_o/I_{inc} for E = 0 and for E = 9.3 V/μm in the liquid-crystal mixture of Fig. 9 at room temperature. The data are plotted vs L in Fig. 10 for a π input polarization. The component of I_o polarized parallel to the input was selected using an analyzer. The detector collected only the on-axis portion of the forward-scattered beam because its 0.8 cm aperture was 160 cm from the sample. As in Fig. 9, the collimated input beam had λ = 6328 Å and $\phi_i = 45°$.

In Fig. 10, we find a significant modulation depth at L = 13 μm, where the transmission doubles from 5% to 10%. Unfortunately, the modulation rates are quite low, the minimum rise and fall times being about one millisecond. The intensity I_o drops sharply with increasing L in Fig. 10. This decrease stems from the exponential growth of diffuse scattering with increasing optical path. The modulation effect in Fig. 10 can be attributed to a field-enhanced macroscopic ordering, i.e. a transition to a quasi-nematic state.

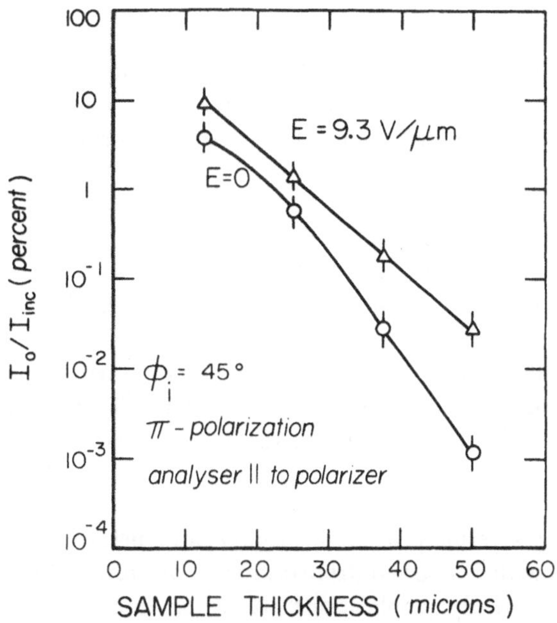

Fig. 10. I_o/I_{inc} vs L for E = 0 and E = 9.3 V/μm. The liquid crystal is the same as in Fig. 8. Other experimental conditions are listed in the text.

LIGHT BEAM DEFLECTION

Beam deflection refers to the voltage-induced change in the diffracted beam's angular position. In the cholesteric liquids mentioned above, wide angular sweeps are possible with a few-hundred-volts drive when $0.3\ \mu m < p < 0.6\ \mu m$. Drive power is low; diffraction efficiency is high (Fig. 9); deflection rates are about 10^3 degrees/sec. One can, therefore, envision a voltage-tuned monochromator based on pitch dilation.

There are, however, many limitations on deflection. When $p \sim \lambda$, the light is diffracted in cones (or portions thereof) centered on the input beam. Also, the cones are not sharp ($\Delta\phi_{S1} > 7°$) and the electro-optic response times are slow.

Deflection conditions may be more favorable when $p >> \lambda$, as in the experiments of Sackmann et al.[12] There the diffracted beam retained the incident beam's circular cross section. Sackmann's photographs also show a narrow linewidth $\Delta\phi_{S1}$, although this may have been due to the five-hour field application.

CONCLUSION

Liquid-crystal light-control effects, both electro-optic and magneto-optic, have been reviewed. New experimental results have been presented on several electro-optic effects: Dynamic scattering and erasable storage in smectics, and helix pitch dilation in cholesterics.

Recent experiments have established the importance of conductivity for field-induced turbulence in nematics; nevertheless, the mechanism of charge production and the charge-carrier species are not yet fully understood.

In smectic liquid crystals, the main features of the turbulence observations (Sec. IV) can be explained using the conductive-torque theory proposed by Carr[14] and Helfrich.[16] A detailed verification of this model in smectics remains to be carried out.

A diffuse light-scattering process was found in undisturbed cholesterics. Although it competes with voltage-controlled diffraction from helix domains, it is not a serious drawback in display applications. However, the diffuse scattering, together with the slow transient response and the diffracted-beam divergence, make other applications of pitch dilation appear less promising, particularly light-modulation and light-beam deflection.

ACKNOWLEDGMENT

The author is deeply grateful to M. J. Rafuse for synthesizing the liquid crystal compounds used in this work.

REFERENCES

1. *P. Chatelain, Acta Cryst., 1(1948) 315.*
2. *W. Haas, J. Adams, and J. B. Flannery, Phys. Rev. Letters, 24(1970) 577.*
3. *E. F. Carr, J. H. Parker, and D. P. McLemore, Liquid Crystals and Ordered Fluids, Plenum Press, New York, 1970, p. 201.*
4. *R. A. Soref, Laser Focus, 6, No. 9 (1970).*
5. *R. A. Soref, Appl. Opt., 9(1970) 1321.*
6. *J. W. Shelton and Y. R. Shen, Phys. Rev. Letters, 24(1970) 23.*
7. *H. Baessler, T. M. Laronge, and M. M. Labes, J. Chem. Phys., 51(1969).*
8. *I. G. Chistyakov and V. M. Chiakovskii, Soviet Phys. - Cryst., 12(1968) 770.*
9. *J. O. Kessler, Liquid Crystals and Ordered Fluids, Plenum Press, New York, 1970, p. 361.*
10. *V. K. Freedericksz and V. Zolina, Trans. Faraday Soc., 29(1933) 919.*
11. *G. Durand, L. Leger, F. Rondelez, and M. Veyssie, Phys. Rev. Letters, 22(1969) 227.*

12. E. Sackmann, S. Meiboom, L. C. Snyder, A. E. Meizner, and R. E. Dietz, J. Am. Chem. Soc., 89(1967) 598.
13. P. G. de Gennes, Solid State Comm., 8(1970) 213.
14. E. F. Carr, Phys. Rev. Letters, 24(1970) 807.
15. G. H. Heilmeier, L. A. Zanoni, and L. A. Barton, Proc. IEEE, 56(1968) 1162.
16. W. Helfrich, J. Chem. Phys., 51(1969) 4092.
17. R. Williams, J. Chem. Phys., 39(1963) 384.
18. P. A. Penz, Phys. Rev. Letters, 24(1970) 1405.
19. G. Durand, M. Veyssie, F. Rondelez, and L. Leger, Compt. Rend., 270B(1970) 97.
20. D. Jones, L. Creagh, and S. Lu, Appl. Phys. Letters, 16(1970) 61.
21. H. J. Caulfield and R. A. Soref Appl. Phys. Letters, (to be published).
22. R. A. Soref, J. Appl. Phys., 41(1970) 3022.
23. H. Koelmans and A. M. van Boxtel, Phys. Letters, 32A(1970) 484.
24. S. Lu and D. Jones, Appl. Phys. Letters, 16(1970) 484.
25. G. H. Heilmeier, L. A. Zanoni, and L. A. Barton, IEEE Trans. Elec. Dev., ED-17(1970) 22.
26. H. Sackmann and D. Demus, Liquid Crystals, Gordon and Breach, London, 1967, p. 341.
27. L. K. Vistin and A. P. Kapustin, Soviet Phys. - Cryst., 13(1969) 284.
28. J. A. Castellano and M. T. McCaffrey, Liquid Crystals and Ordered Fluids, Plenum Press, New York, 1970, p. 293.
29. P. G. de Gennes, Solid State Comm., 6(1968) 163.
30. R. B. Meyer, Appl. Phys. Letters, 12(1968) 281.
31. F. J. Kahn, Phys. Rev. Letters, 24(1970) 209.
32. M. G. Cohen and E. I. Gordon, Bell System Tech. J., 44(1965) 693.
33. D. W. Berreman and T. J. Scheffer, Bull. Am. Phys. Soc., 15(1970) 286.
34. J. Adams, W. Haas, and J. Wysocki, J. Chem. Phys., 50(1969) 2458.
35. R. A. Soref, Proc. IEEE, 58(1970) 1163.
36. C. V. Raman and N. Nath, Proc. Ind. Acad. Sci., 2(1935) 406.
37. G. H. Heilmeier and W. Helfrich, Appl. Phys. Letters, 16(1970) 155.
38. G. H. Heilmeier and J. E. Goldmacher, Proc. IEEE, 57(1969) 34.
39. J. Wysocki, J. Adams and W. Haas, Molecular Cryst. & Liquid Cryst., 8(1969) 471.
40. G. H. Heilmeier and J. E. Goldmacher, J. Chem. Phys. 51(1969) 1258.
41. R. B. Meyer, Appl. Phys. Letters, 14(1969) 208.
42. J. E. Goldmacher and G. H. Heilmeier, (1970), U. S. Patent No. 3,499,702.
43. R. Williams, J. Chem. Phys., 50(1969) 1324.
44. R. P. Twitchell and E. F. Carr, J. Chem. Phys., 46(1967) 2765.
45. E. F. Carr, Adv. Chem. Ser., 63(1967) 76.
46. E. F. Carr, Molecular Cryst. & Liquid Cryst., 7(1969) 253.
47. G. H. Heilmeier, J. A. Castellano and L. A. Zanoni, Molecular Cryst. & Liquid Cryst., 8(1969) 293.
48. H. Baessler and M. M. Labes, J. Chem. Phys., 51(1969) 1846.
49. J. R. Hansen and R. J. Schneeberger, IEEE Trans. Elec. Dev., ED-15, (1968) 896.
50. J. L. Fergason, (1968), British Patent No. 1,123,117.
51. W. Haas, J. Adams, Chem. Abstr. 84933d, (1969), French Patent No. 1,573,335.
52. W. Haas, J. Adams and J. Wysocki, Appl. Opt., Suppl. 3(1969) 196.

OPTICS OF SOLID STATE PHASE TRANSFORMATIONS

DAVID ADLER

Department of Electrical Engineering and Center for Materials Science and Engineering, Massachusetts Institute of Technology, Cambridge, Massachusetts*

and

JULIUS FEINLEIB

Energy Conversion Devices, Inc., 1675 West Maple Road, Troy, Michigan

ABSTRACT

The optical spectra of Mott insulators and amorphous semiconductors are discussed in detail, and are contrasted with the spectra of ordinary crystalline semiconductors. Particular emphasis is placed on the transition-metal oxides and the chalcogenide glasses. The major difference between the spectra of such materials and those of the more common semiconductors arise from the breakdown of the k-conservation selection rules and the presence of intrinsic localized states.

The striking changes in the optical properties of materials that exhibit conductivity anomalies are analyzed. In particular, temperature-induced insulator-metal transitions and photo-induced amorphous-crystalline transitions are discussed. The former often yield striking changes in optical properties, transforming from a material transparent in the infrared with an absorption edge to one of high reflectivity with a plasma edge. The latter transformations can shift the absorption edge to lower energy, resulting in a transition from a material transparent to light of a given frequency to one opaque at the same frequency. The reversibility of such transformations and the resolution obtainable using modern laser technology suggest the application of photon-induced amorphous-crystalline transitions as optical memories of extremely high packing density. In addition, write, erase, and read operations can be accomplished with the same laser.

References pp. 252-253

INTRODUCTION

Many types of solid-state phase transitions exist.[1,2] Since we shall primarily be concerned with those transitions which produce striking changes in the optical properties of the material, we shall restrict discussion to three classes of phenomena: Mott transitions, semiconductor-metal transitions, and amorphous-crystalline transitions. In addition to the discontinuous nature of the optical properties, these transitions also have in common sharp anomalies in the electrical conductivity, which can change by a factor of up to 10^8. The conductivity jumps have been the object of considerable attention in the past, but only recently has much interest been focused on the optical transformations.

The optical properties of ordinary metals and semiconductors are very well known and need not be rehashed here. However, the properties of Mott insulators and amorphous semiconductors differ sufficiently from those of ordinary crystalline semiconductors that they will be carefully analyzed. We shall conclude with a detailed discussion of the optics of the reversible, laser-induced amorphous-crystalline transitions that have recently been discovered in certain chalcogenide glasses.

OPTICAL PROPERTIES OF MOTT TRANSITIONS

Mott[3,4] was the first to discuss the main reason why a large class of transition-metal and rare-earth compounds are insulating when a straightforward band approach suggests that they should be metallic. It had previously been taken for granted that the outer electrons in a crystalline solid delocalized and spread into Bloch states, this delocalization, in fact, being responsible for the cohesive energy of the material by reducing the electronic kinetic energy. Mott called attention to the fact that we must pay a price for this kinetic energy reduction. Delocalization leads to the occasional simultaneous presence on a single ion core of two outer electrons initially on spatially separated ion cores. Since two electrons in the same vicinity Coulombically repel each other, delocalization can bring about an increase in potential energy. Since the decrease of kinetic energy is of the order of half the bandwidth, or approximately 10 eV or more for outer electrons, while the increase of potential energy is screened down to small values in metals and even semiconductors, the Bloch state is the lowest energy state of the system. But Mott pointed out that for narrow-band insulators, the potential energy increase could well outweigh the kinetic energy decrease, and delocalization then does not occur. The ground state is that in which the electrons remain localized on their respective ion cores. Such a material is now known as a Mott insulator.

The optical properties of Mott insulators are very different from those of ordinary band insulators.[5] It is very clear that when the electrons are localized, the periodicity of the lattice is irrelevant, and **k** is not a good quantum number. Each localized state, in fact, is made up from all **k** states in the Brillouin zone, and **k**-conservation is not a factor in optical transitions. Furthermore, since the localized electrons are generally either d or f electrons, higher multiplet and crystalline-field split states exist for each configuration. In addition, it is possible to excite higher configurations optically, as the following transitions from a d^n or f^n ground state:

$$d^n + d^n + \hbar\omega \longrightarrow d^{n+1} + d^{n-1}$$

$$f^n + f^n + \hbar\omega \longrightarrow f^{n+1} + f^{n-1}$$

In each case, higher crystalline-field and multiplet split final states can be obtained. For example, in NiO, which has a $3d^8$ ground state, the following optical transitions are possible solely within the 3d bands:[6]

$$3d^8 + \hbar\omega \longrightarrow 3d^{8*} \quad (1)$$

$$3d^8 + 3d^8 + \hbar\omega \longrightarrow 3d^7 + 3d^9 \quad (2)$$

$$3d^8 + 3d^8 + \hbar\omega \longrightarrow 3d^7 + 3d^{9*} \quad (3)$$

$$3d^8 + 3d^8 + \hbar\omega \longrightarrow 3d^{7*} + 3d^9 \quad (4)$$

$$3d^8 + 3d^8 + \hbar\omega \longrightarrow 3d^{7*} + 3d^{9*} \quad (5)$$

In (1) - (5), the notation $3d^{n*}$ refers to all higher crystalline-field and multiplet split states, so that these equations represent an enormous number of possible transitions.

But, in addition, other bands always exist in the vicinity of the Fermi energy. If this were not true, the Mott insulator would not bind. For example, in the case of EuS, there is a filled anion 3p band below E_F and empty cation 5d and 6s bands above E_F, in addition to the localized $4f^n$ states. In NiO, the oxygen 2p band is just below E_F and the nickel 4s band is somewhat above E_F. Since s and p orbitals are generally quite spread out relative to d and f orbitals of approximately the same energy, we should expect ordinary Bloch-type s and p bands to form, even in Mott insulators. It is not yet clear whether **k**-conservation need be a valid selection rule for optical transitions between such bands. For NiO, an example of such a transition is:

$$\text{(electron in 2p band)} + \hbar\omega \longrightarrow \text{(electron in 4s band)} \quad (6)$$

References pp. 252-253

Thus, (6) represents normal interband transitions. However, these are the *only* normal transitions in the material.

It is also possible to excite an electron from a localized state to a Bloch state and vice versa. In our example of NiO, these possibilities introduce four new sets of transitions:

$$3d^8 + \hbar\omega \longrightarrow 3d^7 + \text{(electron in 4s band)} \qquad (7)$$

$$3d^8 + \hbar\omega \longrightarrow 3d^{7*} + \text{(electron in 4s band)} \qquad (8)$$

$$3d^8 + \text{(electron in 2p band)} + \hbar\omega \longrightarrow 3d^9 \qquad (9)$$

$$3d^8 + \text{(electron in 2p band)} + \hbar\omega \longrightarrow 3d^{9*} \qquad (10)$$

In (7) - (10), **k**-conservation does not apply.

The complexity of the optical spectrum of a Mott insulator is evident. A schematic plot of the absorption of pure, stoichiometric NiO as predicted[5] from the free ion spectra of Ni^{2+} and O^{2-} is shown in Fig. 1. This sketch is in qualitative agreement with experiment.[7] Were it not for the suppression of optically induced

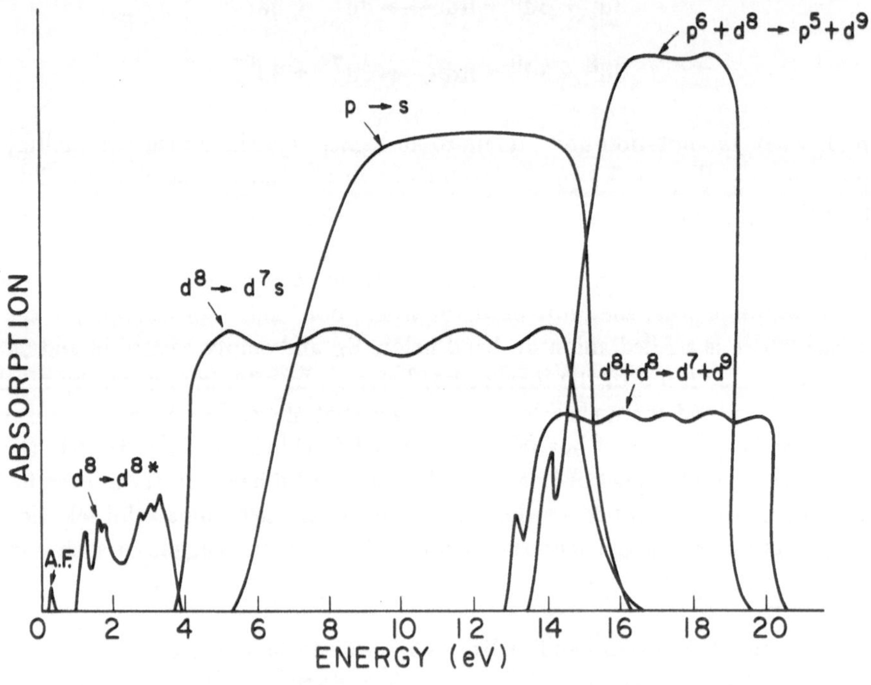

Fig. 1. Sketch of the optical absorption of pure, stoichiometric NiO below 20 eV, as predicted from the free ion energies.

$d \rightarrow d$ transitions, NiO would be virtually opaque throughout the entire spectrum. As it is, a trough in the crystalline-field absorptions between 2.0 eV and 2.8 eV leads to intrinsic NiO being green in color.

However, NiO has never been prepared in pure, stoichiometric form. When non-stoichiometry exists or heterovalent ions are present in the lattice, configurations other than $3d^8$ exist in the *ground state*. This introduces a number of additional transitions equivalent to all those in (1) - (10). These are not too important optically, since the density of, say, $3d^7$ ions is small compared to the density of $3d^8$ ions. However, the fact that the crystalline-field spectrum of $3d^7$ contains absorptions in the 2.0 - 2.8 eV range causes non-stoichiometric or doped NiO to be black rather than green.

To summarize, the optical spectra of Mott insulators are extremely complex, and many different types of transitions are possible. Generally speaking, **k**-conservation is inapplicable, and most materials are highly absorbing throughout the entire optical energy range.

In Mott's original papers,[3,4] he suggested the possibility that a Mott insulator could be converted to a metal under the application of high pressures, and presented arguments that such a transition should be sharp. The metallic state should come about because pressure increases the nearest-neighbor overlap and thus the bandwidth, and eventually the kinetic energy reduction from delocalization must overcome the potential energy increase. The sharpness results from the fact that self-consistent screening leads to a reduction of the effective Coulomb repulsion between electrons as the critical energy balance is approached. Such a pressure-induced insulator-metal transition is called a Mott transition.

On the insulating side of the Mott transition, the optical spectrum is that described previously - relatively highly absorbing for an insulator in the visible and near i.r. because of the intraionic transitions. On the metallic side, the spectrum should be that characteristic of ordinary metals, albeit electronic correlations remain important. At low energies, free carrier absorption must occur, and the material becomes highly reflecting. However, at high energies, interband transitions determine the spectra of both the insulating and metallic states, and the possibility exists that these are essentially the same. But this can be the case only if the spectra in this energy range are dominated by transitions between Bloch-type bands, such as those given in (6) for NiO. The interband transitions represented by (7) - (10), on the other hand, will be very different once the 3d electrons delocalize. If we envision a metallic NiO, produced by a pressure-induced Mott transition,[8] a 3d band must form. This band may be split in two by the exchange energy, and each of these bands may be further split by the crystalline field into a lower t_{2g} sub-band and an upper e_g sub-band. The electron density in NiO is such that the 3d band is

80% filled. Many new 3d→4s and 2p→3d interband transitions are now possible, whether or not **k**-conservation is an important selection rule. Since it is clear from Fig. 1 that the spectrum of NiO in the 14-20 eV maximum absorption region is dominated by processes (9) and (10), a Mott transition in NiO should sharply change the high-energy optical spectrum as well as the spectrum below the 4 eV absorption edge. As we shall see in the next section, this predicted difference in the interband spectra through the transition provides a simple experimental means of differentiating a Mott transition from an ordinary semiconductor-metal transition.

OPTICAL PROPERTIES OF SEMICONDUCTOR-METAL TRANSITIONS

Ordinary semiconductor-metal transitions are induced by changes in the electronic structure of the material.[1,2] The semiconducting state is characterized by an energy gap separating a filled valence band from an empty conduction band. The optical spectrum is dominated by an absorption edge which sets in at the lowest energy allowed transition. If **k**-conservation is a valid selection rule, only vertical transitions are allowed. Below the absorption edge, transmission is high.

The metallic state is characterized by the collapse of the energy gap. If the Brillouin zone remains the same as in the semiconducting state, there must be at least two partially filled bands in the metallic state. However, if a change in crystalline symmetry occurs at the transition, a single partially filled band can occur.[9] If **k**-conservation is not important, this distinction is not optically observable.

Clearly, the major change in the optical spectrum of a material that undergoes a semiconductor-metal transition is a sharp drop in transmission at energies below the absorption edge. A typical example of this is shown in Fig. 2, which shows the transmission at 0.31 eV of a thin film of crystalline VO_2, which undergoes a sharp temperature-induced transition at 68°C.[10] The energy gap of VO_2 is approximately 0.6 eV.[10,11]

Generally bands other than the semiconducting valence and conduction bands exist in the vicinity of the Fermi energy. For example, in VO_2, the vanadium 3d band is the relevant one in determining the semiconducting or metallic nature of the material. However, the filled oxygen 2p band is only about 2.5 eV below E_F. Thus, interband transitions from the 2p band to unfilled parts of the 3d band will contribute to the optical spectrum above 2.5 eV. Small changes in the energies of the Bloch states near E_F will produce only slight shifts in the high-energy spectrum, despite the fact that they induce drastic effects on the dc conductivity and the low-energy spectrum. At still higher energies, the 2p→ 4s transitions begin to contribute, and in this range, the low and high temperature spectra may be indistinguishable.

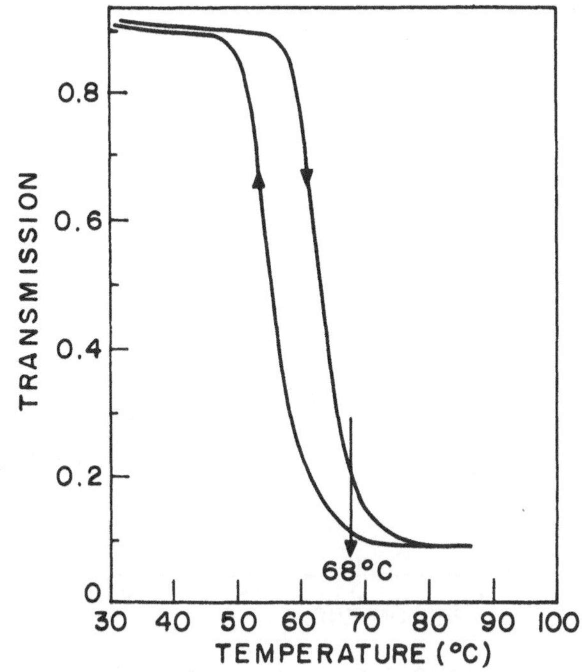

Fig. 2. Temperature dependence of the optical transmission at 0.31 eV of a 1000 Å film of crystalline VO_2 (Verleur *et al.*[10]).

Experimentally, the reflectivity of VO_2 as a function of energy above and below the transition temperature is as shown in Fig. 3.[10,12] At low energies, the spectra are entirely unrelated; the metallic reflectivity exhibits a Drude peak, while the semiconducting reflectivity falls sharply. However, above 2.0 eV, the spectra are essentially identical. Similar results hold for the transmission spectra.[10] We can conclude from these data that the transition in VO_2 is not likely to be a Mott transition, but is rather a Bloch-type semiconductor-metal transition.[13] Similar results appear to hold for the transition in $FeSi_2$.[14]

With respect to the modulation of light, clearly large changes occur only in the energy range below the semiconducting band gap. But the gaps in the semiconducting phase of materials that exhibit temperature-induced transitions to a metallic state appear to be rather small. Of the materials investigated to date, the band gaps range from 0.06 eV ($Ti_2 0_3$) to 0.6 eV (VO_2). However, a large number of ordinary semiconductors exhibit pressure-induced transitions to a metallic state,[15] and these include Si, GaAs, Se, CdS, and ZnS, materials whose gaps run the gamut from the i.r. through the u.v. The gaps decrease gradually with increasing pressure until they suddenly collapse to zero at the point of a first-order phase transformation.

It should also be noted that several common semiconductors, such as Ge, Si,

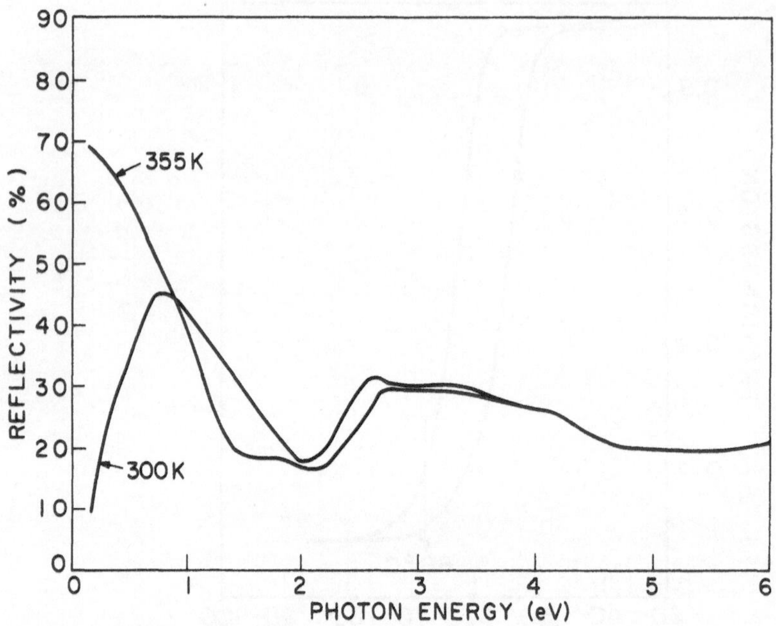

Fig. 3. Reflectivity as a function of energy for VO_2: (a) 90-300°K; (b) 355°K. (Verleur *et al.*[10]; Mokerov and Rakov[12].)

and InSb, transform to metals upon melting.[16] In each case, a change in the local coordination number occurs at the melting temperature, and it is this effect which leads to the sharp collapse of the energy gap.

OPTICAL PROPERTIES OF AMORPHOUS SEMICONDUCTORS

Amorphous semiconductors have no long-range periodicity, as do crystalline materials. However, they generally have a very high degree of short-range order, and the nearest-neighbor environment of a particular atom in an amorphous solid may be precisely the same as that in the corresponding crystal. It is this short-range order which is of major importance in determining the optical properties of amorphous semiconductors.

The first point to make with respect to disordered systems in general is that the fundamental translation group of the crystal is of no significance. Thus, we must give up entirely some of our time-honored concepts, such as the Brillouin Zone, **k**-conservation, and Bloch states. Although these losses are tantamount to a many-body theorist giving up his Green's functions, it is important not to overreact. We no longer have plots of E vs **k**, but we can retain density-of-states diagrams. We lose Bloch states, but the eigenstates of Schrodinger's equation may still be itinerant.

The simplest approach is to analyze the problem in terms of the tight-binding approximation of energy-band theory. In this limit, the major contributions to the electronic energies are the nature of the atom on which the electron is initially localized and the positions and nature of its nearest neighboring atoms. The effects of second and farther neighbors can be treated as a small perturbation. However, as we already pointed out, the nearest-neighbor environment in amorphous solids is generally the same as that of the corresponding crystal. Consequently, we can expect a strong similarity between the densities of states for amorphous and crystalline solids of the same composition. But we must bear in mind two important differences. Firstly, crystalline solids have sharp band edges, which are a consequence of long-range periodicity. If we introduce long-range disorder in the amorphous materials by means of a perturbation on the crystalline band structure, we find that amorphous solids have band tails, or states which are located in what would be the energy gaps of crystalline solids. Secondly, without Bloch's theorem, we cannot assume that all states are itinerant. Current theory[17,18] suggests that a critical density of states exists, above which the states are itinerant, but below which all states are localized. Thus, band edges, in the usual sense of the term, do not necessarily exist in amorphous materials. Instead, the energies at which the critical density of states for delocalization is reached can be thought of as mobility edges.

It is clear that the extent of the band tails must be directly related to the amount of disorder. For elemental amorphous semiconductors, such as Ge and Si, only long-range positional disorder is present, and the band tails should not be very extensive. Indeed as is shown in Fig. 4, the optical spectrum of amorphous Ge exhibits an absorption edge nearly as sharp as that of the crystal.[19] On the other hand, multicomponent amorphous alloys, such as the chalcogenide glasses, possess both positional and compositional disorder. For such materials, Cohen *et al.*[20] suggested that the band tails could become sufficiently extensive that the valence and conduction band edges overlap deep in the gap. If this occurs, a redistribution of electrons takes place at thermal equilibrium, leading to a large density of positively and negatively charged traps in the gap, even at low temperatures. This postulate is in agreement with a large number of experimental observations peculiar to the chalcogenide glasses.[20,21]

At first glance, it would appear that optical measurements should permit direct determination of the extent of the band tails in amorphous semiconductors. However, it now appears that such is not the case. It is well known that crystalline semiconductors generally exhibit an exponential tail on the low-energy side of the absorption edge, the so-called Urbach tail, which has nothing to do with any structure in the density of states. Although many explanations of the Urbach tail have

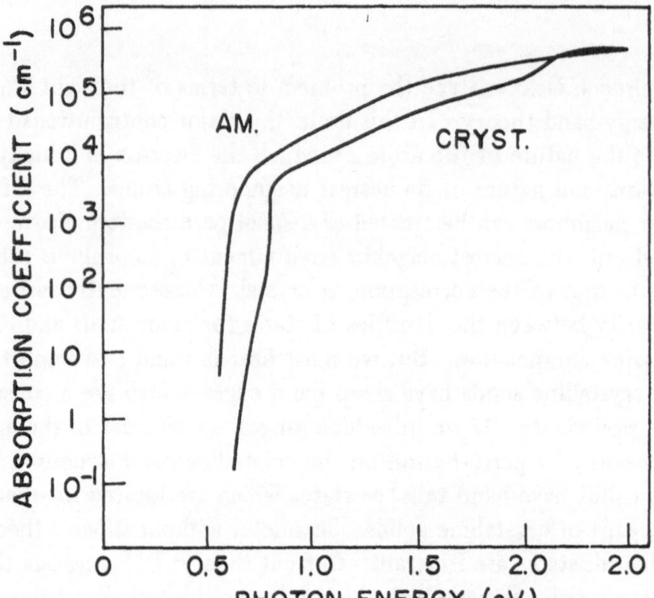

Fig. 4. Optical absorption as a function of energy for amorphous and crystalline Ge (Donovan *et al.*[19]).

been proposed,[22] it is still basically a mystery, particularly with regard to the fact that the edge has essentially the same slope in almost all common semiconductors. The absorption spectra of many amorphous semiconductors exhibit tails with the same characteristic slope. Consequently, it is unlikely that the origin of these tails has any relationship to states arising from the potential fluctuations in disordered systems. This point is underlined by the fact that the absorption edges of trigonal and amorphous Se are virtually identical,[23] as shown in Fig. 5.

One might speculate that for amorphous materials in which the band tailing should not be particularly extensive, such as the elementary amorphous semiconductors, the Urbach edge, whatever its origin, masks the transitions involving tail states. On the other hand, for materials with both positional and compositional disorder, the band tails should be sufficiently strong to dominate the Urbach edge in the absorption spectrum.

One possible method for differentiating between these two situations is by means of the shape of the absorption as a function of energy. If we assume that the absorption spectrum is simply proportional to the joint densities of states of the valence and conduction bands, it can be shown[24] that, if the density of states of the valence band obeys the relationship

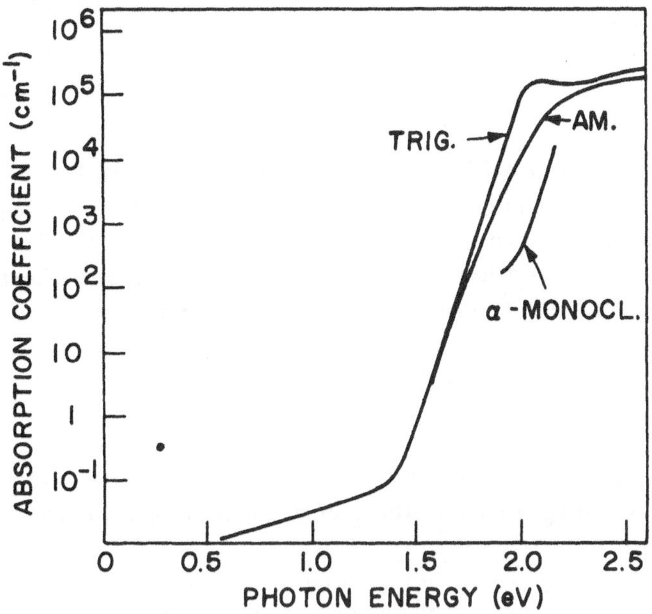

Fig. 5. Optical absorption as a function of energy for the amorphous and crystalline phases of Se (Stuke[23]).

$$g_v(E) \propto (E - E_1)^{n_1} \tag{11}$$

and if the conduction band density of states obeys

$$g_c(E) \propto (E - E_2)^{n_2}, \tag{12}$$

then the absorption constant α should vary as

$$\alpha(\omega) \propto (\hbar\omega - E_o)^{n_1+n_2+1}/\hbar\omega, \tag{13}$$

where E_o is a constant of the order of the gap. For a crystalline semiconductor, $n_1 = n_2 = ½$, but this is unlikely to be appropriate for amorphous materials. A more reasonable assumption[25] is $n_1 = n_2 = 1$. In this case, the relationship (13) can be written

$$(\alpha \hbar\, \omega)^{1/3} \propto (\hbar\omega - E_o) \tag{14}$$

Such a proportionality has been found[26] in a chalcogenide glass, $Te_{55}As_{25}Ge_{15}Si_5$, and this provides direct evidence for the existence of band tails in these highly disordered systems.

References p.p. 252-253

On the other hand, most amorphous semiconductors for which absorption has been studied obey[27,28]

$$(\alpha \hbar \omega)^{1/2} \propto (\hbar\omega - E_0) \tag{15}$$

This relationship follows from (13), for example, with $n_1 = n_2 = 1/2$, but it also can be derived[24] under the assumption that $n_1 = n_2 = 1$, but that the matrix elements for interband transitions between localized states are vanishingly small. This ambiguity prevents any rigorous conclusion from being drawn when (15) is obeyed. An interesting sidelight is that the absorption in amorphous Se varies as[24]

$$(\alpha \hbar \omega) \propto (\hbar\omega - E_0)\ , \tag{16}$$

a relationship which is difficult to justify on physical grounds. However, a convincing argument has been presented[29] that in this region ($E_0 \approx 2.1$ eV) the absorption is excitonic in nature, resulting from excitations of Se_8 molecules in the amorphous solid.

At this point, we might well ask if there is any major experimental difference between the optical properties of crystalline and amorphous semiconductors. Indeed the most significant difference now appears to be not in the shape but rather in the position of the edge. With the exception of the excitonic-type edge in amorphous and trigonal Se, shifts in the absorption edge between the crystalline and amorphous phases generally exist. For example, the edge in amorphous Ge is approximately 0.6 eV,[19] 0.1 eV below that in crystalline Ge, while the $As_2\,Se_3$ edge decreases from 1.8 eV in the crystalline phase to about 1.6 eV in the amorphous phase.[30] On the other hand, the edges in InSb and Te are shifted to higher energies in the amorphous phases,[23] InSb increasing from 0.2 eV to about 0.4 eV, and Te increasing from 0.3 eV to 0.9 eV. Thus, even in the edge shifts, a consistent pattern does not emerge.

OPTICAL SWITCHING OF THIN-FILM AMORPHOUS SEMICONDUCTORS

Much recent interest in amorphous semiconductors has been spurred by the discovery by Ovshinsky[31] of two types of reversible switching phenomena, in which the material undergoes a sharp transition between a high resistance and a low resistance state in an applied electric field. One type of switching is known as threshold switching, in which a minimum field is required to maintain the low-resistance state. On the other hand, materials of a somewhat different composition exhibit memory switching, in which either state is maintained after the applied field is removed. Memory-type materials clearly possess two distinct equilibrium states and thus must exhibit a solid-state phase transformation. It has recently been shown[32] that these transformations are accompanied by first-order optical effects, and thus these materials have useful applications for information processing and modulation of light. In this section, we shall describe the optical properties of the two equilibrium states, and discuss the mechanism for obtaining fast, reversible write/erase capability in an information storage element using optical read-out.

THERMALLY INDUCED AMORPHOUS-CRYSTALLINE TRANSITIONS

An important key to the origin of memory switching can be found from thermal cycling experiments on bulk material.[33,34] Fig. 6 shows the results of a differential thermal analysis, in which a material is heated at a constant rate, and the signal obtained measures any change in the rate of heat absorbed or given off when compared to a standard of constant heat capacity. Such changes occur during phase transitions in the material under investigation. The material, in the case of Fig. 6, is a chalcogenide glass of composition $Ge_{16}Te_{82}Sb_2$. Fig. 6 (a) shows the results when the glass is heated from room temperature. At about 125°C, a drop in the signal indicates the glass transition temperature, at which point the viscosity of the material decreases sharply. With continued heating to approximately 225°C, a large increase in the signal occurs, indicating the occurrence of an exothermic transition. This transition has been associated with the nucleation and growth of crystalline

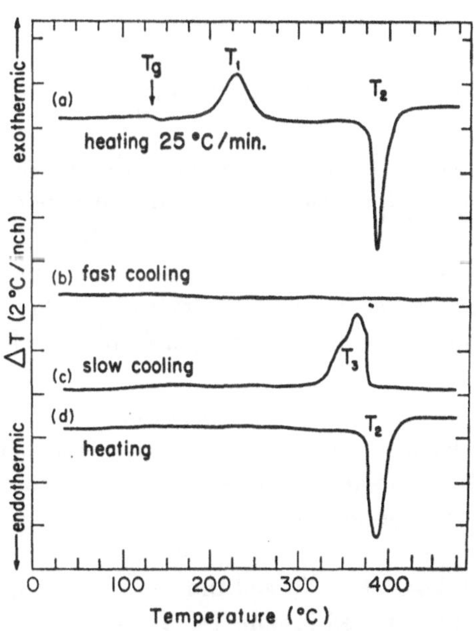

Fig. 6. Differential thermal analysis of a memory-type chalcogenide alloy, $Ge_{16}Te_{82}Sb_2$: (a) heating from the amorphous phase; (b) rapid cooling from the liquid phase; (c) slow cooling 'from the liquid phase; (d) heating from the crystallized phase.

References pp. 252-253

Te.[35] When the material is cooled at any rate from the region between the exothermic peak and the large endothermic peak near 375°C, the low-resistance state of the material is obtained. The endothermic transition represents the melting of the crystallized material.

If the material is cooled rapidly (quenched) from the liquid phase, Fig. 6 (b) is obtained. No transitions at all are found, indicating a return to the glassy state by freezing in the disorder of the liquid. The resulting material indeed has high resistance.[34]

On the other hand, Fig. 6 (c) shows that a slow cooling (anneal) from the liquid phase produces an exothermic transition, corresponding to a solidification to the low-resistance crystallized state. A reheating of this material, Fig. 6 (d), shows only a melting signal.

The thermal cycling of bulk material exhibits the characteristics of the phase transformation necessary to explain field-induced memory switching. However, the amorphous-crystalline transition is greatly accelerated in the electrical device. One reason for this is that only a small filament[36] of material is switched electrically. In addition, the material is already in a conducting state, with large densities of free carriers, when crystallization occurs in electrically switched devices. As we shall discuss, this can be expected to increase the crystallization rate sharply.

OPTICAL PROPERTIES OF THE AMORPHOUS-CRYSTALLINE TRANSITION

As we indicated in Section IV, changes occur in the optical properties of an amorphous material upon transformation to the crystalline state. An example of these effects in Te is shown in Fig. 7. The sharp absorption edge at low-energy in the crystal is shifted to somewhat higher energy and spread out in the amorphous material. Thus, it is clear that large changes in transmission are possible for light in the 0.5 - 0.7 eV range when Te undergoes an amorphous-crystalline transition.

The left-hand curve in Fig. 7 shows the imaginary part of the dielectric constant, which is directly related to reflectivity. It is clear that significant differences exist between the crystalline and amorphous states, largely because of the sensitivity of ϵ_2 to the shift of the absorption edge.

Amorphous Te crystallizes at 10°C, and thus no room-temperature amorphous-crystalline transition can be obtained in pure Te. However, chalcogenide glasses of the form $Ge_{15}\,Te_{81}\,X_4$, where X is some addition of As, Sb, or S, are stable below approximately 200°C and have DTA traces much like Fig. 6. Furthermore, they still maintain the large optical changes at the amorphous-crystalline transition ob-

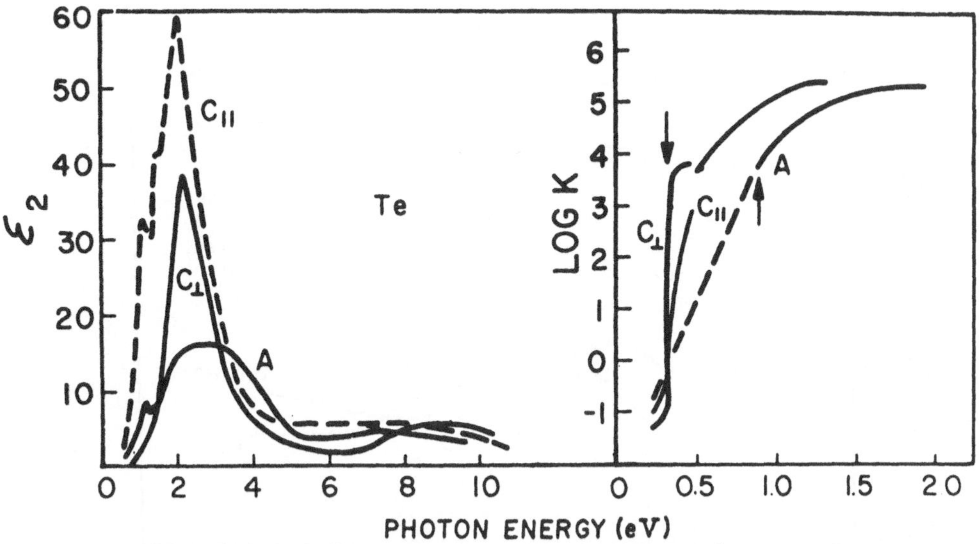

Fig. 7. Imaginary part of the dielectric constant and absorption as a function of energy for amorphous (A) and single crystal Te; $C_{\perp}$ and $C_{\parallel}$ indicate light with electric vector perpendicular or parallel to the crystalline c-axis (Stuke[23]).

tained in Te, as shown in Fig. 8.[32] The reflectance of the crystalline state below 4 eV is approximately 50% larger than that in the amorphous state. The reversibility of the optical changes after several cycles is clear. This is in agreement with NMR[37] and thermal cycling[34] results.

The optical changes are a consequence of band-structure changes that are already apparent from observation on pure Te. The direct absorption edge in crystalline Te has been associated[38,39] with a transition near a Brillouin-zone corner rather than one along the direction of the c axis. But the chain structure of crystalline Te is parallel to the c-axis, and thus the strongest interband transitions are the bonding-antibonding transitions along a given chain. These chains are maintained in the amorphous state, although they are disordered. Because of the disorder, transitions that depend on interchain spacings, such as those leading to the absorption edge, should have their oscillator strengths spread in energy in the amorphous material. This explains why the edge is spread out in amorphous, as compared with crystalline, Te.

As a consequence of this change in the nature of the absorption edge, large discontinuities in the optical properties of Te and Te-rich alloys are associated with the amorphous-crystalline transition. Such a transition can be accomplished ther-

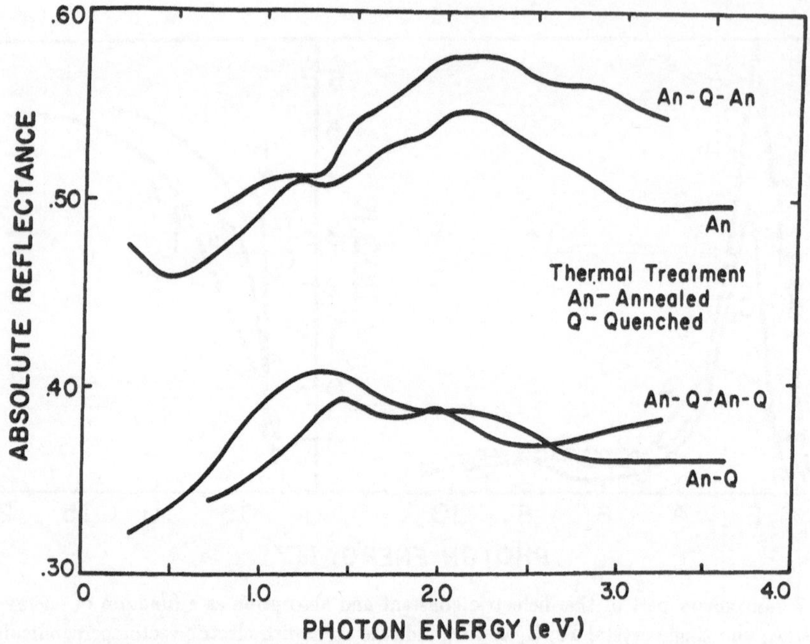

Fig. 8. Reflectance of bulk samples of $Ge_{15}\ Te_{81}\ X_4$. Quenched material is in the amorphous state, annealed material is in the crystalline state (Feinleib and Ovshinsky[32]).

mally, electrically, or by the absorption of light. The last process, as we shall describe, involves both a heating effect and a direct photo-crystallization.

LIGHT-INDUCED AMORPHOUS-CRYSTALLINE TRANSITIONS

When light is absorbed by an amorphous semiconductor, there is a sharp increase in the free carrier concentration. Excitation of electrons across the mobility gap results in a large density of broken covalent bonds in the material. Once these bonds are broken, the metastability of the amorphous state is weakened, and crysallization can proceed at a greatly accelerated rate. This process is called photo-crystallization.

The crystallization rate of amorphous materials is composed first of a nucleation process and then of a growth of crystallites. One of the few photo-crystallization studies that has been reported was carried out by Dresner and Stringfellow.[38] In their experiments on thin films of amorphous selenium, they observed the growth of individual crystallites under the influence of moderate photon fluxes in a configuration in which the light and thermal processes are separable. They found that crystallite growth rates exponentially increase with temperature with an activation energy of about 0.49 eV when low-intensity light impinges on the material and an activation energy of 0.24 eV for high-intensity light. As the intensity is increased, the growth rate is initially directly proportional to the intensity, but becomes proportional to the square root of the intensity at higher values. The transition point between the two regimes decreases when the temperature of the film is raised above the glass temperature. At the highest temperature investigated, 84°C, well below the melting point of Se, they observed an increase in growth rate of about a factor of 20 over the equilibrium rate, and no saturation effects were discernible up to the highest intensities used. Since the maximum growth rates occur near the melting point, no data exist on photo-crystallization when the thermal growth rate is high. However, Dresner and Stringfellow determined that the photo-crystallization rate depends on photon flux rather than total photon energy.

Much higher growth rates were observed by Evans *et al.*[39] in chalcogenide alloys of the type $Ge_{15}Te_{81}X_4$. In these experiments, no attempt was made to separate thermal and direct light effects. Nevertheless extremely high growth rates of about 4×10^5 Å per second, 100 times larger than the highest rates in the amorphous Se study, were obtained. Furthermore, Evans *et al.* reported that light pulses of approximately 6×10^{-4} seconds duration revitrified the films.

In a more recent study,[40] reversible, light-induced optical effects have been demonstrated in a configuration useful for memory storage. In the experiments, a laser was used to achieve high light intensities in small areas of a thin film. Short pulses of argon ion laser radiation at 5145 Å wavelength was focused to a spot about 5 μm in diameter at the interface between a glass substrate and an amorphous Se-Te alloy film 5 μm thick. At this wavelength, the film has an absorption coefficient of about 10^5 cm^{-1} and the light is therefore absorbed in a thickness of about 1,000 Å. By varying the intensity of the pulse or increasing the pulse length, it was found that at a threshold intensity, a sudden darkening of the film can be observed through a microscope, in about a 3 μ diameter area. The observation is made using broad band tungsten illumination where the film in the amorphous state appears to be a deep red color since the band edge is near 7,000 Å. The large contrast between the red background amorphous material and the changed material occurs presumably by both changes in the intrinsic optical properties and the light scattering from small spots. By applying a pulse from the same source but with less amplitude to the spot, a threshold is found at which the spot is reversed back to the original optical state. The effects observed in these experiments typically required pulses of about 1 μsec duration and a peak intensity of about 100 milliwatts to write, and about one third of this intensity to erase. Taking into account the reflection and other optical losses, this comes to about 0.2 Joules/cm^2 absorbed by the active material. With a simple model of heat losses, the temperature in the laser irradiated area will remain above the glass transition temperature for the order of 1 μsec. These times are much shorter than those encountered in thermal phase changes or in the electrically induced effects.[39]

References pp. 252-253

In order to investigate the nature of the light-induced structure changes, a sample was prepared for electron-microscope examination. A film of the $Ge_{15}\,Te_{81}\,X_4$ material was sputtered to a thickness of 700 Å on an amorphous carbon substrate, which was supported on a fine mesh grid. The argon ion laser pulses were adjusted in intensity until the reflectivity of the material, as viewed in an optical microscope, sharply changed. Some of the spots were then erased by means of smaller-amplitude laser pulses. Electron micrographs[41] clearly showed that the transformed material had been very finely crystallized. Either the light created an extremely high density of nucleation centers, or else such centers already exist before irradiation and the light induced the growth of small crystallites from them. Because of this density of nucleation centers, even moderate crystal growth rates lead to a complete crystallization of a 2 μm spot in the order of a microsecond.

When the diffraction pattern of a transformed spot is obtained, sharp lines characteristic of crystallites appear. On the other hand, when a transformed area has been erased by a second laser pulse, the diffraction pattern shows just the diffuse rings characteristic of amorphous material. This demonstrates that the optical transformations are indeed reversible amorphous-crystalline transitions.

As we have already indicated, the physical mechanism for the amorphous-crystalline transformation is a combination of a thermal effect and a non-equilibrium photo-crystallization process. By use of intense radiation such as from a focused laser, it should be possible to obtain an optimum density of electron-hole pairs. As is clear from Fig. 9, the optimum condition for crystallizing the amorphous material at ambient temperature T_A occurs when the pulse intensity and duration are such as to bring the material to the temperature at which the crystallization rate is a maximum, without exceeding the melting temperature T_M. This is shown in curve A. The material is then held at temperatures between the glass temperature T_G and the melting temperature T_M for a time t_1. During this time, thermal growth of crystallites can occur. However, during a fraction t_2 of this time, the light is on, and the crystallization rate is vastly enhanced by photo-crystallization. It has further been shown[42] that a non-equilibrium excited state is preserved for a time, estimated to be of the order of milliseconds at room temperature, after the light is turned off. This is consistent with the band model of Cohen *et al.*[20] and is due to the large density of charged traps which always exist in the gap of the glass. Thus the crystallization rate is also somewhat enhanced even after the light is off. Provided the melting temperature is not exceeded, the material will remain in the crystallized state as it cools.

In the reverse process (curve C), the crystallized material is heated by the absorption of light energy. Since the absorption constant of the crystallized material is larger than that of the amorphous material at the photon energy being used (see Fig. 7) and, in addition, the crystallized material has a lower heat capacity than the

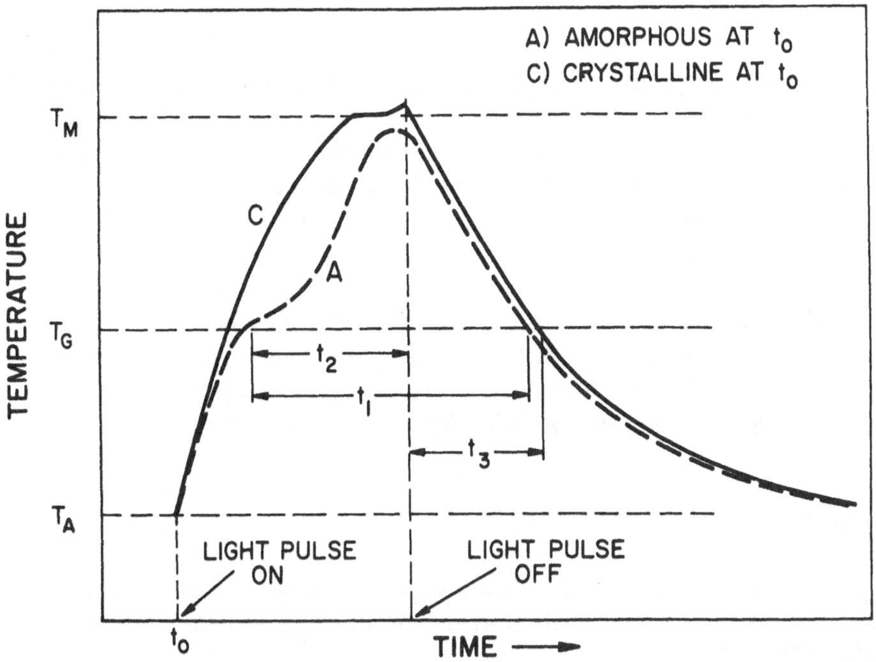

Fig. 9. Temperature-time profile of the active material illuminated by a square light pulse energy above both amorphous and crystalline band gaps.

amorphous material in the temperature range between T_G and T_M, the crystallized material will reach a higher temperature with the same or even less laser energy than will an equivalent amount of amorphous material. If the temperature reached exceeds T_M, as shown in Fig. 9, the material will melt. After the light is turned off, the liquid then cools through the crystallizing range between T_M and T_G in time t_3. Since $t_3 < t_1$, the cooling material spends less time in the range where crystallization can occur than does the material in curve A. But, more important, the liquid in curve C spends all the time in the crystallization range while the light is off, so that the crystallization rate is much lower than in the range t_2 of curve A. By this means, the cooling in curve C quenches the material into the amorphous state.

This large asymmetry in crystallization rates should be associated with all photo-induced amorphous-crystalline transitions, so the reversible optical properties may be expected to be observed as a general phenomenon in these materials. There is insufficient data to make this model quantitative, but the observed high speed reversible electrical and optical memory effects in amorphous semiconductors favor the qualitative features of the model.

CONCLUSION

We have reviewed the effects on the optical properties of three distinct types of solid state phase transformations-Mott transitions, semiconductor-metal transitions, and amorphous-crystalline transitions. In each case, the nature of the optical absorption and reflectivity spectra change strikingly at the point of the phase transformation. Mott transitions have not yet been observed unambiguously, although the promise of significant changes in optical absorption throughout the entire photon energy range make them potentially very interesting. Ordinary semiconductor-metal transitions can be extremely useful for modulating light by means of temperature and pressure variations, particularly at long wavelengths. Finally, reversible amorphous-crystalline transitions produce sharp changes in optical properties, and have the added advantage that they can be induced entirely by optical means.

REFERENCES

* *Research supported by the Advanced Research Projects Agency of the Department of Defense and monitored by the U.S. Army Research Office - Durham, under Contract No. DAHCO4-70-C-0048.*

1. *D. Adler, Rev. Mod. Phys., 40(1968) 714.*
2. *D. Adler, Essays in Physics, 1(1970) 33.*
3. *N. F. Mott, Proc. Phys. Soc. (London), A62(1949) 416.*
4. *N. F. Mott, Phil. Mag. 6(1961) 287.*
5. *D. Adler and J. Feinleib, Phys. Rev. B2(1970) 3112.*
6. *D. Adler, IBM J. Res. Develop. 14(1970) 261.*
7. *R. J. Powell and W. E. Spicer, Phys. Rev. B2(1970) 2182.*
8. *No Mott transition has been observed in NiO, despite the application of 500 kbar pressure. See S. Minomura and H. G. Drickamer, J. Appl. Phys. 34(1963) 3043.*
9. *D. Adler and H. Brooks, Phys. Rev. 155(1967) 826.*
10. *H. W. Verleur, A. S. Barker, Jr., and C. N. Berglund, Phys. Rev. 172(1968) 788.*
11. *L. A. Ladd and W. Paul, Solid State Commun. 7(1968) 425.*
12. *V. G. Mokerov and A. V. Rakov, Soviet Phys. - Solid State 10(1968) 1231.*
13. *The possibility that the transition in VO_2 is a Mott transition can also be eliminated by noting the absence of local moments in the low-temperature phase.*
14. *U. Birkholz and J. Naegele, Phys. Stat. Sol. 39(1970) 197.*
15. *H. G. Drickamer, Solid State Phys. 17(1965) 1.*
16. *A. F. Ioffe and A. R. Regel, Progr. Semicond. 4(1960) 237.*
17. *N. F. Mott, Festkorperprobleme 9(1969) 22.*
18. *M. H. Cohen, J. Non-Crystalline Solids 4(1970) 391.*
19. *T. M. Donovan, W. E. Spicer, and J. M. Bennett, Phys. Rev. Letters 22(1969) 1058.*
20. *M. H. Cohen, H. Fritzsche, and S. R. Ovshinsky, Phys. Rev. Letters 22(1969) 1065.*
21. *D. Adler, Electronics 43 (Sept. 28, 1970) 61.*
22. *J. J. Hopfield, Comments on Solid State Physics 1(1969) 16.*
23. *J. Stuke, J. Non-Crystalline Solids 4(1970) 1.*
24. *E. A. Davis and N. F. Mott, Phil. Mag. 22(1970) 903.*
25. *N. F. Mott, Phil. Mag. 19(1969) 835.*

26. *E. A. Fagen, personal communication.*
27. *K. Weiser and M. H. Brodsky, Phys. Rev. B1(1970) 791.*
28. *J. Tauc, Optical Properties of Solids, F. Abeles, ed., North-Holland, Amsterdam, 1969, p. 123.*
29. *G. Lucovsky, Proc. Tenth Interm. Conf. on the Phys. of Semicond., Cambridge, Mass., 1970, p. 799.*
30. *J. T. Edmond, Brit. J. Appl. Phys. 17(1966) 979.*
31. *S. R. Ovshinsky, Phys. Rev. Letters 21(1968) 1450.*
32. *J. Feinleib and S. R. Ovshinsky, J. Non-Crystalline Solids 4(1970) 564.*
33. *H. Fritzsche and S. R. Ovshinsky, J. Non-Crystalline Solids 2(1970) 148.*
34. *D. Adler, J. M. Franz, C. R. Hewes, B. P. Kraemer, D. J. Sellmyer, and S. D. Senturia, J. Non-Crystalline Sol., 4(1970) 330.*
35. *A. Bienenstock, F. Betts, and S. R. Ovshinsky, J. Non-Crystalline Solids 2(1970) 347.*
36. *R. Uttecht, H. Stevenson, C. H. Sie, J. D. Griener, and K. S. Raghaven, J. Non-Crystalline Solids 2(1970) 358.*
37. *S. D. Senturia, C. R. Hewes, and D. Adler, J. Appl. Phys. 41(1970) 430.*
38. *J. Dresner and G. B. Stringfellow, J. Phys. Chem. Solids 29(1968) 303.*
39. *E. J. Evans, J. H. Helbers, and S. R. Ovshinsky, J. Non-Crystalline Solids 2(1970) 334.*
40. *J. Feinleib, J. de Neufville, S. C. Moss, and S. R. Ovshinsky, Appl. Phys. Letters 18(1971), in press.*
41. *S. C. Moss, unpublished data, 1970.*
42. *E. A. Fagen and H. Fritzsche, J. Non-Crystalline Solids 2(1970) 180.*

MAGNETO-OPTICAL PROPERTIES OF THE Eu-CHALCOGENDES*

J. O. DIMMOCK

Lincoln Laboratory, Massachusetts Institute of Technology, Lexington, Massachusetts

ABSTRACT

The Eu-chalcogenides, EuO, EuS, EuSe, and EuTe exhibit remarkably large magneto-optical effects in the visible and infrared spectral regions arising from highly polarized europium $4f^7 \rightarrow 4f^6 5d$ optical transitions. Faraday rotations as large as 8.5×10^5 deg/cm have been observed in EuO at 20°K and a wavelength of 0.7 μ, comparable to the value observed in iron. However, in contrast to the metallic ferromagnets, the Eu-chalcogenides are relatively transparent in this wavelength region. For example, at 0.75 μ and 4.2°K EuSe has a Faraday rotation of 1.4×10^5 deg/cm and an absorption coefficient of only 66 cm^{-1}. This results in a specific rotation per unit attenuation of 500 deg/dB, approximately four orders of magnitude greater than that of iron. In the infrared, at energies below the semiconducting bandgap, the absorption is less and the specific rotation per unit attenuation can be even larger. The specific rotation of EuO at low temperatures is in excess of 2×10^4 deg/dB at 2.5 μ and 230 deg/dB at 10.6 μ. These materials also exhibit remarkably large Kerr effects with longitudinal Kerr rotations as large as $2\phi = 2.1°$ being observed in EuO at 12°K, 0.56 μ and 30° angle of incidence. The Kerr effects can also be considerably enhanced by antireflection coating and/or by depositing the film on a mirror substrate. These very large effects are of considerable technological importance and the possible use of the Eu-chalcogenides for magneto-optical memory systems, amplitude and phase modulation of light beams and optical isolators for laser systems will be briefly indicated.

References pp. 270-271

INTRODUCTION

The optical and magneto-optical properties of the magnetic semiconductors EuO, EuS, EuSe and EuTe have been studied by many workers using a number of different experimental techniques.[1-6] The optical and magneto-optical studies have revealed considerable structure not only associated with the absorption edge in these materials but also at considerably higher photon energies.[7] This structure has been identified with the onset of Eu $4f^7 \rightarrow 4f^6\ 5d(t_{2g})$ (absorption edge) and $4f^7 \rightarrow 4f^6\ 5d(e_g)$ (higher energy) transitions split by the octahedral crystal field.[1,7] Large magneto-optical effects including Faraday rotation, longitudinal and transverse Kerr effects and magneto-optical dichroism have been observed associated with the absorption edge. In addition, these crystals show a large shift of the absorption edge to lower energy as the temperature is reduced through the ordering temperature or as a magnetic field is applied.[1-3, 5,8,9] Also, large polarization effects have been observed in photoemission experiments, yielding a total polarization of the photoemitted electrons of up to 25% in the visible region of the spectra.[10] The magnetic and optical properties of the europium chalcogenides have been reviewed by several workers from different points of view emphasizing different aspects of the field.[1-6] Recently, the optical properties of these materials were reviewed and an attempt was made to determine what aspects of their electronic structure had been established by experimental and theoretical investigations.[11] In this survey, we will concentrate more on the magneto-optical properties of these materials and their application in magneto-optical devices.

ELECTRONIC STRUCTURE AND BASIC PROPERTIES

Some of the pertinent magnetic and optical data on the europium chalcogenides are given in Table I. EuO and EuS are both ferromagnetic at low temperatures, whereas EuSe is metamagnetic, ordering antiferromagnetically but becoming ferromagnetic with the application of a small external magnetic field. EuTe appears to be a typical antiferromagnet. The energy gaps for these magnetic semiconductors at 300°K are given in the table. These, however, are a strong function of temperature, as was discussed previously.[11]

Primarily as a result of optical and magneto-optical studies on these materials, the band diagram shown in Fig. 1 has been deduced.[12] The valence band indicated P^6 consists of fully occupied anion p-states. The level labeled $4f^7$ represents the localized Eu $4f^7$ ($^8S_{7/2}$) state which is in fact a multiplet level. The europium 5d levels are shown split by an amount 10 D_q into a lower t_{2g} triplet and an upper e_g doublet by the octahedral crystal field. The approximate positions of the Eu-6s and 6p bands are also shown. The fundamental absorption edge at an energy E_g

TABLE I

MAGNETIC AND OPTICAL DATA OF Eu-CHALCOGENIDES[a]

	Curie Temperature (°K)	Magnetic Order	Ordering Temperature (°K)	Absorption Edge at 300°K (eV)
EuO	76	Ferro-	69.4	1.12
EuS	19	Ferro-	16.5	1.64
EuSe	9	Antiferro-	4.6	1.85
EuTe	- 6	Antiferro-	7.8 to 11	2.0

[a] *After Methfessel and Mattis, Ref. 1.*

occurs due to the onset of Eu^{++} 4f → 5d transitions of the type $4f^7$ ($^8S_{7/2}$) → $4f^6$ (^7F_J) 5d(t_{2g}) from the $4f^7$ level to the lower set of 5d states. This also gives rise to a strong reflectance peak which is shown labeled E_1 in Fig. 2.[13] The higher energy reflectivity peak labeled E_2 is primarily due to $4f^7$ ($^8S_{7/2}$) → $4f^6$ (^7F_J) 5d(e_g) transitions to the upper set of crystal field split 5d states, although anion p-valence

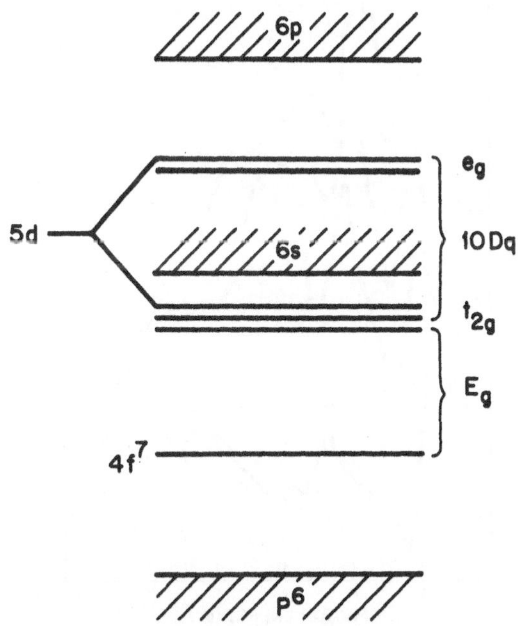

Fig. 1. Qualitative energy level scheme for the Eu-chalcogenides (after Methfessel, Holtzberg and McGuire, ref. 12).

band to Eu conduction band transitions may also be involved.[11] The optical absorption spectra[14] of EuO, EuS and EuSe shown in Fig. 3 indicate structure similar to that observed in reflectivity.

The Eu $4f^7$ levels are believed to lie above the anion p-states and below the conduction band as shown in Fig. 1. However, it is not established whether the $5d(t_{2g})$ states lie below the 6s band as shown or above. In addition, it is not known just how broad the $5d(t_{2g})$ and $5d(e_g)$ levels are. Although the structure in the optical reflectivity data, Fig. 2, indicates that the 5d states form relatively narrow t_{2g} and e_g levels, as shown in Fig. 1, recent energy band calculations[15] yield a 5d band some 10 eV wide composed of overlapping t_{2g} and e_g states. Exciton effects arising from the coulomb attraction between the excited and 5d electron and the Eu^{+++} $4f^6$ configuration could cause a strong enhancement of the optical structure associated with the onset of transitions to the 5d band, giving the illusion of narrow bands. The role of exciton effects in the optical properties of the Eu chalcogenides has been discussed at some length by Kasuya and Yanase.[16] However, the importance of these effects in the 4f to 5d transitions remains one of the more interesting unanswered questions regarding the electronic structure of these materials.

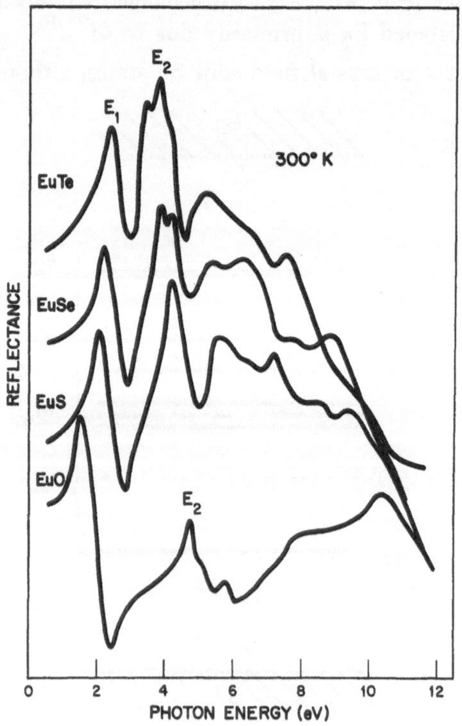

Fig. 2. Room temperature reflectivity spectra of EuO, EuS, EuSe and EuTe single crystals (after Pidgeon *et al.*, ref. 13).

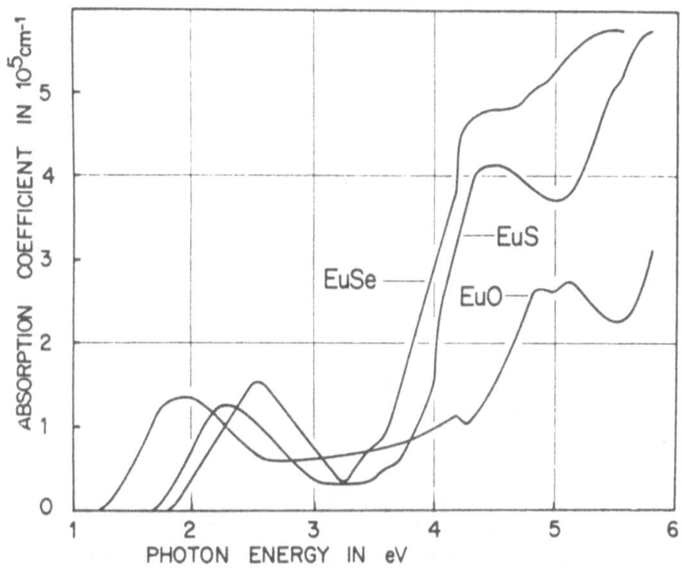

Fig. 3. Absorption coefficient of EuO, EuS and EuSe at room temperature as determined from reflectivity data (after Guntherodt, Schones and Wachter, ref. 14).

MAGNETO-OPTICAL PROPERTIES

The large magneto-optical effects observed arise primarily at low temperatures when the materials are ordered magnetically. In this case the $4f^7$ levels shown in Fig. 1 are fully polarized and in addition, both sets of 5d levels are split into primarily spin up and spin down sub-bands. The situation is shown schematically in Fig. 4.[17] The energies indicated are appropriate for EuO. As a consequence, the optical properties in the vicinity of the two transitions are different for right and left circularly polarized light. The effects of this on the reflectivity spectra of EuS and EuSe are shown in Fig. 5.[7] Similar effects have been observed in the reflectivity spectra of EuO[17] and of EuTe in magnetic fields greater than 80 kG where ferromagnetic order is induced.[13,18] As can be seen in the figure, very large differences occur in the reflectance for right and left circularly polarized radiation. These differences arise from relatively large differences in the oscillator strengths[18a] resulting in large differences in both the real and imaginary parts of the dielectric constant. These reflectivity studies indicate the magnitude of the magneto-optical polarization effects displayed by these materials. These large magneto-optical effects also show up in the absorption spectra of these materials. The absorption edge of EuSe[19] determined using various polarizations of the transmitted light at 4.2°K is shown in Fig. 6. The large shift of the absorption edge as a function of temperature

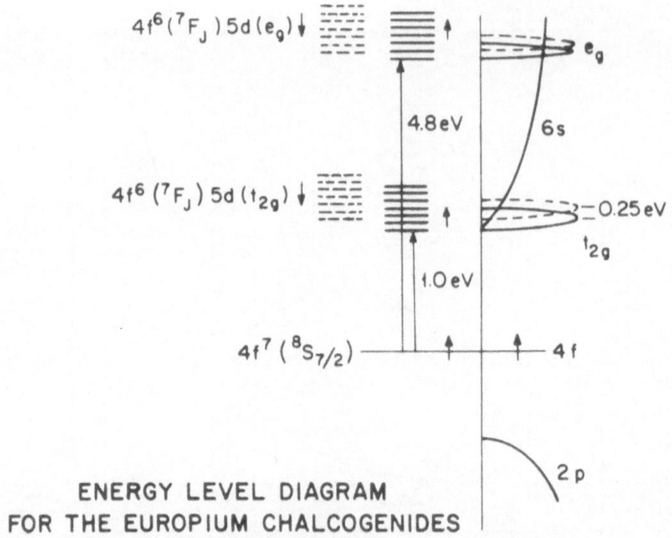

Fig. 4. Schematic band structure of ferromagnetic EuO. The right side shows the fundamental energy gap and the narrow 5d bands. The left side shows the E_1 and E_2 reflectance structure as transitions from the localized $4f^7$ ground state to a final state consisting of the ladder structure of a $4f^6$ multiplet excited state and an electron in a spin-split component of the 5d state (after Feinleib *et al.*, ref. 17).

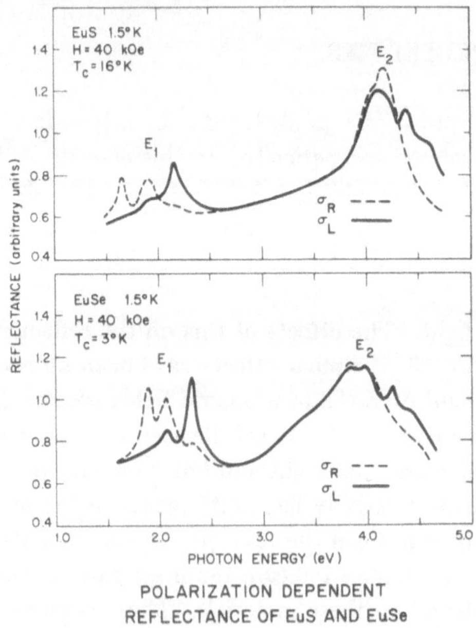

Fig. 5. Reflectance spectra of EuS and EuSe at 1.5°K at near normal incidence using right, σ_R, and left σ_L, circularly polarized light in a 40 Kgauss field, E⊥H, showing E_1 and E_2 polarization dependence (after Scouler *et al.*, ref. 7).

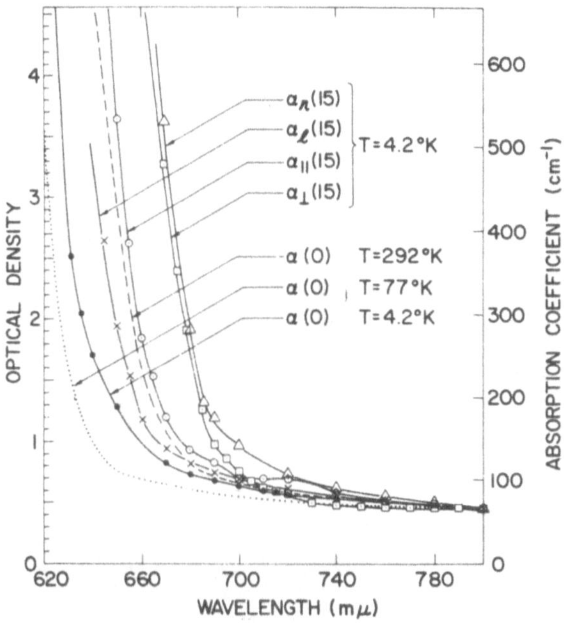

Fig. 6. Absorption edge of EuSe single crystal in the zero field paramagnetic state [$\alpha(0)$ at 77°K and 292°K], in zero field antiferromagnetic state [$\alpha(0)$ at 4.2°K] and in the high-field ferromagnetic state in an external field of 15 Kgauss as a function of polarization (after Argyle, Suits and Freiser, ref. 19).

mentioned above is also indicated by the unpolarized optical density curves taken at 292, 77, and 4.2°K.

MAGNETO-OPTICAL PHENOMENA

Large magneto-optical phenomena including Faraday rotation and Kerr effects have been observed in EuO, EuS, and EuSe. Paramagnetic Faraday rotation was first observed in EuSe at room temperature[20] and a very large Faraday rotation and linear magnetic birefringence was measured at 4.2°K.[21] The Faraday rotation observed in EuO at 5°K in a field of 20.8 kgauss is shown in Fig. 7, along with the wavelength dependence of the optical absorption coefficient.[22] The Faraday rotation for EuS is shown in Fig. 8.[14] Both of these curves, as well as the data on EuSe, indicate that very large Faraday rotations are obtainable in the visible region using EuO, EuS and EuSe. Comparable Faraday rotations have been observed for EuO in the infrared.[23] At photon energies below the absorption edge the absorption coefficient in good crystals of these materials can be quite small, whereas the Faraday rotation is still considerable. The Faraday rotation observed in EuO at 4.2°K and 50 kgauss between 1.5 and 16 μ is shown in Fig. 9, and the measured transmission of a 1.66 mm thick crystal in this wavelength region is shown in Fig. 10.[23] The

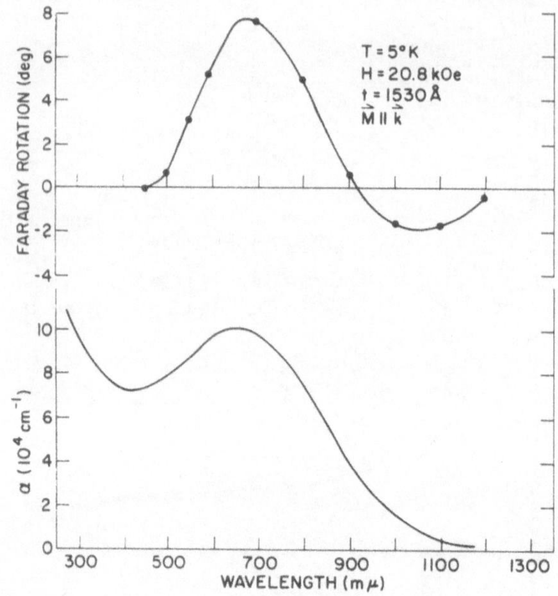

Fig. 7. Wavelength dependence of Faraday rotation at 5°K and absorption coefficient at 8°K for a 1530 Å thick EuO film in 20.8 Kgauss applied perpendicular to the film (after Ahn and Suits, ref. 22).

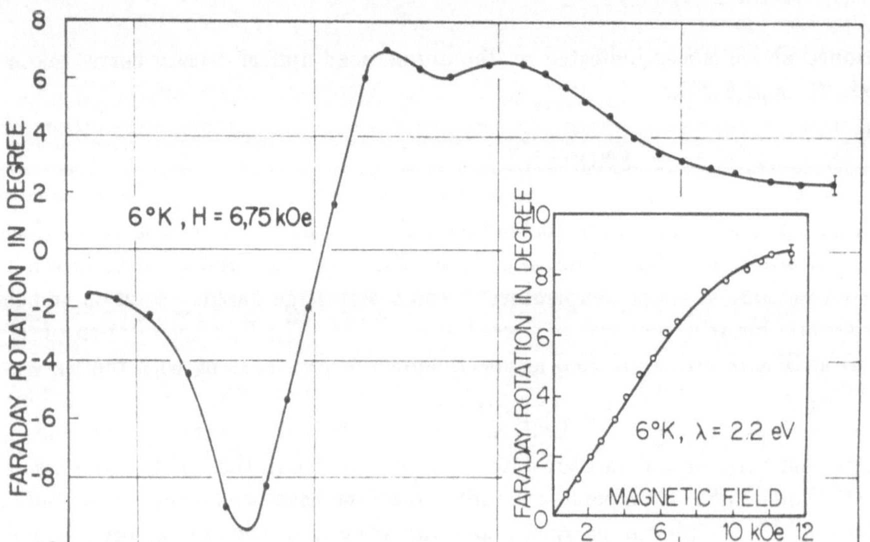

Fig. 8. Faraday rotation of a 1650 Å thick evaporated EuS film at 6°K in a field of 6.75 Kgauss (after Guntherodt, Schoenes and Wachter, ref. 14).

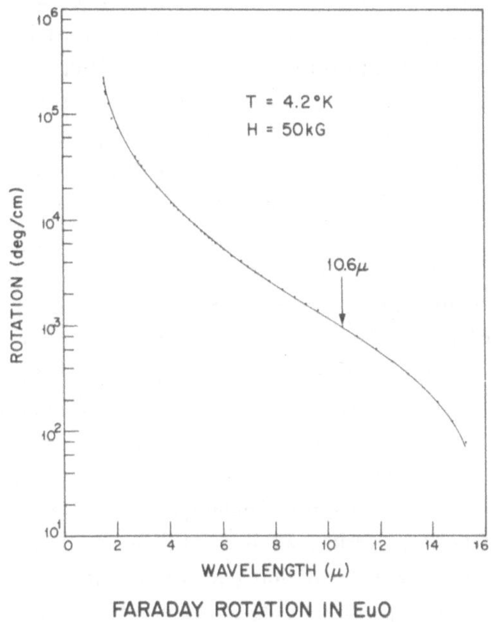

Fig. 9. Infrared Faraday rotation of EuO single crystals at 4.2°K in a saturating field of 50 Kgauss (Dimmock, Ward and Reed, unpublished).

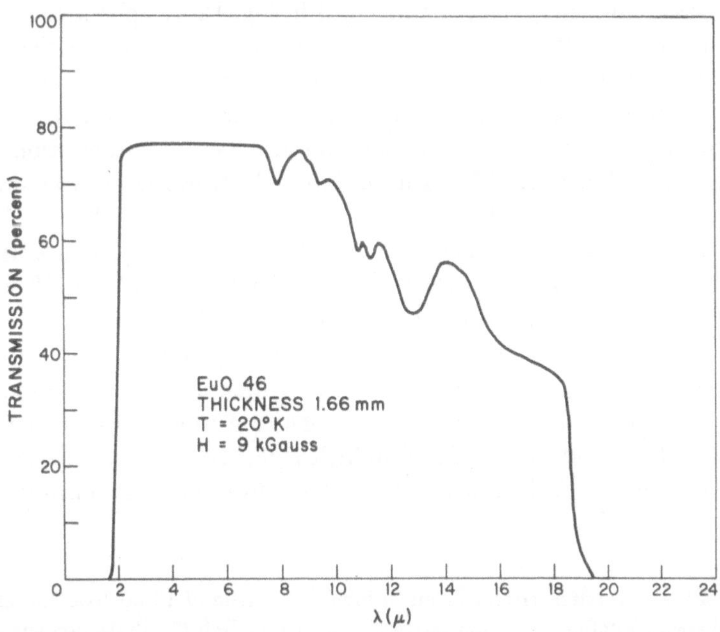

Fig. 10. Infrared transmission of a 1.66 mm thick EuO single crystal at 20°K in a field of 9 Kgauss (after Dimmock, Hurwitz and Reed, ref. 23).

References pp. 270-271

reflection limit for this transmission was determined to be approximately 77% corresponding to the flat region of the spectrum. Throughout the wavelength region between 2.5 and 8 μ the absorption coefficient for this crystal was determined to be less than 0.5 cm^{-1}. At 10.6 μ the Faraday rotation is approximately 1000 deg/cm with an absorption coefficient of 1 cm^{-1} corresponding to a rotation per unit attenuation figure of merit of 230 deg/dB. This figure rises to above 2×10^4 deg/dB at 2.5 μ with a rotation of 5×10^4 deg/cm.

It is interesting to compare these results with those which have been obtained on other materials. Figure 11 shows the Faraday rotations which have been obtained in a number of materials in the visible and infrared. As can be seen, the rotations observed for the europium chalcogenides are larger than those of any other materials except the ferromagnetic metals. However, for most applications the important figure of merit is the degrees rotation per unit attenuation. A comparison of this figure for several of these materials is shown in Fig. 12. In the visible spectral region this figure for the europium chalcogenides is comparable to that of other non-metallic ferromagnetic materials, such as $FeBO_3$ and FeF_2. These latter materials, however, have the advantage of being ferromagnetic at room temperature and thus do not require the cryogenic cooling necessary to observe the effect in the Eu-chalcogenides. This is a tremendous practical advantage. The specific rotation per unit attenuation for the Eu-chalcogenides in the infrared below the band edge is considerably larger than in the visible since the absorption coefficient is much smaller. These results have not been included in Fig. 12 except for those on EuSe, since they are not strictly a fundamental property of the material as the absorption coefficient depends mostly on crystal purity, perfection and surface preparation. It is nevertheless worthwhile comparing the results which have been obtained. In the infrared region YIG at 300°K has a specific rotation per unit attenuation of 800 deg/dB at 1.2 μ and at least 325 deg/dB at 2.5 μ.[29] However, YIG absorbs rather strongly beyond 5 μ and at 10.6 μ the absorption coefficient is about 115 cm^{-1} with a specific rotation of less than 0.25 deg/dB.[30] To our knowledge, the only other magnetic material besides EuO measured at 10.6 μ is $CdCr_2Se_4$[26] which has a rotation of 100 deg/cm and a specific rotation per unit attenuation of 1.77 deg/dB. The value of 230 deg/dB for EuO is thus the largest figure of merit at 10.6 μ for any magnetic material.* However, the high absorption coefficient in $CdCr_2Se_4$ may be due to crystalline imperfections, as it was observed to vary from crystal to crystal, and might be considerably reduced in higher quality crystals, whereas the absorption structure observed in EuO near 10.6 μ (Fig. 10) is likely an intrinsic property

* However, Faraday rotation measurements in InSb[31] in a field of 15 kG have indicated that the intrinsic Faraday rotation is approximately 100 deg/cm at 10.6 μ. Absorption coefficients in this wavelength region for high purity InSb as low as 0.01 cm^{-1} can be achieved giving a figure of merit of 2300 deg/dB, a factor of 10 greater than that of EuO.

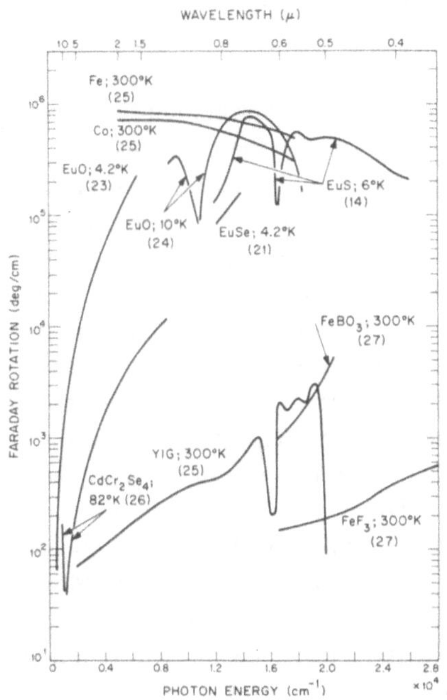

Fig. 11. Comparison of the Faraday rotation observed in various magnetic materials. The numbers in parentheses refer to the references.

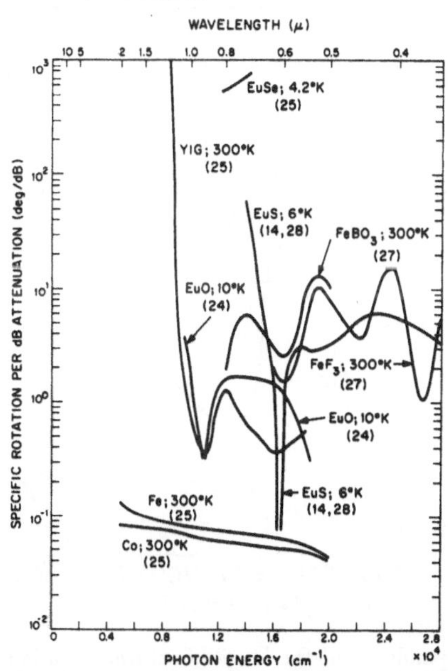

Fig. 12. Comparison of the figures of merit in degrees rotation per dB attenuation observed in various magnetic materials. The numbers in parentheses refer to the references (the absorption coefficient for EuS was taken from Wild, Shinmei and Anderson, Fig. 1, ref. 28).

References pp. 270-271

of the material (as two phonon absorption) since it was reproducibly observed in many crystals.

These materials also exhibit large magneto-optical Kerr effects. The longitudinal magneto-optical Kerr effect spectra for EuO and EuS[32] are shown in Fig. 13. The Kerr coefficients obtained were over an order of magnitude greater than those of iron films, which gives rise to two orders of magnitude increase in the Kerr signal. The Faraday rotation and Kerr effects in thin films of EuO doped with various trivalent rare earth ions have been studied by Ahn and McGuire.[33] A comparison between the Faraday rotation and the longitudinal and transverse Kerr effects for a Gd^{+++} doped EuO film is shown in Fig. 14. It was observed that the trivalent rare earth doping did not have much effect on the magneto-optical properties and that its primary effect was to raise the Curie temperature as had been observed previously.[34] Similar effects on the magnetic properties can be observed by varying the stoichiometry of these materials. The effects of stoichiometric variations on the magnetic properties and Faraday rotation of EuO films have been studied by Ahn and Schafer.[24] Although the Curie point could be increased considerably by the incorporation of excess Eu, the largest Faraday rotation was observed for near stoichiometric films. Eu deficient films had reduced Curie points and considerably less rotation. The largest results obtained for a near stoichiometric film are represented in Figs. 11 and 12.

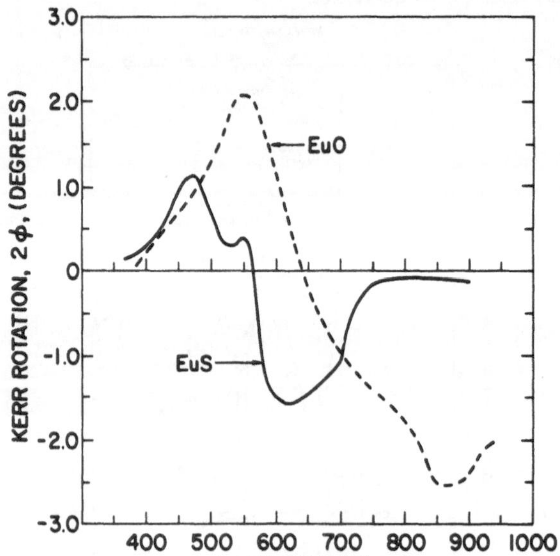

Fig. 13. Wavelength dependence of the longitudinal Kerr rotation in EuO at 12°K and EuS at 8°K in an applied field of 25 Kgauss. The angle of incidence was about 30° from the normal and the incident beam was polarized near the plane of incidence (after Greiner and Fan, ref. 32).

Since the Kerr effects are related to both the diagonal and off-diagonal components of the reflectivity tensor, the magnitude of the Kerr rotations can be strongly influenced by changing the reflectivity properties of the film. This can be done either by placing the film on a substrate with particular reflective properties, by coating the film with dielectric overlayers or both. The Kerr magneto-optic effects in thin films of EuO deposited on reflecting substrates have been studied by Ahn.[35] A considerable enhancement of the transverse Kerr effect was obtained by closely matching the antireflection condition through a variation of the thickness of the semitransparent EuO film deposited on Ag mirror substrates. The transverse Kerr effect for three thicknesses of EuO on a silver mirror is shown in Fig. 15. The peak of curve C corresponds to a minimum in reflectance at 0.52 μ. Ahn[36] has also studied the effects of deposited Eu_2O_3 dielectric films on the transverse Kerr effect of thick EuO films. As the dielectric film thickness increases, the reflectance minimum shifts to longer wavelengths accompanied by decreasing reflectivity. The minimum reflectance of 0.04% occurs at 0.675 μ for a dielectric thickness of ~500 Å. The transverse Kerr effect for three thicknesses of Eu_2O_3 films on EuO is shown in Fig. 16. In each case the maximum effect corresponds to a minimum in reflectance. A comparison of the Kerr effect enhancement in EuO films with a dielectric overlayer, a film on a mirror substrate and the bulk EuO is shown in Fig. 17. These effects are considerably larger than those observed in any other material.

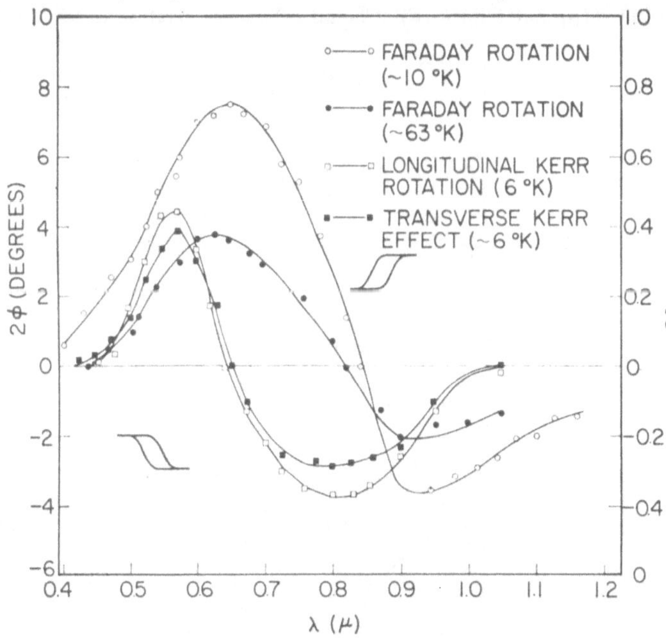

Fig. 14. Variations of the magneto-optic properties with wavelength for a 4000 Å thick Gd^{+++} doped EuO film. Hysteresis loops corresponding to the sign of rotation are also given. The Kerr effect measurements were made at 45° angle of incidence (after Ahn and McGuire, ref. 33).

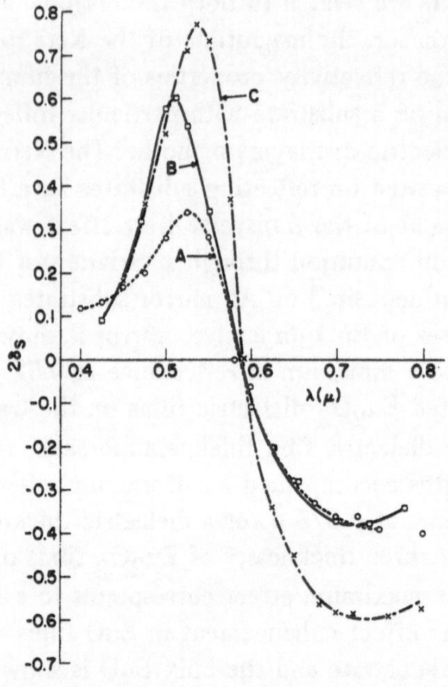

Fig. 15. Transverse Kerr effect for three EuO films on Ag mirror substrates with thickness of 1600 Å (A), 2400 Å (B) and 3200 Å (C) at liquid helium temperature (after Ahn, ref. 35).

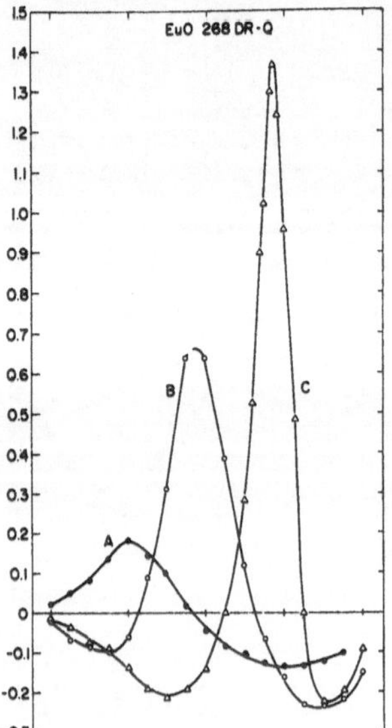

Fig. 16. Dependence of the transverse Kerr effect for a thick EuO film with Eu_2O_3 dielectric overlayers of thickness 150 Å (A), 320 Å (B) and 475 Å (C) at 4.2°K (after Ahn, ref. 36).

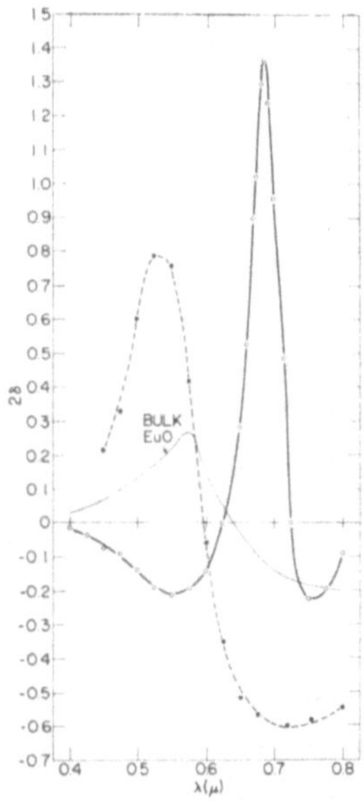

Fig. 17. Comparison of Kerr effect enhancement in EuO films with a dielectric overlayer (heavy solid line), a film on mirror substrate (broken line) and the bulk EuO (extrapolated values of 2δ from longitudinal rotation) (after Ahn, ref. 36).

MAGNETO-OPTICAL DEVICES

The origin and use of Faraday rotation in magnetic crystals has been reviewed by Dillon,[25] and Freiser[37] has recently given a survey of magneto-optic effects and their applications. A number of potentially useful devices are made possible by the magneto-optical effects observed in magnetic materials. Magnetic domains in the Eu-chalcogenides have been observed through optical diffraction[38] and diffuse scattering.[39,40] The possibility of using these magneto-optical effects to produce optically addressed magneto-optical memory systems has been discussed by a number of workers.[41-44] The feasibility of a beam-addressable memory using EuO as a storage medium and a GaAs laser as a transducer has been discussed by Fan and Greiner.[45] In this scheme, the heating produced by an intense laser beam focused to a small spot is used to induce the magnetic switching in the thin magnetic films. The magnetic sense of a particular spot in the film can then be ascertained by addressing with a second beam. The coercive force and consequently the resistance to switching can be reduced by effecting a coupling between films of the Eu chalcogenides and various transition metal ferromagnets.[46,47] In addition to the more standard form of magneto-optical storage Fan, Pennington and Greiner[48] have considered the use of thermographic writing on a thin film of EuO to produce a hologram which can be subsequently reconstructed.

One of the more interesting potential applications of Faraday rotation in the visible and infrared region using transparent magnetic materials is the development of rotators, isolators and modulators of visible and infrared laser beams. This has been discussed by Dillon[25] and Freiser.[37] The figure of merit for an optical isolator is the specific rotation per unit attenuation, although the specific heat and thermal conductivity are important also. Part of our initial motivation for studying the optical properties and Faraday rotation of EuO in the infrared was to develop an isolator for CO_2 laser radar systems. The figure of merit of 230 deg/dB for EuO at low temperatures is the largest obtained at 10.6 μ in any magnetic materials, although as noted above the figure of merit for high purity InSb should be greater. The figure of merit for EuO in the 2.5 to 7.5 μ range is very large, being in excess of 2.3×10^4 deg/dB at 2.5 μ, 4.2×10^3 deg/dB at 5 μ and 1.3×10^3 deg/dB at 7.5 μ. High purity EuS and EuSe should have similar figures of merit in the wavelength range from approximately 0.8 to 1.4 μ and 0.7 to 1.2 μ respectively, based on the differences in low temperature band gaps. For example, high purity EuS should have a figure of merit in excess of 1.5×10^4 deg/dB at 1.06 μ. The value of nearly 100 deg/dB observed in EuS at 0.7 μ is the largest figure of merit obtained in the visible, although as seen in Fig. 12 $FeBO_3$ and FeF_3 are better at shorter wavelengths. The main disadvantage of the Eu chalcogenides is their low transition temperatures which require that magneto-optical devices employing these materials must be operated at cryogenic temperatures. Nevertheless, it is clear that the Eu chalcogenides will find considerable application for magneto-optical devices because of the magnitude of the magneto-optical effects in these materials.

ACKNOWLEDGMENT

I would like to thank J. H. R. Ward for compiling the data represented in Figs. 11 and 12 and constructing these figures.

REFERENCES

* *This work was sponsored by the Department of the Air Force.*

1. *S. Methfessel and D. C. Mattis in Handbuch der Physik XVIII/1 (Springer-Verlag 1968), pp. 389-553.*
2. *G. Busch and P. Wachter, Phys. Kondens. Materie 5, 232(1966).*
3. *G. Busch and P. Wachter, Z. Angew. Physik 26, 1(1968).*
4. *G. Busch, J. Appl. Phys. 38, 1386(1967).*
5. *P. Wachter, Helv. Phys. Acta 41, 1249(1968).*
6. *J. C. Suits, B. E. Argyle and M. J. Freiser, J. Appl. Phys. 37, 1391(1966).*
7. *W. J. Scouler, J. Feinleib, J. O. Dimmock and C. R. Pidgeon, Solid State Commun. 7, 1685(1969); C. R. Pidgeon, J. Feinleib, W. J. Scouler, J. Hanus, J. O. Dimmock and T. B. Reed, Solid State Commun. 7, 1321(1969).*
8. *M. J. Freiser, F. Holtzberg, S. Methfessel, G. D. Pettit, M. W. Shafer and J. C. Suits, Helv. Phys. Acta 41, 832(1968).*

9. G. Busch, P. Junod, and P. Wachter, Phys. Letters 12, 11(1964).
10. G. Busch, M. Campagna and H. C. Siegmann, Solid State Commun. 7, 775(1969); J. Appl. Phys. 41, 1044(1970).
11. J. O. Dimmock, IBM J. Res. Dev. 14, 301(1970).
12. S. Methfessel, F. Holtzberg and T. R. McGuire, IEEE Trans. Mag. 2, 305(1966).
13. C. R. Pidgeon, J. Feinleib, W. J. Scouler, J. O. Dimmock and T. B. Reed, IBM J. Res. Dev. 14, 309(1970).
14. G. Guntherodt, J. Schones and P. Wachter, J. Appl. Phys. 41, 1083(1970).
15. S. J. Cho, Phys. Rev. 157, 632(1967); Phys. Rev. B 1, 4589(1970).
16. T. Kasuya and A. Yanase, Rev. Mod. Phys. 40, 684(1968); A. Yanase and T. Kasuya, J. Phys. Soc. Japan 25, 1025(1968).
17. J. Feinleib, W. J. Scouler, C. R. Pidgeon, J. O. Dimmock, J. Hanus and T. B. Reed, Phys. Rev. Letters 22, 1385(1969).
18. J. Feinleib and C. R. Pidgeon, Phys. Rev. Letters 23, 1391(1969).
18a. J. O. Dimmock, J. Hanus and J. Feinleib, J. Appl. Phys. 41, 1088(1970).
19. B. E. Argyle, J. C. Suits and M. J. Freiser, Phys. Rev. Letters 15, 822(1965).
20. J. C. Suits and B. E. Argyle, J. Appl. Phys. 36, 1251(1965).
21. J. C. Suits and B. E. Argyle, Phys. Rev. Letters 14, 687(1965).
22. K. Y. Ahn and J. C. Suits, IEEE Transactions on Magnetics MAG3, 453(1967).
23. J. O. Dimmock, C. E. Hurwitz, and T. B. Reed, Appl. Phys. Letters 14, 49(1969).
24. K. Y. Ahn and M. W. Schafer, J. Appl. Phys. 41, 1260(1970).
25. J. F. Dillon, Jr., J. Appl. Phys. 39, 922(1968).
26. P. F. Bongers and G. Zanmarchi, Solid State Commun. 6, 291(1968).
27. R. Wolfe, A. J. Kurtzig and R. C. LeCraw, J. Appl. Phys. 41, 1218(1970).
28. R. L. Wild, M. Shinmei and A. L. Andersen, Proc. 9th Conf. on Physics of Semiconductors, Vol. 2, p. 1191(1968).
29. R. C. LeCraw, D. L. Wood, J. F. Dillon, Jr., and J. P. Remeika, Appl. Phys. Letters 7, 27(1965).
30. D. L. Wood and J. P. Remeika, J. Appl. Phys. 38, 1038(1967).
31. B. Lax, Proc. Intl. Conf. on the Physics of Semiconductors, Prague, 1960, p. 321.
32. J. H. Greiner and G. J. Fan, Appl. Phys. Letters 9, 27(1966).
33. K. Y. Ahn and T. R. McGuire, J. Appl. Phys. 39, 506l(1968).
34. F. Holtzberg, T. R. McGuire and S. Methfessel, J. Appl. Phys. 37, 976(1966).
35. K. Y. Ahn, IEEE Transactions on Magnetics MAG4, 408(1968).
36. K. Y. Ahn, J. Appl. Phys. 40, 3193(1969).
37. M. J. Freiser, IEEE Transactions on Magnetics MAG4, 152(1968).
38. J. C. Suits, J. Appl. Phys. 38, 1498(1967).
39. F. Rys and P. Wachter, Solid State Commun. 6, 805(1968).
40. P. Wachter, Phys. Kondens. Materie 7, 1(1968).
41. A. H. Eschenfelder, J. Appl. Phys. 41, 1372(1970).
42. C. D. Mee, IEEE Trans. Mag. 5, 449(1969).
43. D. Treves, R. P. Hunt and B. Dickey, IEEE Trans. Mag. 5, 449(1969).
44. R. P. Hunt, IEEE Trans. Mag. 5, 700(1969).
45. G. Fan and J. H. Greiner, J. Appl. Phys. 41, 1401(1970).
46. K. Y. Ahn and G. S. Almasi, IEEE Trans. Mag. 5, 180 and 944(1969).
47. G. S. Almasi and K. Y. Ahn, J. Appl. Phys. 41, 1258(1970).
48. G. Fan, K. Pennington and J. H. Greiner, J. Appl. Phys. 40, 974(1969).

SYMPOSIUM SUMMARY

W. V. SMITH

IBM Research Laboratory, Zurich, Switzerland

Three things strike me as important when I compare the state of knowledge on opto-electronic materials revealed at this symposium with the situation five years ago--a time when the subject had already achieved a vigorous growth for several years as a result of the discovery of the laser.

My first observation is the addition of two new classes of opto-electronic materials to those then under investigation for control of light. These are the modified lead zirconate titanate and other ferroelectric ceramics and liquid crystals. I think it is significant that both examples represent a response to less esoteric and more traditional light control applications than those evoked by the appearance of the laser. That is, with these materials, one is primarily concerned with the control of "ordinary" incoherent light, at relatively slow response times, for use in displays viewed by the human eye.

My second observation is the persistence of the old forms of light control. Five years ago I, at least, thought of acousto-optic light modulators as relatively slow holdovers of pre-laser technology that could soon be displaced by electro-optic modulators. They have not been so displaced. However, the subject has developed in a way that emphasizes many common features of electro-optic and acousto-optic materials. Indeed, one may see a certain unity in my first and second observations by noting the persistence with which "slow" phenomena (phenomena which involve molecular or macromolecular motion) remain attached to many approaches to the control of light. In the examples cited, these slow motions are macromolecular rotations and crystal vibrations, acoustic or, in the case of ferroelectric, soft mode optic ones. The control of electron motion in injection lasers or back biased p-n junctions (to add examples of opto-electronic materials not covered in this Symposium) are, indeed, exceptions to this statement, but so far they have not replaced slower phenomena. Another exception is the control of light by light. Here electronic motion is usually more important than molecular motion.

My third observation is the major advances that have been made in the past five years in our understanding of the fundamental theory of opto-electronic materials, and the contribution that nonlinear optical experimental techniques have made to this understanding. Before reminding you of the ways in which specific papers presented here lead to these observations, I would like to comment briefly on the beautiful interrelations between the subjects of phase transitions, optical scattering, and lasers as an example of these theoretical and experimental developments. Not only are laser beams used to probe the details of density fluctuations in materials near phase transitions, but additionally, though not the subject of this conference, the theories of such fluctuations and phase transitions have been shown to be closely related to the theories of noise and the threshold transition from incoherent to coherent light emission in a laser. Alternatively put, the laser itself is a nonlinear optical material with properties analogous to those of other nonlinear optical materials.

Let me now refer to the specific papers presented here, and note that a common theme of the importance of light scattering runs through many of them. Let me first list the most obvious titles: "The Scattering (Diffraction) of Light by Elementary Excitations in Solids", by Eli Burstein and M. Pinczuk; "Visible Light Scattering in Ferroelectric Ceramics", by our Chairman, Walter Albers, and M. Kaplit; and "Light Scattering Properties of Nematic Liquid Crystals", by Peter Pershan. We see here first a fundamental interest in scattering as it reveals information about the phonon, plasmon, magnon, and electron-hole pair excitations and their coupling with photons in solids. Near critical points, in particular, the size and frequency spectrum of density or other relevant fluctuations in the material are deduced from the spatial and frequency spectrum of the scattered light. Near ferroelectric phase transitions the scattering, including Raman scattering, gives information concerning the relevant soft modes that maximize electro-optic interactions, thus providing one of the basic physics inputs Stu Kurtz used in his excellent review of "Nonlinear Optical Modulators".*

If the scattering is strong enough, it not only is a useful spectroscopic probe with which to study optoelectric materials, it also becomes a useful mode in which to operate devices. Thus, acousto-optic modulation and deflection (R. W. Dixon) are the consequences of controlled coherent scattering phenomena, and liquid crystal light control devices (P. A. Soref) rely on controlled incoherent scattering. Ferroelectric ceramics can be operated in a similar mode although Albers and Kaplit and Land and Thacher put more emphasis on electro-optic modes. I am glad that Bloembergen in a comment from the floor recalled that acousto-optic scattering can be achieved in two-dimensional configurations where acoustic surface waves deflect

** Ed. note: Regretably, Dr. Kurtz' paper is not included in this volume.*

laser beams travelling down an optical film wave guide, as well as in the more familiar three dimensional configurations.

This symposium treats both linear and nonlinear optical phenomena. Thus, John Hopfield's talk emphasizes the simple relations he, Phillips, and VanVechten have evolved between the chemical bonding, or the covalent and ionic energy gaps, and the dielectric constant (or refractive index) of a material. These are linear phenomena, also treated somewhat differently, by M. Cardona and F. H. Pollak, and these several authors also note the close relation between linear and nonlinear electro-optic coefficients. Most simply summarized - by Kurtz for electro-optic materials and Dixon for acousto-optic materials, although similar relations hold for frequency doubling as well - the nonlinear optical coefficients are all proportional to a fairly high power of the refractive index, and this is the common bond among all these effects which otherwise differ in the crystal symmetries required, selection rules, etc. I remind you of Giordmaine's remark from the floor that ionic size and ionic charge are simply related by Levine to the sign as well as the magnitude of the frequency doubling coefficient.

Perhaps this is a good place at which to return to the persistence of slow phenomena in electro-optic and acousto-optic modulation to which I alluded early in my talk. The point which unites these phenomena is related to the fact that a modulating field - electric or magnetic - which cannot act directly on a photon since the photon carries no charge, must couple effectively with the natural resonances of the system. Piezoelectric resonances of transducers and soft mode optical phonon resonances of ferroelectrics responsible for their large dielectric constant, to which the electro-optic effect is proportional (or macromolecular reorientations in liquid crystals), lie much closer in frequency to the usual applied modulating fields than do the electronic resonances of optical transitions.

Ron Shen's paper on Nonlinear Optical Effects is more concerned with the nonlinear wave equation and unusual boundary condition problems than with material properties as such. However, there is a persistence of slow phenomena in his optical studies since the nonlinear refractive index, important in self-focusing, arises from phenomena averaged over many cycles of the optical wave. Hence, the relevant response time is some envelope of the optical signal rather than an optical period. (You will recall how a precurser optical pulse prepared a filament that could trap the trailing edge of the pulse.)

Electron transitions - this time of free carriers in semiconductors - is also the subject of L. T. Lauder, Jr., and J. G. Gay's paper on Optical Effects of Perturbation of Semiconductor Surfaces. Indeed, this is another exception to my generality that the motion of heavy particles seems relevant to most optical modulators. The efficiency-contrast ratio tradeoffs they quote remind us that direct electron control in this case is difficult to match to optical control.

I come finally to David Adler's paper on the Optics of Solid-State Phase Transformation and J. O. Dimmock's paper on Magneto-Optical Properties of the Eu-Chalcogenides. These papers introduce two different optical phenomena under investigation for high density optical memory and other applications. Amorphous semiconductor-crystalline phase changes, one part of Adler's talk, are of current interest in a number of potential applications, most of which involve change in the resistive state of the device. However, the optical memory application involves a change in transparency. The degree of practical reversibility of these transitions is of course of primary interest in determining their potential utility. The short time constants Adler quotes are most interesting. I hesitate to comment on the exceptional reversibility with equal optical pulses.

Somewhat more is known about magneto-optical memories and devices than about amorphous semiconductor ones. Over the course of the past five years, magneto optical materials of increasingly attractive combinations of high rotation and low absorption have been found. Dimmock's concentration on EuO illustrates the fact that most of these materials work optimally at low temperatures. We will not dwell on these device considerations, but merely note that while magneto-optical phenomena are in some sense weaker than electro-optical ones, this is not necessarily as much of a disadvantage as one might at first guess. The same comment on relative strength of interactions could be made about ferroelectric vs. ferromagnetic phenomena, but the nonexistence of free magnetic charges is a compensating advantage, allowing the achievement of essentially permanent storage more easily in magnetic than ferroelectric materials. Looping back to the interrelation of experiment with theory, you will note the sharper more dispersive resonances involved in magnetooptic compared to electro-optic material.

In conclusion, I would like to compliment Frank Jamerson, Walter Albers and his advisory committee and our host institution, General Motors, for the excellent coverage they have achieved in this continuously dynamic and important research field. Our speakers have all made recent personal research contributions, so that we have observed a snapshot of the moving frontier in our investigations of the physics of opto-electronic materials. While much progress has been made in our understanding of these effects, controversy remains. Witness the stimulating interchanges between Hopfield and Cardona on the relative importance of chemical and band gap properties in determining nonlinear optical properties. (I have translated "accidental properties" to "band gap" here.) Furthermore, new phenomena are continually being discovered. Dixon's reference to Haries' tunable optical filter is one example of a fairly recent advance. At the other end of the frequency spectrum, many of you have seen preprints referring to P. S. Eisenberger and S. McCall's achievement of x-ray parametric fluorescence, so there are more exciting things to come. I want to thank Walt Albers for inviting me to participate in this symposium, and to add that the opportunity for all of us to exchange informal as well as formal information is another very valuable function performed by this symposium.

SUBJECT INDEX